The Everglades,
Florida, and
the Politics of Paradise

———

The
Swamp

MICHAEL GRUNWALD

Simon & Schuster

New York   London   Toronto   Sydney

SIMON & SCHUSTER
Rockefeller Center
1230 Avenue of the Americas
New York, NY 10020

Photography credits are on page 452.

For information about special discounts for bulk purchases,
please contact Simon & Schuster Special Sales at 1-800-456-6798 or
business@simonandschuster.com.

Designed by Ellen Sasahara

Manufactured in the United States of America

5   7   9   10   8   6   4

Library of Congress Cataloging-in-Publication Data

Grunwald, Michael.
The swamp / Michael Grunwald.
p. cm.
Includes bibliographical references pp. 371-432.
1. Everglades (Fla.)—History. 2. Everglades (Fla.)—Environmental conditions.
3. Environmental protection—Florida—Everglades—History.
4. Drainage—Florida—Everglades—History. I. Title.
F317.E9G78 2005
975.9'39—de22   2005056329

ISBN-13: 978-0-7432-5105-1
ISBN-10:   0-7432-5105-9

For Mom and Dad
with love

Contents

And God said unto them: Be fruitful, and multiply, and replenish the earth, and subdue it: and have dominion over the fish of the sea, and over the fowl of the air, and over every living thing that moveth upon the earth.

—Genesis 1:28

Nature is overrated.
But we'll miss it when it's gone!

—Florida golfers, in the 2002 film *Sunshine State*

Introduction

"A Treasure for Our Country"

ON DECEMBER 11, 2000, the Supreme Court heard oral arguments in *George W. Bush, et al. v. Albert Gore Jr., et al.*, the partisan battle royale that would end the stalemate over the Florida recount and send one of the litigants to the White House. The deadlocked election had exposed a divided nation, and pundits were describing Governor Bush's "Red America" and Vice President Gore's "Blue America" as if they were separate countries at war. After five weeks of ferocious wrangling over "pregnant chads" and "hanging chads," hard-liners in both camps were warning of an illegitimate presidency, a constitutional crisis, a bloodless coup.

Inside the Court's marble-and-mahogany chambers, Senator Robert Smith of New Hampshire watched the legal jousting with genuine awe. Smith was one of the hardest of Red America's hardliners, a passionate antiabortion, antigay, antitax Republican, and he believed he was watching a struggle for the soul of his country. Smith was also a former small-town civics teacher, less jaded than most of his colleagues in Congress, and *Bush v. Gore* was a civics lesson for the ages, a courtroom drama that would decide the leader of the free world. "It doesn't get any bigger than this," he thought.

But less than an hour into the proceedings, Smith suddenly walked out on history, squeezing his six-foot-five, 280-pound frame past his perplexed seatmates. *"Excuse me,"* he whispered. *"Excuse me."* A bear of a man with fleshy jowls, a bulbous nose, and a sloppy comb-over, Smith could feel the stares as he lumbered down the center aisle, then jostled through the hushed standing-room crowd to the exit. *"Excuse me. Excuse me."*

Smith's abrupt departure looked like one of his unorthodox protests, like the time he brandished a plastic fetus on the Senate floor, or the time he announced he was resigning from the Republican Party because it was cutting too many big-government deals with the Democrats.

1

Smith was an unabashed ideologue, rated the most conservative and the most frugal senator by various right-wing interest groups. He had voted against food stamps and Head Start, clamored for President Bill Clinton's impeachment, and even mounted his own quixotic campaign for president on a traditional-values platform.

But this was no protest. Smith was rushing to the White House, to celebrate a big-government deal with the Democrats.

At the height of the partisan war over the Florida recount, President Clinton was signing a bipartisan bill to revive the Florida Everglades, a $7.8 billion rescue mission for sixty-nine endangered species and twenty national parks and refuges. It was the largest environmental restoration project in the history of the planet, and Smith had pushed it through Congress with classic liberal rhetoric, dismissing its price tag as "just a can of Coke per citizen per day," beseeching his colleagues to "save this treasure as our legacy to our children and grandchildren." So after his dash from the Court, he headed straight to the Cabinet Room, where he exchanged congratulations with some of the Democratic Party's top environmentalists, like Interior Secretary Bruce Babbitt, the former head of the League of Conservation Voters, and White House aide George Frampton, the former head of the Wilderness Society. And Smith was not even the most surprising guest in the West Wing that day.

That was Florida's Republican governor, another key supporter of the Everglades plan, a former Miami developer named Jeb Bush. As the world waited to hear whether his brother would win his state and succeed their father's successor in the White House, Jeb was already there, staring out at the Rose Garden with the air of a quarterback who had stumbled into the opposing locker room near the end of the Super Bowl. "The last time I was here, your father was president!" one lobbyist told him. Jeb tried to smile, but it came out more like a grimace. One Clinton appointee began babbling about the Cuban Missile Crisis—possibly the last time that room had felt that tense. Jeb even said hi to a Miami congresswoman who had publicly accused him of suppressing black votes. "This," thought Jeb's top environmental aide, "is as surreal as politics can get."

Unless, that is . . . but no, Vice President Gore, a key architect of the Everglades plan, stayed home to listen to the Supreme Court audiotape. "I was really proud of what we accomplished in the Everglades," Gore later recalled. "But I was in a pretty pitched battle that day."

At 1:12 P.M., an ebullient President Clinton invited everyone into the Oval Office, the room that George W. Bush liked to say needed a good scrubbing. If the president was upset about Gore's plight, or Jeb's presence, or the legacy of impeachment, or his imminent move to the New York suburbs, the legendary compartmentalizer hid it well. "This is a great day!" he said. "We should all be very proud." He used eighteen ceremonial pens to sign the bill, graciously handing the first souvenir to Jeb. Senator Smith quipped that it was lucky Clinton's name wasn't Cornelius Snicklefritzer, or else the ceremony might never end. The president threw his head back and laughed. "Wow," thought his chief of staff, John Podesta, "this is like a Fellini movie."

If Florida's political swamp was tearing Americans apart, Florida's actual swamp had a knack for bringing people together. The same Congress that had been torn in half by Clinton's impeachment had overwhelmingly approved his plan for the Everglades, after lobbyists for the sugar industry and the Audubon Society walked the corridors of Capitol Hill arm-in-arm. The same Florida legislature that was in turmoil over *Bush v. Gore* had approved Everglades restoration without a single dissenting vote.

At a press conference after the ceremony, Jeb sidestepped the inevitable *Bush v. Gore* questions to highlight this unity: "In a time when people are focused on politics, and there's a little acrimony—I don't know if y'all have noticed—this is a good example of how, in spite of all that, bipartisanship is still alive." Reporters shouted follow-ups about the Court, but the governor cut them off with a smile. "No, no, no, no, you're going the wrong way on that one. We're here to talk about something that's going to be long-lasting, way past counting votes. This is the restoration of a treasure for our country."

The Test

TODAY, EVERYONE AGREES that the Everglades is a national treasure. It's a World Heritage Site, an International Biosphere Reserve, the most famous wetland on earth. It's a cultural icon, featured in Carl Hiaasen novels, Spiderman comics, country songs, and the opening credits of *CSI: Miami,* as well as the popular postcards of its shovel-faced alligators and spindly-legged wading birds. It's the ecological equivalent of motherhood and apple pie; when an aide on NBC's *The West Wing* was asked the most popular thing the president could do for the environment, he immediately replied: "Save the Everglades."

But there was once just as broad a national consensus that the Everglades was a worthless morass, an enemy of civilization, an obstacle to progress. The first government report on the Everglades deemed it "suitable only for the haunt of noxious vermin, or the resort of pestilential reptiles." Its explorers almost uniformly described it as a muddy, mushy, inhospitable expanse of razor-edged sawgrass in shallow water—too wet to farm, too dry to sail, too unpredictable to settle. Americans believed it was their destiny to drain this "God-forsaken" swamp, to "reclaim" it from mosquitoes and rattlesnakes, to "improve" it into a subtropical paradise of bountiful crops and booming communities. Wetlands were considered wastelands, and "draining the swamp" was a metaphor for solving festering problems.

The heart of the Everglades was technically a marsh, not a swamp, because its primary vegetation was grassy, not woody; the first journalist to slog through the Everglades called it a "vast and useless marsh." But it was usually described as a dismal, impenetrable swamp, and even conservationists dreamed of draining it; converting wet land into productive land was considered the essence of conservation. Hadn't God specifically instructed man to subdue the earth, and take dominion over all the living creatures that moveth upon it? Wasn't America destined to overpower its wilderness?

This is the story of the Everglades, from useless bog to national treasure, from its creation to its destruction to its potential resurrection. It is the story of a remarkable swath of real estate and the remarkable people it has attracted, from the aboriginals who created the continent's first permanent settlement in the Everglades, to the U.S. soldiers who fought a futile war of ethnic cleansing in the Everglades, to the dreamers and schemers who have tried to settle, drain, tame, develop, sell, preserve, and restore the Everglades. It's a story about the pursuit of paradise and the ideal of progress, which once inspired the degradation of nature, and now inspires its restoration. It's a story about hubris and unintended consequences, about the mistakes man has made in his relationship with nature and his unprecedented efforts to fix them.

THE STORY BEGINS with the natural Everglades ecosystem, which covered most of south Florida, from present-day Orlando all the way down to the Florida Keys. For most of its history, it was virtually uninhabited. As late as 1897, four years after the historian Frederick Jackson Turner declared the western frontier closed, an explorer marveled that the Everglades was still "as much unknown to the white man as the heart of Africa."

But once white men got to know it, they began to transform it. A Gilded Age industrialist named Hamilton Disston was the first visionary to try to drain the swamp. A brilliant oilman-turned-developer named Henry Flagler considered his own assault on the Everglades while he was laying the foundation for modern south Florida. And an energetic Progressive Era governor named Napoleon Bonaparte Broward vowed to create an Empire of the Everglades with more canals, declaring war on south Florida's water.

The Everglades turned out to be a resilient enemy, resisting man's drainage schemes for decades, taking revenge in the form of brutal droughts and catastrophic floods, converting Florida swampland into an enduring real estate punchline. In 1928, a hurricane blasted Lake Okeechobee through its flimsy muck dike and drowned 2,500 people in the Everglades, a ghastly foreshadowing of Hurricane Katrina's assault on New Orleans. Mother Nature did not take kindly to man's attempts to subjugate her.

But the U.S. Army Corps of Engineers, the ground troops in America's war against nature, finally conquered the Everglades with one of the most elaborate water-control projects in history, setting the stage for south Florida's spectacular postwar development. Suburbs such as Weston, Wellington, Plantation, Pembroke Pines, Miami Lakes, and Miami Springs all sprouted in drained Everglades wetlands. So did Miami International Airport, Sawgrass Mills Mall, Florida International University, Burger King corporate headquarters, and a vast agricultural empire that produces one out of every five teaspoons of American sugar. Disney World was built near the headwaters of the Everglades. And some people began to wonder whether the creation of a man-made paradise across Florida's southern thumb was worth the destruction of a natural one.

So the story of the Everglades is also the story of the transformation of south Florida, from a virtually uninhabited wasteland to a densely populated Fantasyland with 7 million residents, 40 million annual tourists, and the world's largest concentration of golf courses. "There has never been a more grossly exaggerated region, a more grossly misrepresented region, or one concerning which less has been known than this mighty empire of South Florida," the *Palm Beach Post* said in 1924. That's still about right.

* * *

AMERICA'S WAR ON NATURE has left a tattered battlefield in south Florida. Half the Everglades is gone. The other half is an ecological mess. Wading birds no longer darken the skies above it. Algal blooms are exploding in its lakes and estuaries, massacring its dolphins, oysters, and manatees. And it is now clear that the degradation extends beyond noxious vermin and pestilential reptiles, affecting the people of south Florida as well. The aquifers that store their drinking water are under siege. Their paradise has been sullied by sprawl, and by overcrowded schools, hospitals, and highways. Most of them are at risk from the next killer hurricane—and the one after that. It is now almost universally agreed that south Florida's growth is no longer sustainable.

The Everglades restoration plan that President Clinton signed with Governor Bush at his side is supposed to restore some semblance of the original ecosystem, and guide south Florida toward sustainability. And the Army Corps of Engineers, after decades of helping to destroy the Everglades, will lead the effort to undo some of the damage. "The Everglades is a test," one environmentalist has written. "If we pass, we may get to keep the planet."

On that December day at the millennium's end, Republicans and Democrats described Everglades restoration as the dawn of a new era in conservation—not only for south Florida, but for mankind. Instead of taming rivers, irrigating deserts, and draining swamps, man would restore ravaged ecosystems. Instead of fighting over scarce fresh water—the oil of the twenty-first century—Floridians would demonstrate how to share. The Everglades, Jeb Bush said, would be "a model for the world," proof that man and nature could live in harmony. America's politicians would finally pass the Everglades test.

It was a noble sentiment. But man had been flunking that test for a long time.

PART 1

The Natural Everglades

ONE

Grassy Water

There are no other Everglades in the world.

—South Florida author Marjory Stoneman Douglas

"The Place Looked Wild And Lonely"

THE NATURAL EVERGLADES was not quite land and not quite water, but a soggy confusion of the two.

It was a vast sheet of shallow water spread across a seemingly infinite prairie of serrated sawgrass, a liquid expanse of muted greens and browns extending to the horizon. It had the panoramic sweep of a desert, except flooded, or a tundra, except melted, or a wheat field, except wild. It was studded with green teardrop-shaped islands of tangled trees and scraggly shrubs, and specked with white spider lilies and violet-blue pickerelweeds. But mostly it looked like the world's largest and grassiest puddle, or the flattest and wettest meadow, or the widest and slowest-moving stream. It had the squish and the scruff of an untended yard after a downpour, except that this yard was larger than Connecticut. It wasn't obviously beautiful, but it was obviously unique. "No country that I have ever heard of bears any resemblance to it," wrote one of the U.S. soldiers who hunted Seminole Indians in the Everglades in the nineteenth century. "It seems like a vast sea, filled with grass and green trees."

The Everglades seeped all the way down Florida's southern thumb, from the giant wellspring of Lake Okeechobee in the center of the peninsula to the ragged mangrove fringes of Florida Bay and the Gulf of Mexico, a sodden

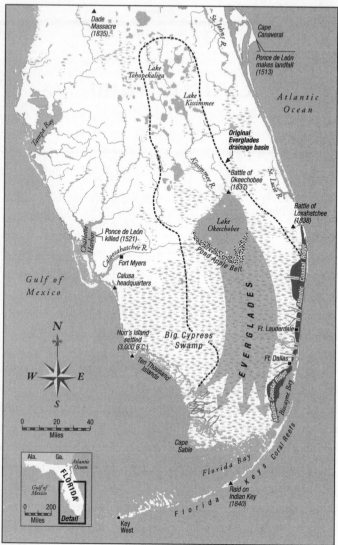

BY GENE THORP—CARTOGRAPHIC CONCEPTS INC.

In its natural state, the Everglades ecosystem covered most of south Florida, and it was all connected by clean, fresh, shallow, gently flowing water. It flowed from the chain of lakes below present-day Orlando into the tortuous Kissimmee River, which emptied into Lake Okeechobee, which spilled into the Everglades. Most of the River of Grass trickled south and southwest into Florida Bay and the Gulf of Mexico, but some of its water gathered into streams that crossed the Atlantic coastal ridge into Biscayne Bay.

savanna more than 100 miles long and as much as 60 miles wide—just grass and water, water and grass, except for the tree islands and wildflowers that dotted the grass, and the lily pads and algal mats that floated on the water. The Seminoles called it Pa-Hay-Okee, or Grassy Water. The American soldiers who trudged through it during the Seminole Wars described it as a grassy lake, a grassy sea, an ocean of grass. The bard of the Everglades, Marjory Stoneman Douglas, later dubbed it the River of Grass. Sawgrass is actually a sedge, not a grass, but the nickname stuck.

The Everglades was relentlessly, remarkably, almost perfectly flat—no majestic canyons, rugged cliffs, or rolling hills, no glaciers, geysers, or craters. Even Everglades National Park's first superintendent admitted that its landscape lacked a certain flair, calling it "a study in halftones, not bright, broad strokes of a full brush," summarizing its attractions as "lonely distances, intricate and monotonous waterways, birds, sky and water." The Everglades was also an incomparably tough slog. It lacked shade and shelter, high ground and dry ground. Breathing its heavy air felt like sucking on cotton. Wading through its hip-deep muck felt like marching in quicksand. Penetrating its dense thickets of sharp-toothed sawgrass felt like bathing in broken glass. And there was something downright spooky about the place, with its bellowing alligators, grunting pigfrogs, and screeching owls—and especially its eerie silences.

"The place looked wild and lonely," one hunter wrote after an 1885 expedition through the Everglades. "About three o'clock it seemed to get on Henry's nerves, and we saw that he was crying, he would not tell us why, he was just plain scared."

The Everglades also teemed with rats, roaches, snakes, scorpions, spiders, worms, deerflies, sand flies, and unfathomably thick clouds of bloodthirsty mosquitoes that flew up nostrils and down throats and into ears. The pioneer Miami naturalist Charles Torrey Simpson loved the Everglades like a son, but he readily acknowledged that "the wilds of Lower Florida can furnish as much laceration and as many annoyances to the square inch as any place I have ever seen."

"My advice is to urge every discontented man to take a trip through the Everglades," another explorer wrote. "If it doesn't kill him, it will certainly cure him."

*   *   *

BUT THE EVERGLADES was more than a river of grass, and it contained more than swarming bugs, slithering reptiles, and lacerating annoyances.

The river of grass was only the most distinctive link of an interconnected ecosystem that once blanketed almost all of south Florida, from its head-waters atop the Kissimmee Chain of Lakes near modern-day Orlando down to the coral reefs off the Keys, an area twice the size of New Jersey. The ecosystem was a watery labyrinth of lakes and lagoons, creeks and ponds, pine flatwoods and hardwood hammocks. It encompassed Biscayne Bay and Florida Bay, the St. Lucie and Miami Rivers. And in addition to its extensive marshlands, it included genuine swamps, most notably the Big Cypress Swamp, a Delaware-sized mosaic of pinelands, prairies, and black-water bogs just west of the sawgrass Everglades.

Sawgrass could be as uninviting to wildlife as it was to people, but the diverse habitats of the broader Everglades ecosystem—also known as the Kissimmee-Okeechobee-Everglades or south Florida ecosystem—supported an astonishing variety of life, from black bears to barracudas, turkey vultures to vase sponges, zebra butterflies to fuzzy-wuzzy air plants that looked like hairy psychedelic squid. The Everglades had pre-historic-looking wood storks that snapped their beaks shut in three mil-liseconds, sausage-shaped manatees that devoured 100 pounds of plants a day, mullet that ran in schools three miles long, and four-foot-tall dwarf cypress trees that looked like skeletal bonsai. The Everglades was the only place on earth where alligators (broad snout, fresh water, darker skin) and crocodiles (pointy snout, salt water, toothy grin) lived side by side. It was the only home of the Everglades mink, Okeechobee gourd, and Big Cypress fox squirrel. It had carnivorous plants, amphibious birds, oysters that grew on trees, cacti that grew in water, lizards that changed colors, and fish that changed genders. It had 1,100 species of trees and plants, 350 birds, and 52 varieties of porcelain-smooth, candy-striped tree snails. It had bottlenose dolphins, marsh rabbits, ghost orchids, moray eels, bald eagles, and countless other species that didn't seem to belong on the same continent, much less in the same ecosystem.

"It is a region so different that it hardly seems to belong to the United States," said the forester Gifford Pinchot, a founding father of American conservationism. "It is full of the most vivid and most interesting life on land, in the air, and in the water. It is a land of strangeness, separate and apart from the common things we all know so well."

For all its mystery and monotony, the Everglades ecosystem did have a few awesome attractions. Charles Torrey Simpson was enthralled by its 100-foot-tall royal palms with trunks like cement pillars, standing guard over its golden ocean of sedge and stream: "It is a picture of unsurpassed beauty set in a wonderful frame. . . . The whole effect is glorious beyond the power of description." Another visitor adored its profusion of wild orchids, "specimens colorless and full of color, scentless and filled with odor that made the surrounding air heavy with their fragrance: some garbed somberly as a Quakeress, others costumed to rival the Queen of Sheba." Early explorers were mesmerized by the millions of ibis, egrets, herons, storks, and other wading birds that seemed to darken the skies; the legendary artist and naturalist John J. Audubon nearly swooned after watching a flock of hot-pink flamingos soar over the Everglades. "Ah! Reader, could you but know the emotion that then agitated my breast. I thought I had now reached the height of my experience." The celebrated zoologist Louis Agassiz was just as fascinated by the luminous coral reefs at the ecosystem's edge, the only living reefs in North America: "Even a brief description of the immense number of shells, worms, crabs, lobsters, shrimps, crawfishes and fishes seen everywhere upon the reef, would be out of place here. In variety, in brilliancy of color, in elegance of movement, the fishes may well compare with the most beautiful assemblage of birds."

For the most part, though, the Everglades was less about beauty than subtlety and originality. It was less ooh or aah than hmm. It was a dainty purple gallinute tiptoeing across a lily pad in a predawn mist, or a vast swath of sawgrass arching in the breeze like a congregation at prayer. It was the vines of a strangler fig slowly choking the life out of a cabbage palm, or a split-tailed, red-eyed hawk called the Everglade snail kite scanning a marsh for the apple snails that made up its entire diet. Everglades vistas seemed to shift like a kaleidoscope with subtle changes in the light or the weather. Guidebooks still warn tourists that the Everglades "takes some getting used to," that it "reveals its secrets slowly," that its appeal "may escape many visitors at first glance."

There was always more to the Everglades than met the eye. Take the golden-brown Everglades goop known as periphyton. It was easy to overlook, clumped around aquatic plants like slimy oatmeal sweaters, floating in sloughs like discolored papier-mâché, crumbling into a snowy powder during droughts. But it was the dominant life-form in much of the Everglades, measured by biomass. It was also the base of the Everglades food chain, providing grazing pastures for small fish, prawns, insects, and snails, which became prey for larger fish and birds. Today, microscopes reveal periphyton mats as action-packed worlds unto themselves, teeming with bacteria, diatoms, and single-cell organisms shaped like candles, spaghetti, bricks, nets, tissues, and tunnels—swimming, splitting, and swallowing one another whole.

If the Grand Canyon was a breathtaking painting, the Everglades was a complex drama, and everything in it had a role. The American alligator, the original Everglades engineer, dug muck out of shallow depressions in the marsh during droughts, creating oases for fish and wildlife like the watering holes of the African bush. The red mangrove, the original Everglades developer, trapped sediments in its spidery prop roots until they formed new spits of swampland, while providing shelter for all kinds of estuarine species. Cauliflower clouds, the mountains of the Everglades, printed their reflections on glittering sloughs as they drifted over the marsh, then funneled and blackened into thunderheads that unleashed spectacular torrents of rain. And that clean, fresh, shallow water was the lifeblood of the Everglades, fueling its flora and fauna, recharging its underground aquifers, keeping its wetlands wet. "A certain kind of lure began to dawn on me," wrote Zane Grey, the best-selling western adventure novelist who was also a record-breaking south Florida snook fisherman. "This is a country that must be understood."

The First 300 Million Years (Abridged)

THE EVERGLADES IS OFTEN described as an "ancient wilderness," a "timeless relic," a "primordial" or "primeval" force of nature that flourished for eons before it was ruined by man. But in geologic time, the Everglades is a newborn. If the history of the earth is condensed to a week, algae started growing Monday, fish started swimming Saturday morning, and birds flew in early Saturday afternoon. The Everglades showed up a half second before midnight, around the time the Egyptians started building pyramids. From the earth's perspective, the story of the Everglades is a rounding error, a momentary blip.

The story of how the Everglades formed stretches back a bit further. It began with a bang about 300 million years ago, after the fish but before the birds, with the cataclysmic shifts of tectonic plates that crunched the planet's major landforms into a single supercontinent called Pangaea. About 100 million years later, plates shifted again, Pangaea split up again and North America dragged away a finger-shaped chunk of northwest Africa. That hijacked appendage became the foundation of the Florida peninsula, the Florida Platform, dangling into the waters of the subtropics like a big toe dipped into a warm bath, dividing the Atlantic Ocean from the Gulf of Mexico.

Then things calmed down. And they stayed calm.

Florida has been geologically stable ever since it was kidnapped from Africa, with none of the seismic upheavals that carved out mountains and canyons elsewhere. South Florida has been especially quiet: It was inundated in the Jurassic Period and spent most of the next 150 million years as a sea floor. Dinosaurs reigned and vanished, and mammals inherited the earth, but south Florida remained underwater, slowly building its limestone backbone from the shells and skeletons of dead marine species, and from microscopic pearls of calcium dissolved in the sea itself. The main result of all those dull millennia of stability was that the region ended up extraordinarily flat. Today, a sign in Everglades National Park announces the towering peak of "Rock Reef Pass: Elevation 3 Feet." One Everglades scientist used to tell the apocryphal story of the cowboy who saw the Grand Canyon and shouted: "Something sure did happen here!" His point was that it wasn't clear from the topography of the Everglades that anything had ever happened there.

South Florida finally emerged from the ocean during the ice ages, when the polar glaciers expanded and retreated, exposing and reflooding the peninsula while man was evolving from *Homo erectus* to *Homo sapiens*. It was only during the last interglacial melt, about 100,000 years ago, that high seas deposited the Biscayne Aquifer, the porous layer of honeycombed limestone that underlies much of the Everglades, and stores much of south Florida's drinking water in its subterranean notches and channels. The same high seas left behind the Atlantic Coastal Ridge, the five-mile-wide ribbon of limestone that became the eastern rim of the Everglades, rising as high as twenty feet above sea level, a virtual Kilimanjaro by south Florida standards. It now supports downtown Miami, Fort Lauderdale, and West Palm Beach.

In the last ice age, low seas exposed the entire Florida Platform, as well as the land bridge across the Bering Strait that Paleoindians crossed to North America 12,000 years ago. Those hunter-gatherers journeyed down and across the continent to discover a supersized Florida, twice as wide as it appears today. It was cooler and drier, with fewer lakes and rivers. But its windswept prairies and oak savannas would have made the modern Serengeti look like a petting zoo, with fourteen-foot-tall mammoths, five-ton mastodons, sloths the size of elephants, jaguars, wild dogs, and saber-toothed smilodons with nine-inch blades for canines. Most of the world's large mammals died out after the ice ages—many scientists suspect their extinctions were related to a certain two-legged predator—but Paleoindians still hunted south Florida's panthers, deer, and ducks, while trolling its waters for oysters, scallops, and fish.

It was only about 5,000 years ago—after prehistoric man was already writing, making pottery, smelting copper, and brewing beer—that seas approached current levels and modern climate conditions prevailed. Sawgrass began to sprout, and dead sawgrass began to decompose into soot-black Everglades muck soils. Geologically, not much had ever happened there, but the ingredients for the Everglades were in place.

THE MOST IMPORTANT ingredient was rain.

South Florida ended up in the Desert Belt, at the latitude of the Sahara and the Arabian. But it is surrounded by cloud-generating water bodies on three sides, with Lake Okeechobee occupying much of the fourth. So instead of a year-round desert climate, south Florida developed a two-tone subtropical climate, a pleasant November-to-May dry season with some of the continent's warmest winters followed by a muggy June-to-October rainy season with the continent's wettest summers. On average, south Florida receives about fifty-five annual inches of rain, significantly more than Seattle, although "average" is a misleading concept when there are such dramatic fluctuations between wet and dry seasons and years. Rain falls in bunches below Lake O, most of it in summer afternoon thunderstorms that can feel like an ocean falling from the sky, dumping a foot in a day. South Florida is also the continent's leading target for tropical hurricanes.

This combination of abundant sunlight and abundant rain was the ultimate recipe for abundant life. Rainy seasons created biological explosions, and the winter dry-downs that followed were just as important, concentrating fish into shallow pools that attracted birds for feeding frenzies. Everglades flora and fauna all adapted to these seesaws between flood and drought; for example, the gambusia's upturned mouth and the gar's oxygen-breathing lung helped both fish survive in low water, which helped the mosquitoes that ate gambusia larvae (also known as mosquito fish) and the alligators that fed on gar (also known as alligator gar) survive extended dry spells as well.

Since south Florida was the subtropical extremity of a temperate land-mass, its mix of flora and fauna was eclectic as well as abundant. Temperate species from the north, including hawks, raccoons, oaks, bobcats, and white-tailed deer, joined tropical species from the south, including roseate spoonbills that flew in, loggerhead turtles that swam in, tree snails that floated in on branches, mahoganies whose seeds blew in during storms, and cocoplums whose seeds were dropped in by birds. Gators came down from the north; crocs came up from the south. They all came together in the Everglades—and nowhere else.

THE OTHER CRUCIAL INGREDIENT was limestone. The rock beneath the Everglades was exceedingly level, declining as little as two inches per mile, so all that water flowed down the peninsula exceedingly slowly. But the Everglades was not quite as flat as a floor or a cracker or a desk, as it is often described, so its water did flow. It trickled ever so sluggishly down the interior of the peninsula, inexorably carving its tree islands into teardrops that pointed the same direction as its almost imperceptible current. It gathered into sloughs and streams that sometimes vanished in the sawgrass, and sometimes found their way to the sea.

South Florida's limestone was also exceedingly porous, so the surface water that accumulated during wet times percolated and recharged the aquifers within the rock, maintaining an unusually high water table. If south Florida's geography made sure it got wet, its geology made sure it stayed wet; in its natural state, 70 percent of the region flooded every year, and 95 percent flooded at least periodically. A man of ordinary height (and extra-ordinary grit) could have walked the entire length of the Everglades without getting his hair wet, but his ankles might have been underwater the whole time.

In such a level landscape, even a few inches of elevation could transform the scenery. Water-lily sloughs that stayed wet all year long were just slightly lower than sawgrass marshes with ten-month hydroperiods, which were just slightly lower than Muhly-grass marshes with four-month hydroperiods. The slightly higher uplands of the coastal ridge rarely flooded at all, so they supported skinny slash pines with fire-resistant bark. And the different plant assemblies all decomposed into different soils, which corresponded almost precisely with the different rock formations beneath them.

Geology and hydrology were not quite destiny in the Everglades, but they were close.

THE EVERGLADES WAS ALSO MOLDED by an ingredient it lacked: phosphorus. It is a common nutrient in nature, critical to plant development. But the Everglades was phosphorus-starved, so its most successful species adapted to a low-nutrient environment. For example, sawgrass is a brutally efficient scavenger of phosphorus, so it outcompeted other marsh plants in the Everglades. Similarly, the microorganisms in periphyton mats were bound together by an intense affinity for phosphorus.

Nature is often called "fragile," which is usually wrong; nature is the essence of resilience. But the Everglades was about as fragile as nature gets, in the sense that even minor changes—in chemistry as well as topography or hydrology—made major differences. Tiny additions of phosphorus could transform the marsh, just like tiny bumps in land elevation or tiny dips in water levels. The Everglades was sensitive that way.

But 5,000 years ago, nothing had been added to the Everglades. Its waters were still pristine, and still flowed south without interruption.

The Lay of the Land

THE NATURAL EVERGLADES ECOSYSTEM, in sum, was an extremely flat drainage basin with extremely poor drainage, an unusually wet watershed that was unusually inefficient at shedding its water. A raindrop that fell in its headwaters in central Florida could have taken an entire year to dribble down to its estuaries at the tip of the peninsula. "The water is pure and limpid, and almost imperceptibly moves, not in partial currents, but, as it seems, in a mass, silently and slowly to the southward," an explorer wrote in 1848. The story of the Everglades, in sum, is the story of that water's journey, and man's efforts to reroute it.

The original journey began with the Kissimmee Chain of Lakes, a sparkling string of shallow potholes down the spine of the peninsula, brimming with bass and bluegill that spawned in the bulrush along their edges. The lakes fed the serpentine Kissimmee River, a kind of riparian Lombard Street, zigzagging like a drunken unicyclist down its narrow floodplain, frequently mutinying its banks to nourish its marshes. The Kissimmee basin attracted hordes of migratory waterfowl each winter, and supported a year-round menagerie of wading birds so abundant that one early visitor mistook the beating of their wings against the water for the churning of an approaching steamboat: "They do not attempt to fly until our boats are among them, and then it seems as if pandemonium has broken loose!"

The Kissimmee emptied into the immense saucer of Lake Okeechobee, Seminole for Big Water. Lake O was only twenty feet deep, but it was the largest lake in the South and one of the largest in America, vanishing in the distance with the curvature of the earth. The lake was crammed with bass, catfish, and trout that literally jumped into the canoes of a U.S. naval force that patrolled the lake in 1842; its shores teemed with deer, turtles, and especially birds. "Feasted sumptuously on wild turkey, broiled and fried curlew, plover and teal, stewed crane, [anhingas] and fried fish, our spoils of the day," the expedition's leader rejoiced. "The Astor House could not have supplied such a dinner."

Lake Okeechobee did not have a traditional outlet. The Caloosahatchee River began three miles west of it and drained west to the Gulf; the St. Lucie River began twenty miles east of it and drained east to the Atlantic; neither river carried water out of Lake O. Instead, during summer storms, the lake swelled until it spilled over its lower lip in a tremendous sheet. That was where the river of grass began, sloshing down the spoon-shaped depression between the Atlantic Coastal Ridge and the Big Cypress Swamp.

There was a washboard pattern to the Everglades, with dense sawgrass ridges alternating with open-water sloughs that carried its waters to sea. The widest strip of water draining the marsh, Shark Slough, curved south-west toward the tidal lagoons of the Ten Thousand Islands, a bewildering green archipelago of mangrove keys at the edge of the Gulf. The second largest Everglades outlet, Taylor Slough, headed due south through the sawgrass before melting into the brackish waters of Florida Bay, a triangular wading pool even larger and shallower than Lake Okeechobee, bracketed by the wide arc of the Keys. There were many smaller outlets as well, including Turner, Harney, and Lostman's Rivers flowing southwest to the Gulf, and "transverse glades" like the Miami, New, and Hillsboro Rivers spilling over or slicing through the coastal ridge on their way southeast to Biscayne Bay.

The Ten Thousand Islands, Florida Bay, Biscayne Bay, and the other coastal estuaries where fresh water from the Everglades mingled with salt water from the sea were the most productive niches of the ecosystem. They sheltered dolphins, manatees, pink shrimp, spiny lobsters, stone crabs, and an almost inconceivable array of fish. "Their number and variety are simply marvelous," another explorer wrote after visiting Florida Bay. "You can at one glance, through this crystal water, see over fifty varieties. The colors would put to blush the palette of an impressionist."

In fact, the Everglades estuaries that ringed south Florida were so full of life that they changed the course of human history.

Native Species

YES, MAN WAS NATIVE to the Everglades, too. In fact, the southwest edge of the Everglades may have been man's first permanent home in North America.

Until 1989, archaeologists believed that all Archaic peoples on the continent were nomadic, that year-round settlements only appeared after the introduction of agriculture. Then a graduate student named Michael Russo excavated Horr's Island, a squiggle-shaped clump of mangrove keys at the head of the Ten Thousand Islands. Russo found evidence of centuries of permanent occupation by a complex society, including traces of wooden posts used in dwellings and huge shell mounds used for rituals and burials. Russo then carbon-dated the site to the late Archaic period—right when the Everglades and its estuaries were taking form.

Russo's findings were archaelogical heresy. It seemed inconceivable that primitive hunter-gatherers with the run of the continent would have settled down on a swampy outpost off the tip of Florida, separated from the mainland by ten miles of tangled mangroves and tidal flats, in a humid archipelago where a twentieth-century entomologist would catch a record-breaking 365,696 mosquitoes in one trap in one night. But that's what happened. Russo revealed why when he dug up the island's food remains, which included seventy-four varieties of fish and shellfish: Horr's Island was an all-you-can-eat seafood buffet. The fishing at the edge of the Everglades was so good that its residents did not need to leave in the off-season to find more food, so good that it forced the archaeology establishment to revise its assumptions about Archaic man.

The people of Horr's Island sometimes ventured offshore to harpoon whales, sharks, marlins, and manatees, paddling canoes they fashioned from hollowed-out cypress logs, but they found most of their food in south Florida's sheltered near-shore estuaries. They hauled in tiny pinfish, catfish, and herring with nets woven from palm fibers, and gathered mollusks that provided raw material for their shell mounds as well as protein for their diets. They harvested different species in different seasons, like reliable underwater row crops: oysters in winter, scallops in summer, fish all year long. Most of the world's ancient societies had agricultural origins, but the bountiful fringes of the Everglades, where mangrove roots and seagrass meadows provided shelter and nutrition for hundreds of estuarine species, proved that cultivation was not a prerequisite for civilization.

BY THE TIME EUROPEANS arrived in the sixteenth century, the people of Horr's Island were gone. The Calusa Indians controlled southwest Florida, and exacted tribute from weaker tribes scattered around the peninsula.

None of these native people were farmers, either. Most were coastal fishermen. Many also maintained hunting camps in Big Cypress uplands or Everglades tree islands, and some may have even lived year-round in drier pockets of the interior—eating more turtles, mammals, and freshwater fish, but thriving just the same. The Europeans marveled at the imposing height, powerful physiques, and rich diets of the Calusa, the ultimate tribute to the bounty of the Everglades. Hernando d'Escalante Fontaneda, a Spanish shipwreck survivor who spent seventeen years as a Calusa prisoner, called his captors "men of strength" in his memoirs. "The people are great anglers, and at no time lack fresh fish," he wrote.

Fontaneda catalogued the marine cuisine of the Glades Indians, including lobsters, oysters, manatees, "enormous trout, nearly the size of men," and eels as thick as thighs. He also noted that the Tequesta Indians, who occupied the high ground of the Atlantic ridge, collected nuts and fruits and made bread out of "coontie," a starchy root abundant in the pinelands. They hunted deer, birds, snakes, alligators, "a certain animal that looks like a fox, yet is not," presumably raccoons, "an animal like a rat," probably

opossum, "and many more wild animals, which, if we were to continue enu-
merating, we should never be through."

The Indians of the Everglades had enough food that they didn't need to
spend every waking moment hunting and gathering; they had plenty of
time for construction, religion, and art. The Calusa built enormous mounds
from their discarded clam, conch, and oyster shells, including the 150-acre
island of Chokoloskee, and topped them with palmetto-thatched homes.
They crafted hammers, bowls, toys, and pendants out of wood, shell, and
bone. They attended rituals in elaborate costumes, and sculpted ceremonial
masks and statuettes depicting turtles, pelicans, panthers, and gators. One
archaeologist was amazed by the "startling fidelity" of the inner ears, hair
tufts, and other details he found on a Calusa deer carving: "The muzzle,
nostrils and especially the exquisitely modeled and painted lower jaw were
so delicately idealized that it was evident the primitive artist who fashioned
this masterpiece loved, with both ardor and reverence, the animal he was
portraying."

The Glades Indians have often been romanticized as wild savages with
hip-length hair and skimpy clothes, worshipping natural creatures and living in harmony with the land. The Calusa certainly were fierce warriors—
"Calusa" meant "fierce"—and it is true that they did not overdress in the
heat. "The men onely use deere skins, wherewith some onely cover their
privy members," gasped a British observer named John Sparke. But the
Glades Indians were sophisticated people, and they did not follow the
Leave No Trace ethic of the outdoors. They built impressive engineering
projects that molded nature to their needs—not just the shell mounds that
still dot the Gulf coast, but seawalls, jetties, weirs, fish traps, and reservoirs.
They dug canals to create canoe routes to their hunting grounds, including
a three-mile cut from the Caloosahatchee River to Lake Okeechobee that
would be reopened centuries later for one of the first Everglades drainage
ditches. The Calusa burned prairies to attract deer, chopped down cypress
trees for their canoes, butchered the animals they idealized in their art, and
preyed on baby fish that would be untouchable under modern catch-and-
release rules. Native people had an impact on the Everglades environment,
just as gators did when they dug their holes, or birds did when they ate
seeds in the tropics and deposited them in south Florida.

But the natives had an extremely modest impact. For one thing, there
weren't many of them—perhaps 20,000 in south Florida at the time of

European contact. They did not slaughter for sport, and their way of life was sustainable without the hunting limits, pollution controls, water restrictions, and wetlands protections associated with modern eco-sensitivity. It wasn't necessarily an admirable lifestyle—Calusa chiefs performed human sacrifices, married their sisters, kidnapped additional wives from conquered villages, and murdered subjects who tried to snoop around their secret meetings with gods—but there is no evidence that it significantly depleted the region's natural resources. "In view of the fact that they lived there for about 2,000 years, the Calusa left surprisingly little impress upon the development of the area," one historian wrote.

The Everglades was still the Everglades before white men arrived.

TWO

The Intruders

We appeal to the Great Father, who has so often promised us protection and friendship, to shield us from the wrongs his white children seem determined to inflict upon us.

—Seminole Indian chief John Hicks

A Hostile Territory

JOHN SPARKE, the British observer who described the near-nudity of Florida's Indians, was right about their deerskins and privy members. But the most illuminating parts of Sparke's account of his 1565 voyage to Florida were the parts he got wrong. He thought Florida was an island, "very scant of fresh water." He reported rumors of "a serpent with three heads and foure feet, of the bignesse of a great spaniel."

"It is thought that there are lions and tygres as well as unicornes, lions especially, if it be true that is sayd, of the enmity between them and the unicornes," Sparke wrote. He revealed the source of this ignorance in a later passage on Florida's birds: "Concerning them on the land I am not able to name them, because my abode there was so short."

This is a common thread in early European accounts of south Florida: They offered dead-on depictions of the coast, and wild speculation about the interior. One Spaniard perfectly described the "islands surrounded by swamplands" at the edge of the Gulf, then lapsed into fantasies of kangaroos, emerald deposits, and mountain ranges further inland, "the emeralds being common near the mountains." That's because he never made it past the Calusa, or the coast.

24

*   *   *

A MIDDLE-AGED SPANISH CONQUISTADOR named Juan Ponce de León was the first white intruder in Florida, and he learned the hard way that it was hostile territory.

Ponce had accompanied Christopher Columbus to the New World, where he had brutally suppressed revolts by Indian slaves and greedily exploited Indian mineworkers. He had become governor of Puerto Rico, then wrangled a charter to colonize Bimini, a Caribbean island rumored to contain fabulous wealth, as well as a magical river that restored youth. Ponce certainly hoped to find the wealth; his charter specified "gold and other metals and profitable things." He may have sought the Fountain of Youth as well; historians tend to scoff at this notion, but at a time when adventurers believed in three-headed serpents and unicorns, it's certainly possible. In any case, Ponce didn't find Bimini. In 1513, in the Easter season—*Pascua Florida*—he found what he thought was another balmy Caribbean island, and named it Florida. Ponce didn't find fabulous wealth or eternal youth, either. He found trouble.

After landing near Cape Canaveral on the Atlantic coast, Ponce sailed around the peninsula to Charlotte Harbor on the Gulf coast, where a Calusa messenger promised him gold for trade—moments before twenty canoes full of war-whooping Indians swooped in to attack his caravels. The Spaniards fought them off with artillery, and Ponce sent two Indian prisoners to the chief with a message of peace. They returned with another promise of trade—and then another ambush, this time with eighty canoes. "The natives of the land [were] a very austere and very savage and belligerous and fierce and untamed people and not accustomed to a peaceful existence nor to lay down their liberty so easily," one Spaniard wrote. This was not the bucolic island of submissive natives that Ponce had envisioned, and he left in frustration.

Ponce returned eight years later with 200 men to try to settle the area, but the Calusa welcomed him with another ambush at Charlotte Harbor. This time, they shot him in the thigh with an arrow dipped in the poisonous sap of the Everglades manchineel tree. Ponce had to retreat to Havana, where he died of his wounds. The Calusa remained untamed, and so did the Everglades.

A Doomed Marriage

THE NEXT CONQUISTADOR who tried to colonize south Florida was Pedro Menéndez de Avilés, who is best known—but still not very well known—for founding America's oldest settlement at St. Augustine in north Florida, fifty-five years before the better-publicized Pilgrims landed at Plymouth Rock. Menéndez was a fearless seaman who rose from modest origins to lead the fabled Spanish armada. He was a muscular, strange-looking man, with oversized ears, an undersized mouth, and a long, crooked nose; in a Titian portrait, his dark eyes give a hint of his fanatic intensity and his considerable self-regard.

In 1564, a group of French Lutherans set up camp in north Florida, threatening Spanish hegemony over the territory and its trade routes. It was the Age of Exploration as well as the height of the Inquisition, and Menéndez persuaded King Philip II that an influx of Spanish Catholics in Florida could serve as bulwarks against colonial rivals and Protestant heretics in the New World. In an early example of Florida boosterism, he also predicted the colony would produce sugar, wheat, cattle, silk, and "endless supplies of fruit" even more valuable than the precious metals Cortés and Pizarro had plundered from Mexico and Peru. Finally, Menéndez also hoped to find a cross-peninsula waterway linking the Atlantic to the Gulf, a shortcut to bypass the reefs, shoals, and pirates that had scuttled so many Spanish ships and sailors—including his only son—in the Florida Straits.

So Menéndez set sail for Florida with supplies that included 3,182 hundredweight of biscuits, eight church bells, and 1,758 cannonballs. Shortly after he dropped anchor in St. Augustine Bay to prepare for the climactic battle, a hurricane destroyed the French fleet on the open seas. God was clearly on Spain's side, and Menéndez showed his gratitude by slaughtering the "evil and detestable" Protestants and burning their prayerbooks. "At them!" he shouted during one rout. "God is helping! Victory! The French are killed!"

His captives begged for mercy, but Menéndez coldly replied that he was waging a war of fire and blood. He spared only twelve musicians and four repentant Catholics out of 150 prisoners, then butchered another 200 Frenchmen at an inlet the Spaniards proudly named Matanzas, for massacre. "They came and surrendered their arms to me, and I had their hands tied behind them, and put them all, excepting ten, to the knife," he told the king. His contemporary biographer praised his mercy, since "by every right he could have burnt them alive."

AFTER SUBDUING THE FRENCH in north Florida, Florida's new *adelantado* turned his attention to the Calusa in south Florida. Menéndez brought gifts to their headquarters on Mound Key near the mouth of the Caloosahatchee: silk breeches, a shirt, and a hat for the strapping young chief Carlos; gowns and mirrors for his wives. Menéndez then invited Carlos to board his galleon for a feast of fish, oysters, and wine. Resplendent in war paint, egret plumes, and a golden forehead ornament, Carlos presented Menéndez with a large bar of silver recovered from a Spanish shipwreck, and grasped his hand in friendship.

But Carlos had foolishly surrendered his manpower advantage by coming aboard with only a few guards; the Spaniards had a strong firepower advantage. Spanish soldiers with matchlit muskets quietly surrounded the Indians, and Menéndez informed Carlos that they would become great friends once he released the Spanish captives he had seized from nearby wrecks—but that he could not leave the ship until then. The captives, including the diarist Hernando d'Escalante Fontaneda, were released.

Carlos wasn't always foolish. He invited Menéndez to celebrate their new alliance in his palm-thatched banquet hall, in front of 1,000 braves and bare-breasted women, and declared that he now considered the *adelantado* his brother. He then offered Menéndez his homely older sister's hand in marriage. To be consummated immediately.

But I can only be with a Christian, Menéndez stammered.

You're my brother, Carlos replied, so we're all Christians now!

It's not so simple, Menéndez said. He began an impromptu lecture about Christian duties, but Carlos was in no mood for theology. We understand, he said: Your food and music are better than ours, and we're sure your religion is, too. But if you don't sleep with my sister right now, Carlos warned, my people will be scandalized.

It was an awkward situation: Menéndez was married to a sister of the expedition's official historian. But according to his understanding brother-in-law, his aides practically forced him to commit bigamy in order to build trust, win converts, and ensure Spanish control of south Florida: "The Ade-lantado showed much [desire] to try some other expedient, but as none

could be found, it was decided that thus it should be done." The chief's
sister was baptized as Antonia, and the new allies all partied together until
2 A.M. Antonia was "very joyful" the next morning, so Menéndez apparently did his part, too.

BUT MENÉNDEZ LEFT WITHOUT HIS NEW BRIDE, dispatching her to
Havana to receive a Christian education while he attended to famines and
mutinies elsewhere in his territory. And the next time he visited Havana, he
did not even intend to see her. To Menéndez, the marriage was pure
realpolitik, but Antonia was heartsick. She had memorized prayers and
mastered Christian doctrine, even while her entourage was dying of colonial diseases.

Menéndez realized that an unhappy Antonia endangered his mission, so
he finally showed up with fancy clothes and necklaces, fibbing that he
hadn't visited earlier because knights of his order were not allowed to sleep
with their wives for eight days after battle. Antonia was not appeased: "She
told him she wished that God might kill her, because when they landed the
Adelantado had not sent for her." That night, Antonia snuck into her
husband's room holding a candle. Please, she sobbed, let me lie in a corner
of your bed. Then my brother won't think you're laughing at me, and he'll
become a true friend of the Christians, or a Christian like me. Menéndez
laughed and sent her off with more baubles. He then sailed her back to Carlos, promising to return to build a home where they could all live together
as Christians. "She was very sorrowful because he didn't stay eight days and
sleep with her," the historian wrote.

Carlos was insulted, too, and Menéndez did not ease his hostility by
badgering him to cut his hair and bring his tribe to Christ. Carlos considered Menéndez a vital ally against the Tocobaga people in Tampa Bay
as well as his internal Calusa rivals, so he agreed. But he said he first
needed nine months to prepare his people to renounce their rituals, or
else they would revolt. Menéndez decided that Carlos could not be
trusted.

The feeling was mutual. By 1567, when Menéndez returned to south
Florida for the last time, the alliance of expedience was as fragile as his marriage of expedience. One missionary complained that Indians only attended
his lectures for the free food, and then had the impudence to question his
Christian logic: "When I showed them clearly and to their face the falsity

and deception of their idols, they threw up to me our adoration of the cross." He said Carlos remained "very much involved in his idolatries and strongly attached to his witchcrafts and superstitions," as well as his multiple wives.

Carlos did agree to accompany Menéndez on a peace mission to Tampa Bay, but they had a rather significant difference of opinion about the purpose of the voyage: As they approached the tribe's village by moonlight, Carlos proposed that they burn it down and kill all the inhabitants. Menéndez refused. Then let me do it myself, Carlos pleaded. The *adelantado* would not budge. Carlos wept with rage, and if Menéndez thought his scorned wife would appreciate his peacekeeping, he was deeply mistaken.

You have two hearts, Antonia cried. One for the Tocobaga, and one for yourself, but none for me or my brother. The marriage was over.

The father of Florida returned to Spain, where he died still clinging to his dreams of colonial grandeur. "After the salvation of my soul, there is nothing I desire more than to be in Florida," Menéndez wrote in his final letter. St. Augustine continued to flourish, but his southwest Florida colony collapsed after he left. Carlos repeatedly attacked the Spaniards, until they killed him and installed a rival on the throne. They soon murdered the rival as well, so the Calusa torched their own village and fled to the sodden interior.

By 1570, the Spaniards in south Florida were missionaries without a mission. They conceded failure and returned to north Florida. Menéndez's nephew reported to the king that south Florida was "liable to be overflowed, and of no use." The cross-Florida waterway had turned out to be as fanciful as the Fountain of Youth. And the Calusa had "a blood lust for killing Christians, for they lose no opportunity they see."

His advice: Stay away from south Florida.

FOR THE NEXT THREE CENTURIES, most white men did.

Florida became a pawn on the global chessboard, a blood-soaked outpost in wars of European conquest. But the fighting was all in north Florida. Spanish settlers fought off the British in the War of the Spanish Succession, the French in the War of the Quadruple Alliance, and the British again in the War of Jenkins' Ear. Nothing changed. Spain did surrender Florida to the British after the French and Indian War, but regained it after the American Revolution. It didn't matter. The Spaniards weren't doing much with it.

Florida was still an expendable backwater, still almost entirely in its natural state. A British visitor noted that Florida's inhabitants, "soldiers and savages excepted, would make but a thin congregation in a small parish-church." And south Florida was its most isolated outpost, the backwater of the backwater. It was obvious that the region was useless, and the Euro-peans were not inclined to try to make it useful. Once again, the Calusa had south Florida to themselves.

But not for long. The Calusa lacked immunity to measles, smallpox, and other Old World diseases, so their occasional exposure to Spanish fisher-men and traders produced virulent epidemics. Raids by British-backed Indian rivals further decimated their numbers. When Spain ceded Florida to the British in 1763, Florida's last eighty Calusa families fled to Havana. The native people of the Glades were gone.

The Wild Ones

IN MODERN SOUTH FLORIDA, where just about everyone comes from somewhere else, it turns out that even the Native Americans are out-of-state transplants. Today, the Seminole Indians and their Miccosukee relatives are known as the people of the Everglades. But they didn't start out anywhere near there. They were driven there.

Seminoles began streaming into north Florida from Georgia and Alabama during the eighteenth century, just as the Calusa were dying out. They had little in common with the Calusa. They were known as *cimar-rones*—"breakaways," or "wild ones"—because most of them split off from the Creek Confederation, and they retained their Creek traditions, wor-shipping the Breathmaker at annual Green Corn harvest ceremonies. They were farmers and traders as well as hunters and fishermen; they were also some of America's first cowboys. They visited the Everglades to hunt, but by 1800, their permanent villages only stretched as far south as Tampa Bay. When a Seminole chief issued his famous vow to remain in Florida—"Here our navel strings were first cut, and the blood from them sunk into the earth, and made the country dear to us"—he meant north Florida.

North Florida was an ideal setting for the Seminoles, a fertile extension of the Deep South. In 1791, the botanist William Bartram described a pros-perous Seminole community near Gainesville in his *Travels,* the classic Romantic narrative that introduced Florida to the world and inspired Coleridge's "Kubla Khan." *Travels* is mostly remembered for its breathless descriptions of flora and fauna, but Bartram was also enchanted by the Seminoles, a people "as blithe and free as the birds of the air." He attributed their rosy outlook to the "superabundance of the necessaries and conveniences of life" in Florida: "They seem to be free from want or desires. No cruel enemy to dread, nothing to give them disquietude, but the gradual encroachments of the white people."

Bartram significantly underestimated the disquietude those encroachments would cause. The Seminoles clashed constantly with whites along the Georgia and Alabama borders, sometimes over cattle, usually over runaway slaves who sought refuge with the tribe. Seminoles kept black slaves, but treated them more like sharecroppers; they raised their own crops, intermarried, and often became full-fledged tribesmen. For white settlers who already coveted Indian land, the threat of a savage tribe becoming a magnet for escaped slaves was an excellent excuse for an invasion. In 1811, Georgia militiamen in the so-called "Patriot Army" began crossing the border to attack the Indians in Spanish Florida, who responded with scalping raids into U.S. territory. Reports of their "uncommon cruelty and barbarism" soon made their way to General Andrew Jackson, America's boldest Indian fighter.

TO OLD HICKORY—or Sharp Knife, as he was known to many Indians after his exploits in the Creek War—the insolence of the Seminoles was unacceptable, and their residence in Spanish territory was irrelevant. In modern terms, Jackson believed Spain was harboring terrorists; he told Secretary of War John Calhoun that as long as Indians were using Florida as a sanctuary to kill American frontiersmen, the U.S. Army had a duty to "follow the marauders and punish them in their retreat." In 1818, Jackson led a scorched-earth march through north Florida designed to "chastise a savage foe, combined with a lawless band of negro brigands." He seized their herds and razed their farms, then commandeered two Spanish forts for good measure. Jackson claimed the right of national self-defense to justify his invasion, but he also described Florida as a natural extension of America, like a paw to a panther. "I view the Possession of the Floridas essential to the peace & security of the frontier, and the future welfare of our country," Jackson wrote to President James Monroe.

Jackson's intrusions sparked international and congressional furors, with critics attacking Jackson as a budding Napoleon, and dismissing Florida as a worthless prize. One congressman argued that Florida was not essential to security, welfare, or anything else: "It is a land of swamps, of quagmires, of frogs and alligators and mosquitoes! A man, sir, would not immigrate into Florida . . . no, not from hell itself!"

But most Americans shared Jackson's disdain for diplomatic niceties, and his faith that Florida was destined to join the U.S. juggernaut. The push to annex Florida was an early flowering of the doctrine that it was America's "manifest destiny to overspread the continent," famously articulated twenty-five years later to push the annexations of Texas, Oregon, and California. One senator called the Florida appendage "a natural and necessary part of our empire . . . joined to us by the hand of the Almighty." A Kentucky jour-nal declared that Florida "as naturally belong[s] to us as the county of Corn-wall does to England." America's right to expand was like the right of a tree to the air above it, or the right of a stream to its channel—a natural right, granted by God, as inevitable as the sunrise. At first, the Monroe adminis-tration tried to distance itself from Jackson's foreign adventure, but Secre-tary of State John Quincy Adams soon negotiated a treaty with Spain that ratified the facts on the ground, transferring Florida to the United States for $5 million. "This rendered it still more unavoidable that the remainder of the continent shall ultimately be ours," Adams wrote.

In 1821, Monroe appointed Jackson to be Florida's first American gov-ernor. Sharp Knife's victory seemed complete. The First Seminole War had extended the frontier to the south, and inaugurated the expansionism that would become standard American policy, outlasting the presidencies of Monroe, Adams, and Jackson himself.

The only hitch was that the Seminoles remained in Florida.

The Drive to the South

EARLY AMERICANS WERE as certain as Menéndez that God was on their side, and against native heathens. The New England colonist John Win-throp saw an Indian smallpox epidemic as proof of heavenly intervention: "God hath consumed the natives with a miraculous plagey." Ben Franklin viewed Indian alcoholism as similar evidence of "the design of Provi-dence to extirpate these savages." In the nineteenth century, these twin beliefs in white supremacy and Manifest Destiny inspired the ethnic cleansing—and often the genocide—of native tribes. For white Americans who believed that they were a chosen people and America their promised land, any fraudulent treaty or land grab could be justified by their moral obligation to overwhelm the red-skinned barbarians who stood in the path of Christian civilization. And if their lofty ambitions happened to coincide with their land hunger and greed, well, what better validation of God's plan?

In Florida, once Americans ousted the Spaniards, it never occurred to them to let the Seminoles keep their land. "To have acquired a territory of such extent, to be left in possession of these Indians, was too absurd to merit one moment's consideration," one officer explained. Whites wanted to cultivate north Florida, so the Seminoles would have to move to the unknown, unwanted south. Letting them remain would have implied that Indians had rights, that America had limits, that primitive hunter-gatherers could take precedence over civilized farmers. The next thing you know, one politician warned, "the progress of mankind is arrested, and you condemn one of the most beautiful and fertile tracts of the earth to perpetual sterility as the hunting ground of a few savages." Actually, the Seminoles were not primitive hunter-gatherers at all; they were successful farmers, which was why their decidedly unsterile land had attracted such ravenous attention in the first place. But Americans weren't about to let such inconvenient facts get in the way of God's design for man's progress. Even the Cherokees, with their literate, agrarian society and their democratic constitution, were forced west on the Trail of Tears.

The task of forcing the Seminoles toward the Everglades was assigned to James Gadsden, a tough-minded Jackson protégé who would later negotiate the Gadsden Purchase in the Southwest as a diplomat. In Florida, though, his job was dictation, not diplomacy. He informed the Indians that if they did not move south, Sharp Knife would crush them again. "The hatchet is buried; the muskets, the white men's arms, are stacked in peace," he warned. "Do you wish them to remain so?" The chief Neamathla begged Gadsden not to force the Seminoles into the Everglades, where the soil was flooded and trees were scarce: "We rely on your justice and humanity. We hope you will not send us south to a country where neither the hickory nut, the acorn, nor the persimmon grows."

The Treaty of Moultrie Creek did not force the Seminoles all the way to the sawgrass Everglades, but it did confine them to four million marshy acres in central Florida, at the fringes of the Everglades ecosystem. Only Neamathla and his subchiefs received tracts in north Florida, thinly disguised bribes to sign the treaty. For surrendering their claim to the rest of Florida, the Seminoles were promised less than a penny per acre. Gadsden candidly described the deal in a letter to Secretary Calhoun: "It is not necessary to disguise the fact to you, that the treaty effected was in a degree a treaty of imposition. The Indians would never have voluntarily assented to the terms had they not believed that we had both the power and disposition to compel obedience."

The Seminoles were now dependent on the Americans, who were stingy with rations, and on the new reservation, an agricultural wasteland. "The best of the Indian lands is worth but little," a Florida official admitted. "Nineteen-twentieths of their whole country within the present boundary is by far the poorest and most miserable region I ever beheld." The chief John Hicks warned a surveyor in the area that his mules might make a tasty meal for hungry braves, since twenty-three Seminoles had already starved to death. "I was in several of their houses & saw nothing except two or three pounds of venison & briar root soup & bread," the surveyor reported. "I am confident of Hicks' statement being true with respect to their starving situation."

Calhoun had predicted that the treaty would pacify Florida for years. But the ink had barely dried before white frontiersmen began seizing slaves from Seminoles, and famished Seminoles began plundering cattle from whites. The settlers began clamoring for the return of all blacks living with the Indians, even though the tribe had purchased many of them legally from the settlers. Florida's legislative council also passed An Act to Prevent the Indians from Roaming at Large, sentencing any Seminole caught off the reservation to thirty-nine lashes. "We were promised justice, and we want to see it!" protested a tribal spokesman named Jumper. "We have submitted to one demand after another, in the hope that they would cease, but it seems that there will be no end to them, as long as we have anything left that the white people may want!"

He was right. There were only 4,000 Seminoles in Florida, but that was 4,000 too many for Florida's settlers. And the new president, one Andrew Jackson, intended to remove them.

Sharp Knife and the Bad Bird

JACKSON HAD COMPLEX ATTITUDES toward Indians. He had raised a Creek orphan as a son, and expressed genuine concern for the red man's survival. But he also saw Indians as inferior beings, and his concern for American security always outweighed his concern for their welfare. He honestly believed that eastern tribes would be overwhelmed by land-hungry frontiersmen unless they moved west of the Mississippi River, but it was no coincidence that the frontiersmen favored his solution as well. His administration would ultimately acquire about 100 million acres of Indian land, driving about 100,000 Indians west—some with high-minded words on parchment, some with guns and bayonets.

In Florida, Jackson's push for Indian removal inspired the same legislative council that had tormented the Seminoles to petition Congress to ease their suffering—by expelling them from the state. The council's pitch brimmed with crocodile-tears compassion for the Indians: "The Treaty of 1823 deprived them of their cultivated fields and of a region of country fruitful of game, and has placed them in a wilderness where the earth yields no corn, and where even the precarious advantages of the chase are in a great measure denied them." The council neglected to mention who had deprived them of those cultivated fields, and who now occupied that fruitful country. Still, Jackson agreed to send Gadsden back to Florida, this time to force the Seminoles to move west instead of south, and to rejoin their enemies in the Creek Nation—once again, for their own good.

Gadsden persuaded several chiefs to sign treaties agreeing to leave Florida, but they were obtained by bribery, intimidation, and fraud, and the tribe renounced them. "My people cannot say they will go," Jumper told the tribe's government agent, Wiley Thompson. "If their tongues say yes, their hearts cry no, and call them liars."

JUMPER WAS ONE of the Seminoles whose tongues had said yes, before a charismatic young brave named Osceola swayed the tribe against emigration with taunts, threats, and patriotic appeals.

Osceola, a mixed-blood Alabama Creek who was driven south by Jackson's army as a child, was not an actual chief. But he had a chief's aura, with a wiry, athletic build, regal cheekbones, feminine lips, and fiery eyes. One U.S. officer, swept up in the Romanticism of the day, marveled at the "indomitable firmness" of Osceola's body, describing his "beautiful development of muscle and power" as "something of the Apollo and Hercules blended." He was a brilliant ballplayer, a superior wrestler, a natural leader. He could be charming in white society, and refused to target women or children in battle, but his resentments burned deep. "I will make the white man red with blood, then blacken him in the sun and rain, where the wolf shall smell his bones, and the buzzard live on his flesh!" he once shouted in council. The legend that Osceola furiously stabbed a treaty of removal may not be true—although there is a slash mark on the document—but he undoubtedly radicalized his fellow warriors and stiffened the spines of his elders.

Thompson, a former general in the Georgia militia, warned Osceola and his fellow Seminoles that their intransigence would not stand. He called them fools, old women, and deluded children, and told them their only choice was whether they wanted to go west over land or water. In March 1835, he read them a typically condescending message from President Jackson, their "Great White Father" in Washington:

> My children: I have never deceived, nor will I ever deceive, any of the red people. I tell you that you must go, and that you will go. Even if you had a right to stay, how could you live where you now are? You have sold all your country. You have not a piece as large as a blanket to sit down upon. What is to support yourselves, your women and your children? . . . Should you listen to the bad birds that are always flying about you, and refuse to remove, I have directed the commanding officer to remove you by force.

When many Seminoles still refused to leave, Thompson theatrically crossed their names off his list of chiefs, withheld their traditional gifts, and barred them from buying rifles or powder. He later clapped Osceola in irons, the ultimate humiliation for an Indian.

At first, Osceola tore his hair, refused to eat, and screamed like a caged beast. But then he calmed down, apologized, and signed a pledge to move west. Thompson set him free after six days, and he seemed like a new man, bantering with U.S. soldiers, even bringing in his followers to register for emigration. Thompson was so grateful that he gave Osceola a silver-plated rifle. The wildest savage had been tamed, and peaceful removal seemed imminent. "The Seminole of the present day is a different being from the warlike son of the forest when the tribe was numerous and powerful, and no trouble in the removal of the remnant of the tribe is anticipated," said the *St. Augustine Herald*. General Alexander Macomb, the commander of the U.S. Army, bragged that he had trounced tribes far more imposing than the Seminoles: "I cannot see that any danger can be apprehended from the miserable Indians who inhabit the peninsula of Florida."

The Passions of a People

OSCEOLA HAD NOT BEEN TAMED. He had only played the penitent to get out of jail.

In November 1835, Osceola and his followers surrounded a Seminole chief named Charley Emathla, who had sold his cattle to move west. When Emathla refused to renounce emigration, Osceola killed him, flinging his cattle money on top of his corpse. The Seminoles fled into the wilderness, and the frontier erupted in panic. "I have no doubt that the object of the whole body of the disaffected is to retire to the wild region of the peninsula of Florida, in the neighborhood of what is called the Everglades," Agent Thompson warned. He was right, if a bit premature.

A month later, while Thompson was smoking a cigar on an after-dinner stroll, Osceola sprang out of hiding and shot him fourteen times with his new silver-plated rifle. The same day, Major Francis Dade was leading a march in central Florida when a band of war-whooping Seminoles leapt out of the tall grass and piney woods, slaughtering Dade and all but three of his 108 men. "The passions of a people, which had been smothered for fifteen years, descending from sire to son, were let loose, and the savage massacres which had appalled the stoutest heart gave undisputed evidence of the character of the contest," an American officer wrote. The Second Seminole War, also known as the Florida War—America's longest, bloodiest, and costliest Indian conflict—was under way.

Osceola brought Thompson's dripping scalp to a drunken victory celebration, and the Seminoles took turns mocking his patronizing speeches. It no longer mattered that their region was poor and miserable, that it lacked hickory nuts and persimmons. They were tired of being pushed around, and determined to fight for their adopted homeland. In Seminole cosmology, west was the direction of death, the path of the setting sun, and Osceola was not going to let his people go voluntarily. He sent a message to General Duncan Clinch, the U.S. commander in Florida: "You have guns, and so have we. You have powder and lead, and so have we. Your men will fight, and so will ours, until the last drop of Seminole blood has moistened the dust of his hunting ground."

THIS TIME, WHITES HAD superior manpower as well as firepower; 40,000 federal regulars and state militiamen would cycle through Florida, while the Seminoles had no way of reinforcing their original 1,000 warriors. But those "miserable Indians" knew their way around the boggy peninsula, especially their Everglades hunting grounds.

For the Americans, by contrast, the Florida peninsula was still terra incognita, known mostly as a protrusion into shipping lanes. In 1835, the United States was a fast-growing nation of more than 15 million people, but fewer than 50,000 of them lived in Florida, and almost all of those pioneers lived in north Florida or the island of Key West. South Florida was still such a mystery that the military distributed maps of the state with the bottom halves blank, "so officers may make additions thereto as they may by their knowledge of the country." One army engineer scoffed at the notion of Lake Okeechobee, insisting that "there is, however, no such Lake in existence, and its position on the maps has been owing to the misapprehension of the Spanish and English geographers." A newspaper described the St. Lucie River, more than 100 miles north of modern-day Miami, as "beyond the ultimate limits of population on the Atlantic border."

The Everglades was still little more than an ugly rumor. In 1823, a British engineer named Charles Vignoles first identified an "extensive inundated Region covered with Pine and Hummock Islands of all sizes and generally called THE EVERGLADES." But Vignoles was frustrated in his own efforts to penetrate that inundated region, admitting that "the dissatisfied traveler has been sent back unable to complete the object of his mission, and confused in his effort to tread the mazes of this labyrinth of morasses." At the outset of the Florida War, U.S. naval forces faced similar confusion. Lieutenant Levi Powell led one foray into the Everglades, but his keelboats got stuck in the sawgrass. "I found it impracticable to navigate the glades," he reported. "We reluctantly commenced our return to camp."

For the Seminoles, that labyrinth of morasses *was* camp, an inaccessible refuge from their persecutors. General Zachary Taylor admitted that "the everglades may be impenetrable to the white man, while they can be penetrated by the hostiles." Taylor's exploits in Florida earned him the nickname "Old Rough and Ready," and helped launch his path to the presidency, but he spoke for many military men when he mused that if the Seminoles wanted the Everglades, they should be allowed to keep it. "I would not trade one foot of Michigan or Ohio for a square mile of Florida swamp," he wrote.

But Taylor's bosses believed America had a duty to drive the Seminoles out of Florida. They vowed to fight for every square inch of the swamp.

THREE

Quagmire

Florida is certainly the poorest country that ever two people quarreled for.

— U.S. Army surgeon Jacob Motte

Catching Water in a Net

GENERAL WINFIELD SCOTT, the big, blustery, impeccably tailored commander who replaced General Clinch a few months after the Dade massacre, figured it would take no more than a few weeks to force the Seminoles to surrender. "Old Fuss and Feathers" had studied infantry tactics in France, and he knew exactly how to fight gentlemanly armies who presented themselves for battle before opening fire. Scott arrived at the front with a large library, fine wines, and a military band that played at his meals, advertising his position to the Indians. He devised an elaborate plan to march three synchronized columns down the peninsula to surround the Seminoles in a pincer attack, tactics better suited to a game of Stratego than the swamps of Florida. "To surround what?" President Jackson snorted when he heard the plan. "The Indians? No."

Scott had been a hero in the War of 1812, and would be again in the Mexican War. But he had no idea how to fight an unconventional war in an unmapped territory that one soldier described as "swampy, hammocky, low, excessively hot, sickly and repulsive in all its features." Scott never imagined that his troops, horses, and wagons would get bogged down in the muck, or that Indians on the warpath would hit and run rather than stand and fight, scattering into the Everglades instead of clumping together in tidy formations.

Scott's columns never did converge on the Seminoles. Instead, his men spent months trudging through trackless wetlands and woodlands in shredded boots and soaked clothes, rarely encountering the enemy except when stumbling into ambushes. The Seminoles, Scott complained, were like the no-see-ums that buzzed his ears at night in Florida. Just when he thought he had his finger on them, they disappeared. "We are not inaptly compared to a prize-ox, stung by hornets, unable to avoid or catch his annoyers," one officer wrote. "Or we are justly likened to men harpooning minnows, and shooting sand pipers with artillery."

"The white man," Osceola sneered, "wants to catch water in a net for fish!"

The Second Seminole War was America's first Vietnam—a guerrilla war of attrition, fought on unfamiliar, unforgiving terrain, against an underestimated, highly motivated enemy who often retreated but never quit. Soldiers and generals hated it, and public opinion soured on it, but Washington politicians, worried that ending it would make America look weak and create a domino effect among other tribes, prolonged it for years before it sputtered to a stalemate. Of the eight commanding generals who cycled through Florida, Taylor was the only one whose reputation was enhanced, when he declared victory after a clash near Lake Okeechobee—a battle that achieved nothing except to confirm the lake's existence. Theodore Roosevelt, a historian before he was president, neatly summed up the conflict: "Our troops generally fought with great bravery, but there is very little else in the struggle, either as regards its origin or the matter in which it was carried on, to which an American can look back with any satisfaction."

At the start of the war, a disgruntled lieutenant named George McCall predicted that it would drag on for seven years and cost $50 million— 10 times what the United States paid for Florida, 100 times Jackson's original budget for driving all the eastern tribes west of the Mississippi. McCall was right about the duration, and just a shade high on price. "Millions of money has been expended to gain this most barren, sandy, swampy and good-for-nothing peninsula," wrote another lieutenant, Amos Eaton. Fifteen hundred federal regulars and hundreds of state volunteers would die in the effort to evict or exterminate every Indian in Florida; hundreds of Seminole men, women, and children would perish as well. "How vastly wide has the earth of Florida opened her grasping jaws, to swallow up human life during this Seminole War!" Eaton lamented.

Privation and Disease

WAR IS ALWAYS HELL, but Florida seemed worse. "Campaigning in Florida," a twenty-six-year-old army surgeon named Jacob Motte wrote in his journal, "was characterized by every species of privation and disease."

Motte had been educated at Harvard, where he had written essays about the nature of genius; he was a pipe smoker, a port drinker, the scion of a notable South Carolina family descended from the French aristocracy. But now he found himself "wading in morasses and swamps waist deep, exposed to noxious vapours and subject to the whims of drenching rains or the scorching sun of an almost torrid climate." Now his companions included snakes, leeches, horseflies, and mosquitoes: "It was intolerable—excruciating!"

Motte had the rare opportunity to experience the Everglades in its natural splendor, and he despised almost every minute. "It was certainly the most dreary and pandemonium-like region I ever visited; nothing but barren wastes," Motte wrote. It teemed with leeches, lizards, and other ugly, slimy creatures, not to mention the Indian "savages" that whites viewed as just another threatening species of the wild: "It is in fact a most hideous region to live in, a perfect paradise for Indians, alligators, serpents, frogs and every other kind of loathsome reptile." It looked nothing like the breathtaking natural landscapes that the Hudson River School and the other artists of the Romantic era were promoting as national symbols of America's greatness. It looked like a sloppy Kansas:

> Before us and on either side of us, the scene presented to our view was one unbroken extent of water and morass, like that of a boundless rice-field when inundated. No obstruction offered itself to the eye as it wandered o'er the interminable, dreary waste of waters, except the tops of a tall rank grass, about five feet or upwards in height, and which harmonized well with the desolate aspect of the surrounding region, exhibiting a picture of universal desolation.

The troops spent so much time slogging through wetlands in the Everglades and Big Cypress that their ankles swelled like balloons, and gruesome inflammations covered their legs. "The doctors at one time thought that the amputation of both limbs would be necessary, and it was more than two years before all the sores were healed," recalled a midshipman who spent two months around Lake Okeechobee. Hundreds of officers quit, and one colonel got so depressed he rammed his sword into his eye and through his brain. One officer suggested that anyone forced to fight in Florida should be promoted for his suffering, and given a year's leave of absence "to polish up and see the ladies."

The Americans had to drag their canoes, rifles, ammunition, and provisions through razor-edged sawgrass that ripped their clothes and sliced their skin, through muck so deep and sticky that one private dropped dead in his tracks from exhaustion. They shredded their boots on jagged pinnacle rocks that Motte compared to "a thick crop of sharply pointed knives," never knowing who or what was behind the next tree. "Every rod of the way swarmed with rattlesnakes, moccasins, and other deadly reptiles," one fighter recalled. Then there were the mosquitoes, the "swamp angels," the bane of every soldier's existence. "Their everlasting hum never ceases," one doctor reported. "One of the sailors swore that they had divided into two gangs, and that one hoisted the net, whilst the other got under and fed, and I verily do believe there were enough of them to have done it." General Alexander Webb's war diaries give a sense of the torment:

April 12: Did nothing but send off express to Fort Deynaud at 4 A.M. and mourn my existence the rest of the day. Mosquitos perfectly awful.

April 13: No peace from mosquitos. . . . Stayed up all night. . . . Mosquitos awful. 1,000,000,000 of them.

April 18: Mosquitos worse than ever. They make life a burden.

April 19: I am perfectly exhausted by the heat and eaten up by the mosquitos. . . . They are perfectly intolerable.

Motte and his fellow medical men did not realize it, because they blamed tropical disease on "swamp miasmas" and the summer "sickly season," but those mosquitoes spread malaria, dengue, and yellow fever. The U.S. troops also suffered from dysentery, tuberculosis, and a kind of collapse one officer

described as a "general sinking of the system a regular cave-in of the constitution." At one point, five battalions could not muster 100 men; after a two-
month trek through Big Cypress, 600 out of 800 troops in one unit reported
unfit for duty. Andrew Humphreys, a young lieutenant who later became one
of the most influential and wrong-headed commanding generals in the history of the Army Corps of Engineers, described the crisis in a letter home:
"My company now left with one sergeant and one corporal—both sick—the
other non-commd officers, 2 sergts and 3 corpls, will be in St. Augustine
sick!" Illness killed four times as many Americans as the Seminoles killed,
with casualty lists dominated by generic diagnoses such as "ordinary disease"
or "fever." One regiment attributed sixty-three of its seventy deaths to "disease incident to climate and service in Florida."

"Oh!" Motte wailed while in the grips of a fever of his own. "That I could
only have escaped from [this] detested soil! That I might once more live like
a human being!"

The Character of the Country

IN THE FRENZY of finger-pointing that followed his campaign, General
Scott called Florida's settlers cowards who "could see nothing but an Indian
in every bush," a jibe that got him hung in effigy in Tallahassee, while a
rival general accused Scott of "folly," "evil genius," and treason. Another
officer complained that President Jackson was "either wholly ignorant, erro-
neously informed or criminally apathetic as to the affairs of Florida," while
Jackson growled that Floridians were such "damned cowards" that they
ought to let the Indians kill them, so that their "women might get husbands
of courage, and breed up men who would defend the country." Trapped in
the White House, Jackson raged over the "unfortunate mismanagement of
all the military operations in Florida," and claimed he could have whipped
the Seminoles with a force of fifty women.

But after all the charges and countercharges, a court of inquiry pinned
the blame for Scott's quagmire on Florida itself, citing "the impervious
swamps and hammocks that abound in the country occupied by the enemy,
affording him cover and retreat at every step." This was an assessment everyone could accept. Scott complained that his men knew Florida about as well
as they knew the Labyrinth of Crete. "The most prominent cause of failure
was to be found in the face of the country, so well adapted to the guerrilla

warfare which the Indians carry on, affording ambushes and fastnesses to them, and retardation to us," another officer wrote. This retardation continued after Scott left Florida; his successor, Richard Call, lost 600 horses to starvation in a brief campaign. Call gave way to General Thomas Jesup, a renowned logistics expert who would serve forty-two years as the army's quartermaster, but was stymied in Florida by logistical snarls that left him with insufficient guns, coffee, and canoes, and beans "utterly unfit for issue."

To the men who fought there, Florida was the enemy, not the prize. In a defense of soldiers stigmatized for their service in Florida—yet another foreshadowing of Vietnam—Motte argued that their problem was morasses, not Indians:

> This is the true secret that so long retarded the victorious termination of Indian hostilities in Florida. It was the character of the country, not the want of valor or persevering energy in our army—notwithstanding the abusive comments of some civilians, who, reclining on cushioned chairs in their comfortable and secure homes, vomited forth reproaches, sneers and condemnation, wantonly assailing the characters of those who, alienated from home and kindred and all the comforts of life, were compelled to remain in this inglorious war.

THEN AGAIN, the Seminoles managed to function in Florida's morasses. Outmanned and outgunned, they repeatedly exploited their superior knowledge of the terrain, luring Americans carefully to prepared battlegrounds with prearranged escape routes, then cutting down a few of them before melting back into the wilderness.

At the Battle of Okeechobee, for example, the Seminoles waited for Taylor in a hammock protected by a half-mile-wide sawgrass marsh, after clearing a narrow path for the Americans. Taylor swallowed the bait, ordering a frontal assault through the clearing into enemy fire. The Seminoles mowed down twenty-six soldiers and wounded 112 others, then vanished before Taylor's troops arrived to seize the hammock. In private letters, Old Rough and Ready sounded quite distraught for a commander who publicly bragged that he had "routed" the enemy: "I may be permitted to say, that I experienced one of the most trying scenes of my life, and he who could have looked on it with indifference, his nerves must have been very differently organized than my own."

Taylor, who hadn't wanted to trade a foot of the Midwest for a mile of Florida swamp, had traded the blood of his men for a hammock in the middle of nowhere. He got so frustrated that he sent an aide to Cuba to buy scent-sniffing bloodhounds, who turned out to be equally inept at tracking Indians in Florida's wetlands. "Their whole object is to avoid coming in collision with us, which the situation of the country enables them to do," Taylor wrote. "It is evident to all acquainted with the country that those people can remain with impunity in the Swamps & hammocks of the everglades . . . until or after the climate draws the white man from the country." The Seminoles weren't fighting a war of extermination. All they had to do was survive.

They were good at that. They knew how to live off the land, even without hickory or persimmon. "They had no difficulty finding plenty of food anywhere and everywhere," wrote a young officer named William Tecumseh Sherman, one of 200 future Civil War generals who saw action in Florida. "Deer and wild turkey were abundant, and as for fish there was no end to them." The Indians gathered bird eggs, turtles, and "swamp cabbage," which trendy eateries now call "hearts of palm." They grew corn, squash, and pumpkins on remote tree islands. They also discovered natural remedies on the run; the Seminole names of Everglades plants included "bitter medicine," "diarrhea medicine," and "ringworm medicine." "There is no country in the world so peculiarly adapted to their wants and habits," one lieutenant wrote.

The Everglades was a rich source of sustenance, but the idea that the Indians thrived while whites suffered was another Romantic myth, fueled by stereotypical comparisons of the Seminoles to wolves, serpents, and other wild creatures. The U.S. fighters could import food and supplies from outside Florida, leave their wives and children home, retreat from the swamps in the summer, and return home themselves after hitches as short as three months. The Seminoles had nowhere else to go. They couldn't herd cattle or grow crops with any certainty on the run; they couldn't visit the coast to fish without risking attacks by naval forces; they couldn't even use their guns to hunt or set fires to cook without exposing their location. Sherman watched one Seminole woman remain mute after she was riddled with buckshot; others suffocated crying babies to avoid detection in the silence of the swamps. At parleys, Indians hid their nakedness with tattered corn bags that U.S. soldiers discarded, and scrounged corn kernels that U.S. horses dropped. Still, they refused to leave Florida.

As the Seminoles fled deeper into the Everglades, it became harder for U.S. soldiers to understand why their government wanted them to drive the Indians out of a "God-abandoned" hellscape where they were unlikely to get in anyone's way anytime soon, when their proposed reservation in Oklahoma lay directly in the path of America's westward expansion. Sherman saw the peninsula as "the Indian's paradise, of little value to us," and suggested that instead of kicking the Seminoles out, America should move the other eastern tribes in. South Florida, General Jesup concluded, was not "worth the medicines we shall expend in driving the Indians from it."

The Betrayal

JESUP, THE GENERAL whose reputation suffered the most damage in the Florida War, was also the general who best understood its folly. He described it as "a Negro war," not an Indian war, fueled by selfish white Floridians who wanted to use his troops as "negro-catchers." Jesup became convinced his nation was running a fool's errand, a "reckless waste of blood and treasure," and he was increasingly unhappy to be the errand boy. "The Indians are a persecuted race, and we are engaged in an unholy cause," he once said.

But removal was the law of the land, and few politicians were inclined to exempt the Seminoles, who had shed American blood and sheltered American slaves. The very existence of the Seminoles in Florida threatened the institution of slavery throughout the South; 800 blacks fled plantations to join the tribe, the largest slave revolt in U.S. history if considered en masse. Secretary of War Joel Poinsett warned Jesup that allowing even one Indian to remain unmolested in Florida would betray weakness and tarnish the national honor, encouraging copycat resistance from other tribes. Jesup thought it was ridiculous to worry about saving face with savages, but he soldiered on.

He tried to negotiate with the Seminoles, but he soon concluded that they could never be trusted to leave Florida voluntarily; they were only coming to parleys for the food. So in October 1837, after Osceola led seventy warriors to another parley under a white flag of truce, Jesup had them surrounded and imprisoned. Jesup was instantly pilloried in the press for his double-dealing, while Osceola was hailed as the patriotic leader of an oppressed people. "We disclaim all participation in the 'glory' of this achievement of American generalship," huffed the *Niles Register.* "If practiced towards a civilized foe, [it] would be characterized as a violation of all that is noble and generous in war." Jesup was by far the most successful U.S. general in Florida, forcing 2,000 Indians to move west, but he would spend the rest of his career writing long-winded justifications of his treachery.

Osceola, already weak with abscessed tonsils and malaria, was shipped to a South Carolina prison, where the frontier artist George Catlin painted his portrait and promoted his legend. The thirty-three-year-old warrior sighed to his prison doctor, Frederic Weedon, that the Seminole birthright had been stolen by "the strong & oppressive hand of the white people." The spirit was draining out of him, this time for real. In a scene immortalized in verse by Walt Whitman, Osceola slowly donned ostrich plumes, leggings, and a turban, as well as a war belt, silver spurs, and red war paint. With his two wives and two children at his side, he clutched his scalping knife in his right hand, lay on his back, and died with a smile. "Thus has a great savage sunk to the grave," Weedon wrote in his diary. Twenty-two towns, three counties, two lakes, two mountains, a state park, a national forest, a snake, and a turkey were later named in his honor, along with the Florida State University mascot, the mistitled Chief Osceola, who plants a flaming spear at midfield before Seminole home games.

But in a final act of white betrayal, the eccentric Dr. Weedon chopped the head off his patient's corpse, later hanging it above his children's beds as a grotesque warning for them not to misbehave. It was never recovered for a proper burial. And for all the popular outrage over Osceola's martyrdom, America kept trying to wipe out his people for four years after his death. Jesup immediately launched a new campaign to hunt the Seminoles in the heart of south Florida, hoping to terminate an interminable war.

IN JANUARY 1838, about 1,500 bedraggled American troops gathered along the Loxahatchee River at the northeast edge of the Everglades. Three hundred Seminoles awaited them in a hammock near today's Jupiter, this one protected by a half-mile-wide cypress slough that would swallow their horses to their saddle girths, much like the marsh that confounded Taylor at Lake Okeechobee. And Jesup ordered a frontal assault on the entrenched Indian position, just as Taylor had done. His artillerymen lobbed shells into the hammock, and infantrymen clawed through the slough on foot, ducking a barrage of bullets. "The Indians yelled and shrieked," Jacob Motte recalled. "The rifles cracked, and their balls whistled; the musketry rattled; the rockets whizzed; the artillery bellowed; the shells burst, and take it all in there was created no small racket for awhile."

The Americans eventually reached the heavily wooded hammock, but by then the Seminoles had retreated to the other side of the Loxahatchee, where they had prepared a second position by cutting notches for their rifles in cypress trees overlooking the river. The Americans were exposed again, and the Seminoles pumped more volleys into their lines; one musket ball shattered Jesup's glasses, slicing open his cheek. Finally, the fearless Colonel William Harney and his dragoons "plunged into the swift torrent, and crossed in the face of a shower of balls which whistled about them." Motte memorably described the Seminoles' response: "They immediately absquatulated."

Jesup claimed victory, just as Taylor had done, bragging that he had "met, beat and dispersed the enemy." But again, it was hard to see what he had won, or why dispersing an enemy he was trying to kill or capture was a good thing. Seven Americans were killed, and thirty-one wounded. "There before us lay death in his most horrible forms; bodies pierced with ghastly wounds, and locks begrimed with gore," Motte wrote. The Americans had taken the field, but again it was not a field they wanted. One Indian had been killed, but the rest had fled to fight another day. Jesup realized that complete removal was a lost cause, and said so in a private protest to Secretary Poinsett:

> In regards to the Seminoles, we have committed the error of attempting to remove them when their lands were not required for agricultural purposes; when they were not in the way of the white inhabitants; and when the greater portion of their country was an unexplored wilderness, of the interior of which we were as ignorant as of the interior of China. We exhibit, in the present contest, the first instance, perhaps, since the commencement of authentic history, of a nation employing an army to explore a country, for we can do little more than explore it.

The War of Exploration

IN THE SUMMER OF 1839, General Macomb cut the deal that Jesup had advocated all along, leaving much of southwest Florida to the Seminoles in exchange for peace. Colonel Harney was assigned to fortify a trading post for a new reservation on the Caloosahatchee River.

William Harney was another Jackson protégé, profane, cocky, and tough as rock. Secretary of War Jefferson Davis called him "physically the finest specimen of man I ever saw," with a muscular chest and astonishing athleticism: "Had he lived in the time of Homer, he would have robbed Achilles of his sobriquet of 'swift-footed.'" Harney could also be stubborn as rock; he was court-martialed twice for insubordination—once for refusing to use a military tailor, once for refusing to drill while recovering from gonorrhea. And he had some anger issues. He once flogged a female servant to death with a rawhide, and beat a dog that had the temerity to cross his vegetable garden. When a private complained that if he were a captain, Harney wouldn't treat him so harshly, Harney told him to consider himself promoted, then thrashed him to a pulp.

But Harney saw no point in fighting the Seminoles. On a stop for provisions at the twelve-acre island of Indian Key, he told a local horticulturalist named Henry Perrine he would go to the Caloosahatchee alone with his hands tied behind his back if that would prove his good faith. "Harney, they are treacherous rascals," Perrine replied. "Don't trust them too much." Harney should have listened. One night after he went to sleep, 160 Indians led by the hulking warrior Chakaika overwhelmed his contingent of thirty soldiers, scalping many of them under their mosquito bars. Harney escaped in his underwear, blackened his face with mud, and raced barefoot through the woods to safety. He hobbled back that night to find half his men dead, some with their entrails ripped out. Nothing had ever infuriated him like the Caloosahatchee double-cross, the duplicity of Indians who had just assured him they were satisfied with Macomb's treaty. "There must be no more talking—they must be hunted down as so many wild beasts," he raged. "Let every one taken be hung up in the woods to inspire terror in the rest!"

BEFORE HARNEY COULD hunt him down, Chakaika led his band across Florida Bay for another raid, this one on Indian Key. Dr. Perrine was one of the first settlers killed, after ignoring his own advice and trying to reason with the Indians. But first he managed to hustle his wife and children out a trapdoor in his cellar. They hid beneath a wharf for hours while the Indians pillaged their property, carrying off trunks, burning books, devouring fresh-baked bread and pies. "After their repast was over they would take first one pile of dishes and then another & throw them upon the floor breaking them to pieces, & they would dance & whoop!" Perrine's daughter wrote. One brave turned their way after hearing a splash, but saw only turtles. The Perrines were forced to dash out of hiding after the Indians set their house on fire, but they miraculously stumbled across a Seminole canoe and paddled it to safety, eluding shots the Indians fired at them across the bay.

It was a cinematic escape, but Chakaika's braves killed thirteen pioneers before retreating to the Everglades. It was becoming clear that the Americans could no longer allow the Seminoles to choose their battlegrounds. It was time to take the fight to the Indians, to flush them out of their hideaways. If the Seminoles can penetrate the Everglades, Harney vowed, so could he.

In December 1840, Harney and a black guide led ninety men in sixteen canoes to hunt Chakaika, ascending the Miami River before crossing the Everglades from east to west. One private noted that its "undying growth of ever-green grass, rising about six feet above the surface of the water, and waving in the breeze, gives it at times the semblance of a vast green ocean." An anonymous soldier wrote in the *St. Augustine News* that the Everglades seemed "expressly intended as a retreat for the rascally Indian."

On their fifth day in the marsh, while lingering on a tree island, the Americans spotted canoes in the distance. For once they ambushed the Indians, capturing two warriors and a woman, accidentally killing another woman with one of the newfangled "revolvers" that the manufacturer Samuel Colt had given Harney to test in the swamp. The next day, the surviving widow led them to Chakaika's island, where they found the chief chopping wood. Chakaika howled and tried to flee, but the Americans shot him, scalped him, and hung his corpse from a tree. Harney's detachment then left the island and stumbled across Shark Slough, which they rode to the mangroves of the southwest coast. "We have now crossed the long fabled and unknown Everglades. . . . We have accomplished what has never been done by white man," the anonymous soldier bragged.

Harney killed or captured only two dozen Indians, but he showed the Seminoles that south Florida was no longer their safe haven, that Americans were determined to penetrate and dominate the Everglades.

*   *   *

THE U.S. TROOPS SPENT the rest of the war on search-and-destroy missions in the Everglades, but they did a lot more searching than destroying. The Everglades was still unmapped and only sporadically navigable, and there were only a few hundred Indians hiding in its vast expanses. "The commands in canoes penetrated every part of the Everglades, finding abandoned fields, villages and trails, but not an Indian or a track was seen," Lieutenant John McLaughlin wrote after three weeks in the Everglades with his 200-man Mosquito Fleet. Still, he argued, the fleet's time was well spent: "If our labors have not been rewarded with the capture of any of the enemy, they have at least gained us information of an extensive country which had never hitherto been explored, and exhibited an imposing force in the heart of a country hitherto deemed impenetrable."

As Jesup had predicted, the most important thing the U.S. military did in the Everglades was explore it, revealing a slice of Florida that had been as mysterious as the interior of China. Most of the explorers concluded the Everglades should be avoided at all costs. In letters and diaries, they denounced it as a "horridly gloomy-looking," "bewildering," "Stygian," "monstrous," "unredeemable," "diabolical," "tiresome" wasteland. "At night as we lay down the uproar around us was fearful," wrote an army captain named Abner Doubleday, who was later credited (falsely) with inventing baseball and (accurately) with firing the first Union shot in the Civil War. "Birds of all kinds were making the night hideous with discordant sounds."

But on occasion, some of the men betrayed a grudging admiration for this singular wilderness. Even Jacob Motte—when he wasn't venting about oppressive heat, poisonous critters, hellacious storms, or the ungrateful politicians who had dispatched him to the Everglades—had his moments of Romantic reverence. He was enchanted by wild ducks that seemed to blot out the sun, by mangroves covered by so many egrets they looked like cotton fields, by radiant roseate spoonbills that would "hover over our heads, looking like the leaves of a rose that had been broken and given to the streaming air." There was something to be said for the "solitary grandeur" of the Everglades, its "savage and undisturbed communion," its magical nuances of color and texture and light. The Everglades was the essence of raw nature, "sacred from the invading plough." It was the opposite of genteel civilization, and even a Charleston aristocrat could appreciate that:

Nothing, however, can be imagined more lovely and picturesque than the thousand little isolated spots, scattered in all directions over the surface of this immense sheet of water, which seemed like a placid inland sea shining under a bright sun. . . . As we threaded this maze of countless islets, studding the unbroken surface of water in loneliness and silence amid the wild romance of nature—far secluded from the haunts of civilized man and marked only by the characteristics of wildest desolation . . . we felt the most intense admiration, and gazed with a mingled emotion of delight and awe.

THE SEMINOLES NEVER DID SURRENDER. In 1842, President John Tyler finally agreed to let the last 300 Indians remain in Big Cypress Swamp and the western Everglades. "The further pursuit of these miserable beings by a large military force seems to be as injudicious as it is unavailing," Tyler told Congress.

White Floridians continued to press for the removal of the Seminoles, and Army officers continued to conduct surveys near their lands. Hostilities flared up again in 1855, when a surveying party at the edge of Big Cypress vandalized some banana plants belonging to the mercurial chief Billy Bowlegs, who retaliated with an ambush. That started the Third Seminole War, a series of skirmishes that achieved nothing, except for more accurate maps and descriptions of the Everglades. The sporadic conflict ended when Bowlegs agreed to move west with 163 supporters for $44,600 in bribes.

But 100 or so unconquered, unbribed Seminoles remained in the Everglades, and few of the Americans who had chased them there had a problem with that. "This country should be preserved for the Indians . . . and if the fleas and other vermin do not destroy them they might be left to live," declared Alexander Webb, the general with the mosquito-obsessed diary. "I could not wish them all in a worse place."

FOUR

A New Vision

Its being made susceptible of cultivation—and instead of being, as now, a waste of waters, fit only for the resort of reptiles—would be a happy epoch for Florida.

—U.S. Army General William Harney

"The Most Desirable District in the Union"

B EFORE THE SECOND SEMINOLE WAR, when south Florida was still just a blank spot on maps of the peninsula, one man had a plan for the Everglades: the U.S. consul to the Yucatán.

The consul had collected seeds of Mexico's tropical plants and trees, and he was sure they would flourish in south Florida if the Everglades could be drained of its excess water. He was an unhealthy and unlucky man—his lab assistant accidentally poisoned him with arsenic, and a Mexican soldier bayoneted him when he approached the buffet without identification at an embassy banquet—but he was a true visionary. He published a paper on the medical benefits of an obscure bark called quinine, and helped secure funds for a fledgling scientific institution called the Smithsonian. He believed just as fervently in south Florida. He knew it was considered "a sickly and sterile territory," but insisted it could produce the world's most valuable crops, "with the least possible labor, and at the least possible price." He foresaw a tropical paradise, supplying fruit, vegetables, and sugar for America's stomachs and hemp for America's ropes, while attracting invalid tourists who might otherwise winter in

France or Italy and pioneers who might otherwise settle in Cuba or Texas. "How many years have I fruitlessly labored to convince the American people that the most slandered section of their immense domains is the most desirable district in the union for the physical enjoyments of the human race?" he wrote.

The consul's labors never did bear fruit—or hemp—in south Florida. In 1838, when the war seemed to be winding down, Congress authorized him to select thirty-six square miles in the Everglades for the introduction of tropical plants. He requested a federal survey around Cape Sable at the peninsula's southern tip, the "sheltered seashore of an ever-verdant prairie in a region of ever-blooming flowers in an ever-frostless tropical Florida." But the survey never took place. That's because the consul was the horticulturalist Henry Perrine. His grandiose plans to develop the Everglades ended with his murder at Indian Key.

But Perrine's impact is still felt in south Florida. He imported the region's first mangoes and avocados. His hemp plants, *Agave sisalana Perrine,* still grow wild in the Keys. His land grant now includes the town of Perrine. And by the war's end, Perrine's lonely vision of a transformed south Florida was no longer so lonely.

THE EVERGLADES WAS STILL a distant wilderness after the Florida war. South Florida's white population had dwindled to about fifty. But the region was no longer unmapped or unexplored, and it was no longer completely undeveloped. It had received an infusion of roads, bridges, and other wartime infrastructure, with new military posts on the Kissimmee River, Lake Okeechobee, and Cape Sable, as well as the future sites of Fort Myers, Fort Lauderdale, and Miami, which was then known as Fort Dallas. And after the war, the Armed Occupation Act of 1842 offered 160 acres to any homesteader willing to settle in the region for at least five years. South Florida no longer seemed like quite so foreign a territory.

It also seemed like an agricultural diamond in the rough. Farming requires rainfall, sunlight, and soil, and south Florida had all three in abundance. It was a region of year-round sunshine and warmth, a potential winter garden that could produce crops while the rest of America's fields lay fallow. It received three times as much annual rainfall as the frigid Great Plains region that was being promoted as America's future breadbasket.

And the deep, ink-black soils that lay beneath much of the Everglades seemed too good to be true. They were some of the world's most highly organic peats, as deep as twenty feet in some areas of the marsh, the accumulation of 5,000 years of decaying vegetation. An early British visitor had proclaimed them "rich as dung."

Some of America's Indian-hunters had noticed them, too. For example, a roguish soldier in Harney's brigade named Edward Judson—who later adopted the name Ned Buntline and became the promoter of Buffalo Bill's frontier show, the father of the dime novel, and an influential nativist demagogue—described an abandoned Indian cornfield he found on an Everglades tree island as "the richest land I have ever seen." He predicted that south Florida's organic muck would one day produce sugar, cotton, coffee, and hemp, as well as pineapples, guavas, oranges, and plantains. "Florida is now, or is soon destined to be, a very important portion of our Confederacy, both in a commercial and a general view," he wrote.

There was just one problem: The spongy flatlands of the Everglades were too wet to support crops. South Florida's excess water would have to be removed.

The Spaniards had abandoned southern Florida after deeming it "liable to be overflowed, and of no use." But retreat was not the American way, not of a time when talking about limits on American prosperity and expansion, in the words of the future president James Buchanan, "was like talking of limiting the stars in their courses, or bridling the foaming torrent of the Niagara." The end of the Florida War coincided with the height of Manifest Destiny, and a new age of internal improvements. "America is a land of wonders, in which everything is in constant motion and every change seems an improvement," Alexis de Tocqueville observed. It was a time for spread-eagle visions of an empire of progress, for white frontiersmen to extend civilization into barren wildernesses, converting prairies, forests, and swamps into money. Inventions like the steel plow and mechanical reaper were making farms more productive, while new roads, canals, and railroads were linking farms to distant markets.

Meanwhile, the U.S. Army Corps of Engineers—which had started as a tiny engineering regiment in George Washington's revolutionary army, building fortifications at Bunker Hill—was evolving into an all-purpose public works brigade for the federal government. The Corps ran the academy at West Point, so it was America's only formally trained engineering organization; its motto was "*essayons*," French for "let us try." Congress was happy to oblige, dispatching Corps engineers to survey the West, build lighthouses, custom houses, and the first national road, and "improve" rivers and harbors for navigation; in Louisiana, for example, a snagboat invented by a Corps captain named Henry Shreve cleared a longstanding logjam on the Red River, allowing steamboat traffic to reach a new inland port called Shreveport. The Corps was the bureaucratic embodiment of the nation's land ethic, refusing to accept natural conditions as inevitable, determined to conquer the wilderness to serve people. Many members of Congress believed that man's control of nature was God's work on earth, justifying American expansion across the continent. During the debate over U.S. annexation of Oregon, John Quincy Adams asked the House clerk to read aloud God's command to mankind: "Be fruitful and multiply, and replenish the earth, and subdue it; and have dominion over the fish of the sea, and over the fowl of the air, and over every living thing that moveth upon the earth."

"That," Adams said, "is the foundation not only of our title to the territory of Oregon, but the foundation of all human title to all human possessions."

SO IF FLORIDA was of no use, Americans would make it useful.

As soon as Americans began to see the Everglades, they began to fantasize about getting rid of its surplus water. "Could it be drained by deepening the natural outlets?" one explorer wondered. "Would it not open to cultivation immense tracts of rich vegetable soil?" In 1845, after Florida became the twenty-seventh state, one of its legislature's first acts was to urge Congress to "survey the Everglades, with a view to their reclamation." The resolution acknowledged that the Everglades had always been considered "wholly valueless," because draining it had always been considered impractical. But attitudes had changed:

Recent information derived from the most respectable sources has induced the belief, which is daily strengthening, that these opinions are without foundation, and, on the contrary, that at a comparatively small expense the aforesaid region can be entirely reclaimed, thus opening to the habitation of man an immense and hitherto unexplored domain perhaps not surpassed in fertility and every natural advantage by any other on the globe.

Most of those "respectable sources" were Seminole War officers. During their grueling treks through the Everglades, they had been struck by the swamp's potential for drainage and improvement.

"I entertain no doubt of the practicability of the measure," General Thomas Jesup wrote in a hastily scribbled letter to Florida senator James Westcott.

"The results of such a work as this are beyond mere speculation," Lieutenant Levi Powell wrote in his own missive.

"I do not know of a project that I regard as more calculated to benefit the country than this," General William Harney agreed.

Draining the Tub

JAMES K. POLK dedicated his presidency to the ideals of Manifest Destiny, adding a million square miles to the Union in his term. By the time Florida requested a survey of the Everglades, the Polk administration was already fighting a war over Texas, plotting the annexation of California, and negotiating the partition of Oregon. Florida was already part of America, and the Everglades was so wet that it could not even be surveyed; one man who had tried had concluded that his effort "might as well have been expended in surveying the moon." So reclaiming south Florida's swamp was not a high administration priority.

Still, Senator Westcott pestered Polk's aides to study Everglades drainage, suggesting that three million acres could be reclaimed for just $250,000. Draining the Everglades, he argued, would unlock America's most productive land for rice, sugar, and tropical fruits, and attract an army of pioneers to a dangerously underpopulated region at the edge of the nation's defenses. Wasn't Manifest Destiny supposed to extend American civilization from sea to shining sea? The administration eventually relented, appointing a Harvard-educated lawyer named Buckingham Smith to investigate the Everglades.

The rumpled, scholarly Smith was not an obvious choice to explore the River of Grass. He was not an engineer, surveyor, or scientist; he was a lawyer, politician, citrus grower, and historian, an accomplished but eccentric aristocrat, a slaveowner who set up a foundation to benefit St. Augustine's blacks after his death. He had served as the speaker of Florida's territorial legislature, and as an American diplomat in Mexico and Spain, but he is best remembered as the fastidious antiquarian who practically discovered Florida's early history, unearthing and translating twenty-five volumes' worth of documents in Madrid's archives, including Fontaneda's memoirs.

Smith had never seen the Everglades before his five-week examination of the swamp, but he described its subtle charms with a lyricism rarely found in Treasury Department documents:

> Imagine a vast lake of fresh water, extending in every direction, from shore to shore, beyond the reach of human vision; ordinarily unruffled by a ripple on its surface, studded with thousands of islands of various sizes, from one-fourth of an acre to hundreds of acres in area, and generally covered with dense thickets of shrubbery and vines. . . . Lilies and other aquatic flowers of every variety and hue are seen on every side, in pleasant contrast to the sawgrass; and, as you draw near an island, the beauty of the scene is increased by the rich foliage and blooming flowers of the wild myrtle and the honeysuckle. . . . The profound and wild solitude of the place, the solemn silence that pervades it . . . add to awakened and excited curiosity feelings bordering on awe.

But Smith had no interest in preserving this awe-inspiring wild landscape. To the contrary, he wanted it destroyed as soon as possible: "The first and most abiding impression is the *utter worthlessness* to civilized man, in its present condition, *of the entire region*." Smith's goal, America's goal, was to get rid of all that unruffled water, to convert a wasteland "as useless as the deserts of Africa" into something useful for mankind, to expand American settlement to the edge of the continent. "The statesman whose exertions shall cause the millions of acres they contain, now worse than worthless, to teem with the products of agricultural industry . . . will merit a high place in public favor, not only with his own generation, but with posterity," Smith wrote.

SMITH MADE A METICULOUS CASE for draining the swamp, backed up by the Seminole War officers, and a meticulous case for how to go about it. His report provided much of the intellectual framework for south Florida's next century. And as hydrological treatises go, its key points were simple.

The first point seems blindingly obvious today, but in 1848 it seemed as radical as the revolutionary movements that were sweeping across Europe. The elevation of the Everglades, Smith declared, was higher than the elevation of the Gulf or the Atlantic. That meant the Everglades could be drained, because water flows downhill. Smith did not know the precise elevation of the Everglades, but he was sure it was at least a few feet higher than sea level. The Miami River, which spilled over the Atlantic ridge in rapids so strong that canoes had to be portaged upstream, was the most striking evidence. The other transverse glades also seemed to drop into Biscayne Bay. "The elevation of the Ever Glades . . . proved the capability of their being drained," wrote James Gadsden, who had visited the area after the negotiations at Moultrie Creek. "And the number of small rivers and creeks, which at seasons relieved the overflowing of the interior basin of Florida, showed that by deepening these natural outlets . . . the whole country, at times submerged, might be reclaimed and brought into profitable cultivation."

This was the second point, which literally flowed from the first: The way to start draining the Everglades was to draw more water down its natural outlets—by deepening them, widening them, and extending them further into the marsh. Just bust open a few more gaps in the already porous Atlantic ridge, so that the transverse glades could flow unimpeded out to the coast, and the Everglades would drain like an unplugged tub. Powell suggested that God had created the transverse Glades for the express purpose of showing man how to drain the swamp. "A bountiful Providence has already pointed out the way, and has partially accomplished it," he wrote. "The surplus waters of the great lake have, at several points, worn down the narrow rocky girdle and opened a deep and ample channel beyond it to the sea. We have only to follow up the work, and break down the barrier to the proper level at these natural outlets, to empty out the basin."

But even if all the plugs were pounded out of the ridge, Smith doubted the little transverse glades could whisk enough water out of the Everglades to drain the entire swamp, even if they were all widened and deepened. The basin was too big, and there was too much water spilling into it from Lake Okeechobee and the sky. Smith believed the key to drainage would be lowering the lake, to prevent it from overflowing into the Everglades in the first place. He proposed one canal east from the lake to the St. Lucie River, and a shorter canal west from the lake to the Caloosahatchee River, following the path of a Calusa canoe path. (Smith wrongly attributed the silted-in ditch to the Spanish, calling it "too considerable to have been undertaken by the Indians of Florida.") His correspondents proposed additional outlet canals from the lake southeast to the Miami River, and southwest to Shark Slough. "That such work would reclaim millions of acres of highly valuable lands, now utterly valueless because incapable of use, I have no doubt," Harney wrote.

The final point of Smith's report was that draining the Everglades would achieve spectacular benefits for minimal costs. Smith estimated a maximum price of $300,000, but to be safe, he said $500,000 would "beyond question, defray all outlay necessary for the successful accomplishment of the work." It was impossible to say exactly how much swampland would be sucked dry, but Smith expected to reduce water levels in Lake Okeechobee and the Everglades by five feet, reclaiming one million acres south of the lake and hundreds of thousands more in the Kissimmee River floodplain. Harney was even more optimistic, and was so entranced by the commercial prospects of the Everglades that he began planning his own tropical fruit and sugar plantation in Cape Sable, following Henry Perrine's example. He was sure that such fertile lands in such a desirable climate would be "the best sugar land in the south," while Jesup predicted sugar plantations "as valuable as any in the world." A drained south Florida could produce just about anything capable of growing anywhere, while offering the only hope of reducing America's dependence on the West Indies for high-priced lemons, limes, and pineapples. "That the results must be of inestimable value to the whole confederacy will be so clearly manifest as to render comment wholly superfluous," Smith proclaimed.

The potential bonanza extended beyond agriculture. The canals tapping the lake to the east and west could double as a cross-peninsula shipping lane, achieving the Menéndez dream of a Gulf–Atlantic connection for the steamboat era, providing the long-awaited alternative to the treacherous Florida Straits. A habitable south Florida could also develop a fishing industry, a maritime industry, and a timber industry; Smith thought it would become a hub for "the making of salt by solar evaporation." All these opportunities would attract settlers, bolstering the security of America's most exposed region, ratcheting up the pressure on Seminole holdouts to join their exiled tribesmen out west. "In less than five years that region will, I

have no doubt, have a population of a hundred thousand souls, and more,"
Harney predicted.

Smith suggested that post-drainage south Florida's white population
would reach 250,000, a five million percent increase, and that it would split
off from north Florida to form a new slave state within a decade. "To be
identified, even in a secondary position, with the commencement of an
undertaking that must be so eminently beneficial to my country is a privilege of no mean consideration," Smith wrote.

SMITH DID INCLUDE ONE VOICE of caution in his otherwise exuberant report.
Stephen Mallory, a customs official who later became a U.S. senator and the
Confederate navy secretary, knew the Everglades like his backyard in Key
West. "I have ate of its fish, drank of its waters, smelt of its snakes and alligators and waded through its mud up to my middle for weeks," he wrote. Mallory was the would-be Everglades surveyor who had scoffed that he might as
well have surveyed the moon, and he had prescient doubts about Everglades
reclamation as well: "My own impression is that large tracts of the Glades are
fully as low as the adjoining sea, and can never be drained; that some lands
around the margins may be reclaimed by drainage or by diking, but that it
will be found wholly out of the question to drain all the Ever Glades."

But while Mallory clearly was not as enthusiastic a drainage booster as
Smith or his military correspondents, he was not as pessimistic as some
historians suggest. He agreed that fruit would grow well in Everglades
hammocks, and that small farms would thrive at the swamp's edges.
Around New River, he had seen settlers growing coconuts, lemons, limes,
and coontie, while supplementing their diets with abundant fish and game.
"The most indolent man I ever knew prospered there," he recounted.

That was the original Perrine dream for South Florida: vast production
and vast profits "with the least possible labor." In the heady days of Manifest Destiny, it did not take a prophet to see that this Panglossian vision was
certain to become reality, and that it was God's vision, too. The Everglades
had the best of all possible soils in the best of all possible climates. It had
no value whatsoever in its current form—only 400 of the eight million
acres below Lake Okeechobee had been sold—but the road to reclamation
was clear. *I entertain no doubt . . . beyond mere speculation. . . . A bountiful Providence has already pointed the way. . . . I am also convinced . . . beyond question . . . so
clearly manifest as to render comment wholly superfluous.*

The drainage of the Everglades, like the annexation of Texas or the sub-jugation of the Indians, seemed as inevitable as the torrent of the Niagara. Give America five years and half a million dollars, and a soldier's hell would surely become a pioneer's paradise. Just get rid of the water, and money and people would pour into the blank spot on the map. The bugs, the snakes, the "solemn silence"—surely it would all be gone soon.

FIVE

Drainage Gets Railroaded

*Draining of the Everglades is a subject of too great magnitude
to be idly dismissed or permanently abandoned.*

— South Florida pioneer John Darling

Yulee's Road

AMERICA WAS A NATION on the move, extending its frontier, taming its wilderness. But in the words of its own governor, "the progress of Florida, if it deserves that name, has no parallel within the limits of the Union in feebleness and insignificance." Nowhere was that truer than south Florida, where the undrained Everglades seemed to mock the advance of civilization.

The Buckingham Smith report of 1848 was the first ray of hope for the backwater's backwater, a plea for progress by the U.S. government. Senator Westcott distributed 5,000 copies, and drafted legislation to give Florida eight million acres of the Everglades for reclamation. His bill imposed strict conditions to make sure the land grant was "exclusively and sacredly" devoted to drainage, to prevent state officials from using it as a "corruption fund." But Westcott's bill was unexpectedly scuttled on the Senate floor— not by senators from other states who thought it too generous, but by his fellow Florida Democrat, David Levy Yulee, who thought it too stingy. Yulee argued that Florida was entitled to all its wetlands, not just the Everglades, and said the conditions tying the state's hands "would make the grant utterly valueless." The rift in the Florida delegation spelled doom for the bill.

A powder keg of a man with bushy eyebrows, a broad forehead, and dark hair that curled around his ears, Yulee did not believe that Florida should be forced to promote the drainage of the Everglades. His belief had nothing to do with environmental concerns. He thought his state should focus instead on promoting railroads—especially his own railroad.

IF THERE WAS EVER a typical path to power in America, Yulee's wasn't it. His ancestors were driven out of Spain during the Jewish expulsion in the fifteenth century. His grandfather was grand vizier to the emperor of Morocco before being burned alive during a palace revolt. His father, Moses Levy, rebuilt his fortune in the lumber industry in the West Indies, where the future senator was born David Levy on the island of St. Thomas in 1810.

Moses later bought 60,000 acres in north Florida, where he created a utopian colony called New Pilgrimage for persecuted Jews. He became increasingly ascetic in his faith, retreating into the Talmud, dabbling in abolitionism, eventually cutting off his teenage son's allowance and tuition in order to avoid showing favoritism over other children of God. David had to drop out of boarding school, and became increasingly estranged from his father. He moved to St. Augustine, making his own way as a self-educated lawyer, defending the rights of the southern slave-holders his father hated, attacking the capitalist class his father represented. When he was only thirty-two, he was named Florida's territorial delegate to Congress, where he skillfully debated more experienced legislators who criticized the Florida War, especially those who dared to express sympathy for blacks and Seminoles. "So far from the practice of cruelty and oppression towards the Florida Indians," he thundered, "the great fault has been a too great and almost criminal kindness, moderation and forbearance!"

Levy saved his highest dudgeon for Florida's moneyed elites, declaring himself "unchangeably" opposed "from the very innermost depths of my soul" to "the spirit of monopoly" and "the supremacy of the money power":

A swarm of mercenary and greedy speculators have settled themselves upon our infant country, and by a system of corruption and deception have selfishly sought to wield the industry and resources of the whole community to their own peculiar profit. . . . Now and forever, we must decide whether we will basely yield our limbs to the chains that have been forged for us, and be content to bear with the insolence, the arrogance, the frauds and oppressions of CORPORATE PRIVILEGE.

Levy was the father of Florida statehood, persuading skeptics at home that statehood would help the territory develop without having to rely on money-grubbing corporatists. At the time, the state's only railroad was a rickety twenty-mile link between Tallahassee and St. Marks that used mules as locomotives, but Congress was awarding all new states 500,000 acres to jump-start their internal improvements. Levy argued that Florida could use those lands to finance a line from the Atlantic to the Gulf, then use the profits to pay the entire cost of state government. In an era when a railroad lawyer named Abraham Lincoln was pushing for 2.6 million acres in grants for the Illinois Central, Levy insisted that Florida's railroad must be state-owned, to keep it free of the "impositions and exactions which a private chartered monopoly would impose." Levy's case for railroad socialism eventually carried Florida into the Union. The day statehood was approved in 1845, St. Augustine's nine-pound cannon fired for David Levy, "Florida's favorite son." He soon became the first Jewish member of the U.S. Senate.

BUT LEVY DID NOT keep his faith. He changed his name to Yulee, married a politician's daughter, and became a devout Christian. His politics also changed once he tasted power.

Senator Yulee still believed a cross-Florida railroad could provide a quicker shortcut than a cross-Florida canal, and help "spread a belt of civilization across the continent through Mexico, and girdle her with American & civilized influence." But instead of a state-owned railroad, he now envisioned a Yulee-owned railroad. He chartered the Florida Railroad, and used his public office to seek federal land grants, surveys, contracts, and rights-of-way for his line. Yulee instructed one Army engineer to write a report concluding that a cross-peninsula railroad would be superior to a cross-peninsula canal, a report "which will be very useful to our company to establish in the public mind from a disinterested official source." A year later, that disinterested official source became the chief engineer of the Florida Railroad.

But Yulee still thought of his railroad as a gift to Floridians. In a letter to his wife, he made his profiteering sound like sacrifice: "You suffer the penalty of having a husband involved in an undertaking too heavy for one person, which yet, now entered upon, must be borne through." He told his six-year-old son he was fulfilling his duty to serve his fellow men, "so they can visit each other easier and get more good from what they labor to make from the earth."

Yulee also saw it as his duty to block Westcott's bill, with its land grant limited to the Everglades, and its conditions requiring the grant to be used to drain the Everglades. Instead, Yulee helped push the Swamp and Overflowed Lands Act, which transferred all federal lands deemed "wet and unfit for cultivation" to the states—still supposedly for the purpose of reclamation, but with fewer strings attached. The swamplands act eventually granted more than 20 million acres of wetlands—about 60 percent of Florida's landmass—to the state to dispose of much as it pleased. The Everglades was just part of the bounty.

NOW THAT IT OWNED much of its land, Florida began giving itself away.

In 1855, the Florida legislature passed a law approving lavish land grants to stimulate railroad construction, creating an Internal Improvement Fund that would give railroads 3,840 acres of swampland and state-backed bonds of up to $10,000 for every mile of track they laid. The law, a later investigation found, was "the product of the brain of Hon. David Yulee."

The internal improvement law wasn't exclusively about railroads. One clause did authorize the fund's board of trustees—the governor and four other elected officials—to promote the drainage of the swamp and overflowed lands. But the trustees consistently ignored proposals to dredge rivers, dig canals, and drain swamps. More than a year after the law was enacted, when the trustees were asked how they intended to handle reclamation projects, they admitted they had "not sufficiently considered the subject to form any definite ideas." Their mandate was to create railroads, which would open the interior and attract settlers, who would buy land and replenish the fund, which could perhaps be used to finance drainage ditches someday in the future. If the state had to give away swampland to attract railroads now, it seemed a small price to pay for growth and development. "The rapid enhancement of the general wealth and population certain to follow their construction would be ample recompense for the surrender of the whole fund," Yulee said. He had a pretty good idea whose wealth would be most rapidly enhanced.

* * *

AT FIRST, THE INTERNAL IMPROVEMENT FUND worked as planned. Rail-roads built tracks. The fund gave the railroads massive land grants. The rail-roads issued bonds backed by the fund, which they sold to build more tracks. And Florida finally enjoyed a mild railroad boom.

The centerpiece was "Yulee's Road," stretching from the Gulf port of Cedar Key to the Atlantic port of Fernandina, which Yulee envisioned as the new hub of the South. (The senator and his partners had secretly bought the prime real estate in both towns, at times using his slaves as collateral.) The *American Railroad Journal* predicted that the Florida Rail-road's economic impact on the South would rival the Erie Canal's up north, and might have said so even if Yulee hadn't recruited its editor as an investor.

Yulee used his chairmanship of the Naval Affairs Committee to score dredging funds for Fernandina's port, and his chairmanship of the Post Office Committee to secure lucrative mail contracts for his railroad. But the line still foundered in the financial panic of 1857. And Governor Madison Perry refused to approve its bonds, accusing the railroad of suckering the state into risky guarantees by hiding its debts and exaggerating its capital. "Railroads are useful, but State credit is a pearl above all price," he warned.

The Florida Railroad was not finished until March 1861, an inauspicious time to launch a new freight line in the South. The Union was disintegrat-ing, and Yulee and Florida had backed secession. "I remember him in the House, standing there and begging us—yes!—begging us to let Florida in as a State!" one senator fumed. "Well! We let her in, and took care of her, and fought her Indians, and now that despicable little beggar stands up in the Senate and talks about HER rights!" Yulee caught the last train out of Fernandina as federal troops occupied the town; the passenger in the seat next to him was killed by Union gunfire as the train crossed the bridge to the mainland.

For all his public rhetoric about southern prerogatives, Yulee's main con-cern during the Civil War was his railroad. He pleaded with the Confeder-ate commander, Robert E. Lee, for troops to protect his line, and filed lawsuits to prevent Lee's army from confiscating his iron and locomotives. In a letter to Florida governor John Milton—who had proclaimed death preferable to defeat, and would later keep his word by shooting himself— Yulee argued that the war was not the kind of emergency to justify suspending his property rights: "I humbly trust I may not be wanting at any time in necessary & dutiful sacrifice & contribution to the great cause in which all citizens are engaged. But I have not the right to make myself free with the property of others, nor to seek merit for a generous patriotism at another's cost." Yulee even sought an injunction to stop the Confederacy from seizing sugar from his plantation.

Yulee's effort to save his railroad was another lost cause. The Civil War left all of Florida's railroads battered and bankrupt. The Internal Improvement Fund, which had guaranteed their bonds, limped out of the war on the brink of ruin. And in the chaotic Reconstruction era, the fund's trustees drove it over the brink, fueling a land-grabbing frenzy that investigators described as "a wild run for all that was in sight." Florida nearly gave itself away completely before the wild run skidded to a halt.

The Great Giveaway

RECONSTRUCTION WAS SUPPOSED to revive the South and uplift blacks. Instead, northern carpetbaggers and southern opportunists joined forces to line their pockets and preserve white supremacy. Florida's antebellum Democratic power brokers—led by Yulee, who reestablished himself as a kingmaker after a brief stint in federal prison for treason—engineered the elections of "moderate" Republicans who agreed to defend the old order's financial interests and block radical Reconstruction in exchange for the keys to the treasury. In an era of scandal—the Tweed Ring in New York, the Crédit Mobilier on the transcontinental railroad, carpetbaggers and scalawags around the South—the Internal Improvement Fund became the "corruption fund" that Senator Westcott had feared. Businessmen eager for contracts descended on Tallahassee with suitcases full of cash, just as they were doing in President Ulysses S. Grant's Washington. If the Wright brothers had come along earlier, one official later said, the Internal Improvement board probably would have given away the air above Florida.

After the war, the fund's trustees decided to seize the state's bankrupt railroads, then sell them to raise cash to pay their bondholders. But in a classic insider deal, Yulee and his partners repurchased the Florida Railroad for a song, foisting its debts on the state while looting its landholdings, raking in windfall profits while the state was begging their bondholders to accept 20 cents on the dollar. "The manipulations have, in all cases, proved successful, and while no honest man can or will approve their course, creditors have nevertheless to do the best they can," one land agent wrote. The board sold Florida's other major railroad to a pair of crooks who made Yulee look like Honest Abe; they had just swindled North Carolina out of millions of dollars in railroad bonds, and now swindled Florida out of millions more, after plying politicians with whiskey and bribes. The two men simply pocketed the cash raised by their bonds, leaving the fund with heaps of new railroad debt and no new railroads. Florida's only remaining asset was its swamp and overflowed land.

Desperate for cash, the trustees put that land up for grabs, and no longer just for railroad schemes. They now entertained all kinds of improvement schemes, including reclamation projects. One young midwesterner, a bushy-bearded pioneer named William Henry Gleason, offered to drain the entire Everglades with canals in exchange for the right to buy up to six million acres for less than 7 cents per acre. The trustees eagerly accepted.

GLEASON WAS NOT THE ONLY wheeler-dealer who sought his fortune from the fund after the war. Others offered to drain the Caloosahatchee River and Kissimmee River floodplains, develop a colony of Alsatians, and start a coffee plantation in the Keys. But Gleason was Florida's archetypal carpetbagger, the cunning antihero of the Reconstruction era.

He got his start on the Wisconsin frontier, where he worked as an engineer and a lawyer, founded the lumber town of Eau Claire, opened a bank, became a Democratic Party functionary, and pestered the legislature for internal improvements. But he was caught using reserves from his bank to speculate in real estate, and accused of writing bad checks to cover his losses. So he fled to Pennsylvania, where he was again attacked for "irregular practices," and then to Virginia, where he profiteered during the war by selling provisions in the South.

After the war, Gleason received a federal appointment to study whether Florida should be converted into a colony for black freedmen. He soon concluded that its sun-kissed lands were far too valuable to turn over to blacks, particularly when they could be turned over to, say, William Henry Gleason. He moved his family to the future site of downtown Miami on Biscayne Bay, an uninhabited area he found "so beautiful and healthful that it must one day become the resort of the invalid, the tourist and the lover of adventure," and began plotting his takeover of south Florida.

Gleason became the Internal Improvement Fund's most aggressive suitor, proposing a multitude of reclamation and navigation projects in exchange for the right to buy swampland on the cheap. The trustees were not supposed to sell land for less than $1.25 an acre, but politics trumped law in Reconstruction Florida, and the former Democrat had the good sense to pursue a new career in Republican politics. In 1868, he was elected lieutenant governor on a ticket with Harrison Reed, another Wisconsin carpetbagger. And when Reed exhibited a few unscripted flashes of conscience—foiling a scam Gleason had concocted to buy bonds with worthless scrip, then vetoing a bill to boost politician salaries—Gleason led a movement to impeach him, declared himself governor, and set up a rival governor's office in the hotel across the street. It was a chaotic time in Florida politics. When Gleason stopped by Reed's office to retrieve his papers, a Reed loyalist shoved a revolver in his face. "Gleason wore a fine beaver hat, which went one way while he went the other, retreating in quick time to the seat of his hotel government," according to one contemporary account. After several weeks of this anarchy, a state court voided Reed's impeachment and ruled Gleason ineligible for high office on a technicality. But a defiant Gleason continued to preside over the Senate, and participated in three more efforts to impeach Reed. Gleason also seized control of Dade County's new government, and stole an election for the state assembly from a hog farmer named Pig Brown.

Meanwhile, the fund continued its fire sale, and Gleason—now calling himself Governor Gleason—continued to float new proposals: navigation canals across the peninsula and down the east coast, a rice plantation in the Everglades, a timber deal in the Panhandle. He signed up a senator, a congressman, and several trustees of the fund as business partners, and secured one pledge of 1.36 million acres from the fund in exchange for nothing in particular. One angry legislator compared Gleason to robber barons like Cornelius Vanderbilt and "political tricksters" like Boss Tweed. A newspaper mocked him in verse:

Far better for Gleason if he had remained
In Wisconsin, where he so much glory attained.
For his talents, peculiar, are out of their sphere.
There are no "wild cat" banks for his management here.

If Gleason had followed up his brassy proposals and political shenanigans with actual drainage work, he might have locked up millions of acres of land and reshaped the face of Florida. But he achieved almost nothing except a modest dredging job that created Lake Worth Inlet. And before he could cash in, the great giveaway was shut down—not by an outraged public or an honest politician, but by a rival capitalist who wanted his share of the loot.

HIS NAME WAS FRANCIS VOSE, and he ended up with a lien on half of Florida. Before the war, Vose's New York factory had supplied iron to the Florida Railroad in exchange for state-backed bonds. After the war, Vose rejected the board's offer of 20 cents on the dollar, and secured an injunction barring the board from disposing of the fund's land for anything less than the official price—and for anything other than cash—until he received full payment plus interest. Vose was recalled as "a little grasping fellow, with a heart no bigger than a mosquito's gizzard," but his greed helped stop the fund's trustees from squandering Florida's only remaining asset. When they tried to defy the injunction by awarding more land to Gleason and other insiders, a federal judge forced the fund into receivership.

David Yulee then persuaded Governor Reed to stop fighting the lawsuit—not because he felt guilty about stiffing his creditors, but because he figured the receiver would shut down the Internal Improvement Fund, protecting his railroad from new competition. He was right. Gleason lost his dubious claim to 1.36 million acres, and was even denied 3,840 acres he had actually earned—3,840 acres that would become the heart of Palm Beach. There would be no more giveaways of pristine wetlands—and no more improvements—until the lawsuit was settled. Soon the fund was so broke that its trustees had to borrow $150 to pay a lawyer.

America was building more railroads than the rest of the world combined, but Florida was once again stuck in limbo. "Our development from internal improvements was stagnant, idle and motionless, and no gleam of light seemed to penetrate the gloom," Governor William Bloxham later recalled. Until it could find a white knight to pay off its debts, Florida's wetlands would remain wet—and its growth would remain stunted.

"How Far, Far Out of the World It Seems!"

BY 1880, FLORIDA RANKED thirty-fourth of the thirty-eight states in population. It had less than one-fifth of Iowa's population and less than half of West Virginia's. It was forty-five times larger than Rhode Island but had fewer people. At a time when New York City had 1.2 million residents, Florida did not have a city with 10,000. Its 530 miles of railroads ranked thirty-sixth in the nation, and its $20 million worth of farmland ranked thirty-seventh. In a bustling industrial era of immigration, urbanization, and innovation, Florida still had 7.6 million acres of unsurveyed land; no other eastern state had any. Americans were conquering time and distance with the telegraph and telephone, the transcontinental railroad and transatlantic cable, the steam engine and suspension bridge. But Florida was still America's hinterland, supplying one-sixth of one percent of its tax revenue.

While Florida was the South's emptiest and poorest state, north Florida at least showed signs of life. Harriet Beecher Stowe, the author of *Uncle Tom's Cabin,* moved to Mandarin on the St. Johns River, and her boosterism helped attract thousands of visitors and settlers. Early Yankee snowbirds began flocking to Jacksonville, to escape frigid northern weather and foul urban air. Henry Flagler, John Rockefeller's right-hand man at Standard Oil, visited with his ailing wife in 1878, never imagining that he would become the father of modern Florida. A young riverboat captain named Napoleon Bonaparte Broward began shepherding tourists from Jacksonville to resort towns like Enterprise, Palatka, and Green Cove Springs, "The Saratoga of the St. Johns," never dreaming that he would become the father of Everglades drainage. Yulee built a luxury hotel in Fernandina, which began calling itself "The Newport of the South." Even Stowe began to resent the stampede of "idle loungers" from the North, especially the ones who snuck into her orange groves, hoping for a glimpse of an abolitionist heroine.

But south Florida was still a watery wilderness. In 1848, General Harney had predicted 100,000 new arrivals in five years; thirty-two years later, the census reported just 257 white residents in southeast Florida. Henry Sanford, Lincoln's ambassador to Belgium, promoted an eponymous town near the center of the peninsula as "The Gate City of South Florida," but Sanford was still a gate to nowhere. South Florida in 1880 was as empty as it had been when Jacob Motte's unit was chasing Seminoles, and emptier than it had been when the people of Horr's Island were gathering oysters. "How still it is here!" Iza Hardy exclaimed in *Oranges and Alligators: Sketches of South Florida Life.* "How far, far out of the world it seems!" And Hardy never even made it to the real south Florida. She described Orlando, north of the peninsula's geographical midpoint and 250 miles north of its tip, as the "extreme south."

It was an understandable mistake. Back then, hardly any whites ventured into south Florida. A few enterprising cattlemen—led by Francis Hendry, the father of Hendry County, and Jacob Summerlin, the "King of the Crackers"—grazed wild herds of wiry "scrubs" in the Kissimmee and Caloosahatchee basins, running them to Fort Myers and Punta Rassa with buckskin whips that cracked like rifle shots. Hendry founded LaBelle and helped develop Fort Myers, while Summerlin founded Bartow and helped develop Orlando. A few pioneers ventured further south, starting villages like Coconut Grove on the high ground of the Atlantic ridge, or homesteading in remote outposts like Chokoloskee on the Gulf coast. But very few. "We call this God's country, because He could not give it to anybody," a Chokoloskee pioneer wrote. The pioneers subsisted by fishing, hunting, making charcoal, salvaging shipwrecks, and growing fruits, vegetables, sugarcane, and coontie. But they had few local customers, and no way to ship their products to outside markets. When Fort Lauderdale's lighthouse keeper had to list the men within three miles of his station, he wrote: "None." In 1879, a visitor named James Henshall counted only twenty-five residences on Biscayne Bay. Even the area's most exuberant booster, William Gleason, moved north to the Indian River town of Eau Gallie, defeated by "mosquitoes and sand flies, recurring hurricanes and a depressing sense of isolation."

Life on the Everglades frontier was not for wimps. Cottonmouths were as common as doctors were scarce. Pioneers shared their shacks with two-inch roaches, horseflies whose bites felt like stab wounds, and "swamp angels" so thick they put out fires. "They are unbearable by anyone not endowed with rhinoceros hide," one settler wrote. "The incriminating mosquito, the nimble and microscopic sand fly, the familiar flea, the industrious warlike ant, with many others, each have their day—and it is frequently a long one!"

The lonesome wilderness tended to attract outcasts and outlaws, self-reliant folks who didn't care for human society. The first act of the first minister to visit Everglades City was to bury a fellow passenger who had been beaten to death on their voyage from Fort Myers. Boss Tweed passed through the Everglades on the lam, and the notorious Edgar "Bloody" Watson—a fugitive suspected of murdering several westerners, including the legendary outlaw Belle Starr—resurfaced as a sugar grower in the Ten Thousand Islands. Watson's farmhands tended to vanish around payday, and he may have killed a few fellow homesteaders as well, before he was gunned down by an impromptu firing squad of his neighbors.

THE SURVIVING SEMINOLES remained scattered around south Florida's interior. In 1880, a minister named Clay MacCauley found thirty-seven Indian families living in five camps around Lake Okeechobee, Big Cypress, and the Everglades. They were doing much better, now that they could hunt, fish, grow crops, raise livestock, and otherwise exploit their environment without fear of attack. "The Seminole, living in a perennial summer, is never at a loss when he seeks something, and something good, to eat," MacCauley observed. They cultivated 100 acres of corn, beans, and melons, raised hogs, chickens, and cattle, and shot deer, birds, and even manatees with their Kentucky rifles. They built a rudimentary mill to turn sugarcane into juice, although they only made enough to serve themselves. "They are the producers of perhaps the finest sugar cane grown in America; but they are not wise enough to make it a source of profit," MacCauley sniffed.

The Seminoles did not share the white man's obsession with profit. They kept their distance from whites, except to swap pelts and plumes for pots, beads, cloth, and ammunition. They taught their children to turn their backs on whites; women were not even allowed to look at whites. Seminoles could be severely punished for learning to read or write, or discussing tribal matters with outsiders. The tribe remained so isolated from white society that long after the Civil War, a chief named Tiger Tail brought a black slave to Fort Myers for sale. As one paper recounted: "When informed that the negroes were free he ejaculated: 'White man's nigger mebbe free, but Indian's nigger, no.' Whereupon Tiger Tail grasped the darkey by the nape of the neck, pushed him into the canoe and paddled back to the Everglades." As late as the 1920s, a group of Seminoles visiting Madison Square Garden were still so suspicious of whites that they nearly rioted after a speaker joked about shipping them to Oklahoma.

It is easy to understand the mistrust, considering the past betrayals and enduring racism. Most white accounts portrayed the Seminoles as dignified and industrious—except when their "wild natures" were "heated up by the crazing liquor"—but hostility often lurked beneath the surface. "If the native Floridian does not extend his encroachments further the Seminole will continue to live in peace and harmony with mankind, asking nothing, needing nothing," one author wrote. A nice sentiment, except that it suggested Indians were not part of "mankind," and that it was up to the natives to avoid "encroachments" on white intruders.

"Nature Reigns Here Undisturbed"

FOR THE TIME BEING, though, most whites stayed away from south Florida.

For example, Lake Okeechobee was still so inaccessible that writers spun yarns of a 170-foot-deep lake surrounded by 150-foot-tall cliffs, overrun by monkeys and gigantic spiders. "It has slept," one writer said, "in a sort of poetical fog of mystery." Only the most dedicated nature lovers tried to pierce that fog.

Kirk Munroe—a noted nature writer and the founder of the American Canoe Association—did try to explore Lake Okeechobee after he moved to Coconut Grove but nearly starved to death after getting lost in "horrible" cypress swamps and "terrible" sawgrass marshes. "Am in despair," he scribbled on his fourth day in the wilderness. "Mosquitoes and lizards abound in numbers I have never seen equaled," he wrote on day six. "God help me!" he wrote on day seven. Monroe finally abandoned his canoe and thrashed through the sawgrass, passing out just as he reached the mainland.

Angelo Heilprin, a Hungarian geologist, had a more pleasant experience around Lake Okeechobee. There were no 150-foot cliffs, no spiders the size of his head—just nature and solitude. Never before had he "so keenly appreciated the insignificance of my own humble being in the sea of life by which I was surrounded." Rushes and reeds bearded the lake's boggy perimeter, along with cypress stands dripping with Spanish moss and a pond apple jungle overgrown with flowering vines. "It would be vain to attempt to depict by word the solemn grandeur of these untrodden wilds, the dark recesses, almost untouched by the light of day, that peer forbiddingly into a wealth of boundless green—or to convey to the mind a true conception of the exuberance of vegetable life that is here presented," Heilprin wrote.

The Everglades also remained a mystery to most Americans, inevitably compared to unexplored lands such as Tibet, Timbuktu, Quintana Roo, and Antarctica. James Henshall tried to dispel some of its myths after a rare visit:

> The singular and wonderful region known as the Everglades is not, as is popularly supposed, an impenetrable swamp, exhaling an atmosphere of poisonous gases and deadly miasma, but a charming, shallow lake of great extent, with pure and limpid waters from a few inches to several feet in depth, in which grow curious water-grasses and beautiful aquatic plants; while thousands of small islands, from a few rods to a hundred acres in extent, rise from the clear waters, clothed with never-ending verdure and flowers.

In these days before drainage, Everglades fauna was still as abundant as flora. Charles Torrey Simpson, the pioneer naturalist who noted the unmatched lacerations of the Everglades, was nevertheless overwhelmed by the explosion of life he encountered on his first visit:

> Bear, deer, otter, mink, raccoons, various wildcats and the opossum were abundant, while every swamp and stream was full of alligators. Vast numbers of roseate spoonbills, snowy herons, American egrets and the great white heron . . . winged their way over the pineland as they visited the swamps for food. And food was everywhere abundant, for the waters were swarming with small fish and the lowlands contained unnumbered millions of pond snails.

The Everglades, in other words, was still the Everglades. In the late nineteenth century, with the Industrial Revolution in full swing, America was rapidly annihilating its natural landscapes—chopping down forests, wiping out buffalo, replacing wild lands with farms and cities that spewed fertilizers and sewage into rivers and streams. But a *Harper's* writer who visited south Florida observed that "nature reigns here undisturbed." He watched a loggerhead turtle lay eggs in the moonlight, a laughing gull steal a fish in midair from a pelican, a panther leap over a shrub. He saw starfish, sawfish, bald eagles, "gorgonias brilliant with iridescence," and eleven species of herons.

The Everglades ecosystem was no longer purely "natural." Its wild boars descended from hogs imported by Spanish explorers; its hemp descended from Dr. Perrine's seedlings; Gleason's ditches had transformed Lake Worth into a salt water lagoon. Seminoles set fires to attract deer, while white pioneers dug wells, cut timber, and burned buttonwoods for charcoal.

But south Florida was more natural than just about anywhere else in America. With the Vose lawsuit freezing the Internal Improvement Fund, plans to drain the Everglades had been shelved. The Everglades was still the haunt of vermin and reptiles, and it looked like it would stay that way. "The Everglades will always retain its present state of wildness, and thus furnish a safe retreat for game animals, where they will multiply and increase in spite of the advance of civilization," Henshall wrote.

Respectable Americans considered this a deplorable state of affairs, a disturbing glitch in the march of progress. The *New Orleans Times-Democrat,* a booster publication dedicated to the development of the South, sponsored an expedition to the Everglades, hoping to promote its commercial potential. But by the time its correspondents escaped this "desolate sawgrass desert," they had reached the same depressing conclusion about the Everglades: "They are nothing more than a vast and useless marsh, and such they will remain for all time to come."

PART 2

Draining the Everglades

SIX

The Reclamation of a Kingly Domain

A radical and recent change has taken place in that section of Florida.

—Florida State engineer H. S. Duval

ALMOST EVERYONE AGREED that the Everglades was a vast and use-less marsh. But not everyone agreed that it would remain that way for all time. Governor William Bloxham, for one, believed that the right capitalist could drain it, transform it, and launch a new age of growth and development for Florida. The backwater era had dragged on long enough.

In 1881, Bloxham found his man in Hamilton Disston, a thirty-six-year-old Philadelphia saw manufacturer who had just inherited his factory and his fortune. Bloxham was a southern planter, a former Confederate captain, a Democratic politician, a lifelong Floridian; Disston was a Yankee indus-trialist, a former Union volunteer, a Republican operative, a tourist who had fished a bit in Florida. But Bloxham decided to place his state's future in Disston's hands. "We want immigration and capital, come from whatever source it may," the governor explained.

Today, Disston is often recalled as a feckless failure who shot himself in his tub after squandering his fortune on an outlandish drainage scheme. But the myth of Hamilton Disston bears little resemblance to reality.

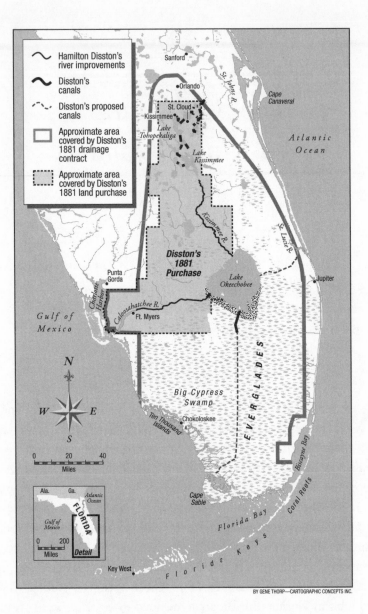

Hamilton Disston's river improvements

Disston's canals

Disston's proposed canals

Approximate area covered by Disston's 1881 drainage contract

Approximate area covered by Disston's 1881 land purchase

Sanford

Orlando

St. Cloud

Kissimmee

Lake Tohopekaliga

Lake Kissimmee

St. Johns R.

Cape Canaveral

Atlantic Ocean

Kissimmee R.

St. Lucie R.

Disston's 1881 Purchase

Punta Gorda

Charlotte Harbor

Lake Okeechobee

Jupiter

Gulf of Mexico

Caloosahatchee R.

Ft. Myers

N

W   E

S

0   20   40
Miles

Big Cypress Swamp

Ten Thousand Islands

Chokoloskee

E V E R G L A D E S

Biscayne Bay

Coral Reefs

Ala.   Ga.

Atlantic Ocean

FLORIDA

Gulf of Mexico

0   200
Miles   **Detail**

Cape Sable

Florida Bay

Florida Keys

Key West

BY GENE THORP—CARTOGRAPHIC CONCEPTS INC.

South Florida was virtually uninhabited until 1881, when Hamilton Disston signed contracts to drain as much as 12 million acres of the Everglades and buy 4 million acres outright. Disston liberated Florida from its debts while attracting a few newcomers to company towns such as Kissimmee and Fort Myers, but he didn't drain the Everglades. His dredges reclaimed some of his own land in the Kissimmee and Caloosahatchee basins, but they only began one canal south of Lake Okeechobee, and they never finished it.

Out of His Father's Shadow

HAMILTON DISSTON'S ANCESTORS were French noblemen of the D'Isney clan, but his father was a self-made millionaire.

Henry Disston was fourteen when he moved from England to America with his own father, who dropped dead just three days after they arrived in Philadelphia. Orphaned and alone in a strange country, Henry apprenticed himself to a sawmaker, and eventually built his own sawmaking empire. He endured countless setbacks and losses on his road to riches—evictions, fires, the death of his first wife in childbirth—but he never lost faith in himself or his products. He'd walk into a hardware store, ask to see a competitor's saw, then break it over the counter: "My name is Henry Disston, and here is a saw that I defy you or any other man to break with similar treatment." He opened his own steel mill at a time when American steel was considered so inferior he had to conceal its use in his saws, but soon his saws became so synonymous with quality that he had to warn customers to look out for counterfeit Disstons. And Henry created more than a business; he created a community. He built a paternalistic company town called Tacony on the outskirts of the city, providing schools, libraries, and churches for his workers, helping them buy coal and medicine, even organizing their baseball team.

Born in Philadelphia in 1844, Hamilton shared his father's restless spirit and confidence, but he could never replicate his father's up-by-his-bootstraps journey. By the time Hamilton quit school to start an apprenticeship at the saw works, Henry was already running a $500,000-a-year operation, with customers as far away as Australia. So Hamilton tried to escape his father's shadow in other ways. He joined a volunteer fire department and bolted from work to fight so many infernos that Henry threatened to sack him. He twice ran off to enlist in the Union army, and after Henry twice paid bounties to get him discharged, he organized a company of his fellow employees during the Gettysburg emergency. Henry finally let him serve, and even bankrolled the "Disston Volunteers."

But when the war was over, Hamilton returned to work as an executive in his father's factory, and immersed himself in his father's Republican Party, serving as a ward leader in Philadelphia and running Tacony politics behind the scenes. He hobnobbed with the protectionist congressman William "Pig Iron" Kelley and the future president Benjamin Harrison, and lent his yacht to Senator Matthew "Boss" Quay. The *New York Times* described Disston as a wealthy young dilettante who "amuses himself with politics," but his political interests were no lark; they invariably promoted his business interests. He founded the Protective Tariff Club to push for duties on foreign steel, and later lobbied his friends in Congress to protect his Florida sugar interests.

Disston was a fun-loving socialite with a yacht named *Mischief,* but he was also a married father of three, a Presbyterian, a Mason, a shrewd executive who knew how to separate work from play. "He can drink plenty of champagne between 11 o'clock and midnight, and be at the saw works at 7 o'clock with cool and capacious brain," one reporter noted. Disston was easy to underestimate. He looked more like a psychiatrist than an industrialist, with a bushy mustache and a gentle smile. The only hint of his inner drive was his narrow, unflinching eyes, which the reporter described as "like that of the great eagle in the cage at the Tampa Bay Hotel, that can look straight at the sun without a tear, or even a blink."

HAMILTON TOOK OVER the saw works after Henry's death in 1878 and steadily expanded production, churning out 1.4 million hacksaws and three million files a year. The firm's 2,000 workers still called him Ham, even though he now walked the factory floor in striped trousers and a morning coat, with a silk hat cocked at a rakish angle. He added political and public relations flair to a staid operation, giving President Rutherford B. Hayes a tour of the plant just a month after Henry's death. At the start, he showed Hayes a hunk of steel; forty-two minutes later, the hunk was a twenty-six-inch handsaw etched with the president's name. The exhibition had the desired effect on the press, which saluted "the extensive and world-renowned Keystone Saw and Tool Works, which has scarcely a rival in the Old World or New World."

But Ham yearned to be more than a caretaker and promoter of his father's company. He invested in a chemical firm, a Chinese railroad syndicate, Atlantic City real estate, and Wild West mines. He finally found his calling after Ambassador Sanford, another wealthy Republican, invited him to Florida to fish black bass. In the peninsula's fens and bogs, Disston saw what Buckingham Smith and his Seminole War correspondents had seen decades earlier: countless acres of fertile soil rendered worthless by water. He was sure he could remove that water, and create millions of acres of sugar fields.

For Disston, remodeling this forbidding wilderness was more than a business venture. It was a chance to create something new, and succeed on his own merits. It was also "a vast amount of fun," a diversion from the drudgery of furnace malfunctions and annealing costs. But it was a business venture, too. When Disston isn't mischaracterized as a sucker or a crook, he is often mischaracterized as a head-in-the-clouds romantic. In fact, he was a visionary capitalist. He saw the Everglades as more than an opportunity for self-actualization; he saw it as an underpriced commodity, just waiting for someone to exploit its potential.

Stupendous Schemes

IN JANUARY 1881, DISSTON cut his first deal with Governor Bloxham and the Internal Improvement Fund, agreeing to drain up to 12 million acres of the Everglades ecosystem in exchange for half the swamp and overflowed land his dredges successfully reclaimed. He planned to improve one-third of Florida's landmass, an area larger than New Jersey, Connecticut, Delaware, and Rhode Island combined. One engineering magazine declared his a challenge more daunting than any drainage project in modern Holland or ancient Rome: "If this remarkable enterprise is . . . carried to a successful conclusion, it will prove to be by far the greatest work of its kind." Pig Iron Kelley hailed Disston's "Napoleonic instinct and foresight," predicting that all Americans would benefit from his "reclamation of a more than kingly domain."

William Gleason and other Floridians with more moxie than money had cut similar deals that never amounted to much, but Disston demonstrated his seriousness by posting a bond, hiring the region's top engineers, and pledging to start dredging within six months. The *New York Times* decreed that his "stupendous scheme" had "every prospect of success," and the *Weekly Floridian* agreed: "All know the value of the lands if reclaimed and the immense benefit that would accrue to the State. Now men of capital and energy have taken hold of the matter with an earnestness that convinces us they mean to carry out the great work."

The only problem was the Vose lawsuit, which still cast a shadow over Florida, paralyzing its Internal Improvement Fund. Vose had died, but his heirs and other creditors were threatening to foreclose on their lien and seize the fund's swampland. Then Disston would have no incentive to drain the Everglades, railroads would have no incentive to build tracks, and set-tlers would have no incentive to come to Florida. The Everglades would remain a vast and useless marsh, probably for all time to come. But Bloxham believed that if the fund could be restored to solvency, Florida's pent-up development potential would explode, and pioneers would flock south. "This growing cancer, in the shape of a rapidly accumulating debt, had to be arrested," he said. "However painful the remedy, the operation had to be performed, or the disease would soon have been fatal to the entire body."

So the governor quietly visited Philadelphia, and soon announced an even bigger deal with his new Republican friend: In addition to his drainage efforts, Disston would pay the fund $1 million to buy four million acres of swampland outright, erasing its debt and freeing up the rest of its land to promote new improvements. The *Times* called the transaction "the largest purchase of land ever made by a single person in the world," and the *Floridian* exulted that it would "give to the State the most vigorous push forward on the road of progress that it has yet received." A Fort Myers telegraph operator nearly fainted after the news came over his wire. "One million dollars in good cold cash!" he gasped. "Imagine that!" His excitement was widely shared in Florida. "Both Democrats and Republicans are in ecstasies over the sale," one paper reported.

Not everyone was ecstatic. As Disston selected his land—most of it in the center and southwest of the peninsula, including most of the Kissimmee and Caloosahatchee basins—some of Governor Bloxham's enemies attacked the 25-cents-per-acre sale as a giveaway. But this was a rather flimsy complaint, considering how many acres the cash-poor fund was literally giving away; it awarded one man 98,000 acres just for surveying Disston's lands. It was true that other potential buyers were sniffing around the fund, but none of them had put up "good cold cash" before Disston came along. With Florida falling further behind its neighbors every day, and Disston offering to retire the fund's crippling debt, it's easy to see why the governor refused to waste time trolling for a slightly better deal. There was more griping six months later when Disston flipped two million acres to a British shipbuilder for a $100,000 profit, but that profit only reflected the surge in property values he had created by freeing the fund from federal custody. As Bloxham boasted, "when this great incubus of incumbrance was lifted, Florida rose up and at once bounded forward more rapidly than any other State in the Union."

That was no exaggeration. The Disston sale sparked Florida's first land boom, bursting the dam of debt that had held back immigration and capital. In the four years following the sale, Florida added 800 miles of railroad tracks; in the previous two decades, it had built fewer than 200 miles. Florida's taxable property doubled in value during Bloxham's term, and land sales to settlers increased sixfold. An estimated 150,000 tourists visited the state in the winter of 1884, prompting one newspaper to complain that New York's businessmen had all fled to Florida. "Would that Florida had a thousand Disstons interested in her future!" the *Floridian* gushed.

The Border of Southern Civilization

ONE OF THOSE NEW YORK BUSINESSMEN, the Standard Oil tycoon Henry Flagler, developed a particularly keen interest in Florida's future after honeymooning in St. Augustine with his second wife. Flagler saw that with first-rate resorts and railroads, America's oldest city could become a real Newport of the South, a winter playground for the Gilded Age leisure class. So he started building luxury hotels, including the 540-room Ponce de Leon Hotel, a Spanish-style castle that was the world's largest concrete structure. Flagler also consolidated and upgraded the area's antiquated railroads into a single east coast line, then extended it south to Daytona Beach. "The scheme has outgrown my original ideas," he admitted to a friend. And while Flagler was remodeling the Atlantic coast, a Connecticut entrepreneur named Henry Plant was building resorts and railroads on the Gulf coast, including the grand Tampa Bay Hotel with the eagle in the lobby.

Most of the development unleashed by the Disston sale was in north Florida, but Disston personally lured some settlers further south. He imported 250 New Yorkers to Orlando's Lake Conway, selling them small farms for $5 an acre. (With a dash of his father's Tacony paternalism, he only sold to families with enough cash reserves to tide them over in case of hard times, and provided experts to educate them about pioneer challenges.) He later founded the coastal resorts of Tarpon Springs and Disston City, which is now Gulfport, and sold the land that became Sarasota and Naples. He also helped a Russian immigrant named Peter Demens extend his Orange Belt Railroad to the Gulf, to a tiny village that Demens named after his hometown of St. Petersburg.

Meanwhile, Disston's drainage operations began to attract pioneers to the edges of the Everglades. The cow town of Fort Myers, the base for his dredging work on the Caloosahatchee, tripled in population almost overnight. Thomas Edison bought a winter home there, prophesizing that "there is only one Fort Myers, and one day 90 million people are going to find it out." The *Times-Democrat* correspondents who stopped in before their Everglades expedition mournfully predicted that this "little Eden in the wilderness" would soon be overrun by hordes of sun-starved northerners:

> We will always think of this little town as we first knew it, and although it may be best for its commercial interest that money and men should crowd to the wall and rob it of its present village simplicity, purity and sweetness; yet it seems to us like trampling to earth the roses which bloom before each door and putting an ax in the beautiful palm and stately coconut trees which grow and thrive on every side.

Disston also transformed a tiny trading post on Lake Tohopekaliga into his corporate headquarters of Kissimmee. Within two years, "The Tropical City" had 700 residents, several hotels, a shipyard, two sawmills churning out lumber for the construction boom, a station on Henry Plant's new South Florida Railroad, and a mayor named Rufus Rose, a steamboat captain who was now Disston's drainage superintendent. In 1883, Disston arranged a visit to Kissimmee by President Chester A. Arthur, a Republican, who spent two days fishing with Captain Rose, landing a ten-pound bass on his first cast. Kissimmee enjoyed a blast of national publicity as "the border of Southern civilization," and Disston began selling downtown lots for $100 an acre.

Disston later helped launch the nearby towns of Southport, Runnymede, Narcoossee, and St. Cloud, where he developed an 1,800-acre sugar plantation, and the fifteen-mile Sugar Belt Railroad. He also sold 7,000 acres in the vicinity for a short-lived Shaker retreat called Olive Branch. One devout Shaker, Andrew Barrett, complained that even the trembling celibates of his ascetic community were succumbing to the lure of Florida real estate. "When I see the greed of money step in and engross our whole attention, I begin to think we have forgotten the primary object of our exit into Florida," Brother Barrett wrote. "To me, this was not intended as merely a speculative scheme. . . . If God is in it, I don't believe He wants any such business."

* * *

PERHAPS NOT, but Disston did. He was the first developer to market Florida swampland on a global scale, opening real estate offices in England, Scotland, Sweden, Denmark, Germany, Italy, and throughout America, distributing maps of the peninsula dominated by lands his company owned (in red) or intended to drain (in green), peddling sun-bathed "lands of inexhaustible fertility without fertilizing," where "heavy downpours are exceptional events" and the summers were cooler than Cincinnati's or Bismarck's. "The immigrant from Europe, or settler from other states, can find no more favorable location, and the capitalist no better investment," his brochures trumpeted.

Disston promoted his domain as America's new winter playground and breadbasket, a frost-free, illness-free, bug-free paradise where 20 acres were worth 100 up north: "You secure a home in a garden spot of the country, in an equable and lovely climate, where merely to live is a pleasure, a luxury heretofore accessible only to millionaires." He depicted a cornucopia of "inexhaustibly rich lands" producing the world's finest fruits, vegetables, rice, timber, and tobacco, as well as sugarcane that would regenerate without replanting:

ATTENTION!!
FARMERS!!
Why stand you idle Six Months in the year, eating up in winter all you make in summer, and saving nothing for old age, or your children?
STOP! CONSIDER!
Have you heard of DISSTON'S purchase of 4,000,000 acres of upland in
FLORIDA
The country where you can raise crops ALL THE YEAR!

"Upland" was as audacious a stretcher as Disston's claim that Florida's downpours were exceptional events, since every one of those four million acres had been deeded to the state as low-lying wetlands, "swamp and overflowed." But Disston was determined that they would not remain overflowed for long. He had pledged to reclaim central and south Florida from its underwater limbo, and he intended to keep his pledge.

Delivering Florida from Evil

FOR CENTURIES, SOUTHERN FLORIDA had been dismissed as "liable to overflow, and of no use." But Disston and his engineers intended to end its overflows for good—by preventing the unruly Kissimmee River from overflowing its banks and soaking its floodplain, while preventing Lake Okeechobee from overflowing its southern rim and flooding the Everglades. In just a few years, they bragged, the Everglades would be "as dry as a bone." Disston's corporate prospectus described the operation as "not only a sure and safe investment, but offering probabilities of greater returns in the future than any enterprise that has been brought before the public in years."

Disston's drainage strategy was straightforward: Move the excess water in the Kissimmee valley down to Lake Okeechobee, then move the excess water in Lake Okeechobee out to sea. In the upper basin, his engineers proposed to link the Chain of Lakes with a series of canals and straighten the serpentine Kissimmee River. In the lower basin, they adopted Buckingham Smith's plan to lower Lake Okeechobee: one canal east to the St. Lucie River and out to the Atlantic, one canal west to the Caloosahatchee River and out to the Gulf, and at least one canal south through the Everglades. "Okeechobee is the point to attack," one Disston associate explained. The key to the plan was to make the outflow from the lake through the Caloosahatchee and St. Lucie Canals "equal to or greater than the inflow from the Kissimmee valley, which is the source of all the evil."

By "evil," of course, he meant "water."

The theory behind the engineering was simple: South Florida was higher than sea level, and water flows from higher elevations to lower elevations. As long as the law of gravity remained in effect, excess water would flow downhill from the Kissimmee River to Lake Okeechobee, and from Lake Okeechobee out to tide. And as long as the canal levels remained lower than the surrounding water table—which in south Florida was always near or above the surface—they would simultaneously suck water out of the surrounding marshes. This was basic hydraulics: Water would seek its own level.

Disston's chief engineer, James Kreamer, proposed to launch the attack on the lake by shunting its water east through a massive canal to the St. Lucie River, the steepest and most direct route to sea; he warned that starting with a short canal west to the lazy and level Caloosahatchee would "undoubtedly inundate" Fort Myers and its surrounding lowlands. So Kreamer designed a 21-mile-long, 220-foot-wide eastern canal with an estimated cost of about $450,000, and recommended "the immediate adoption of measures for an energetic prosecution of the work." He calculated that its current would be strong enough to scour out its own bottom, and move pebbles as large as eggs.

But Disston was too low on cash to follow Kreamer's advice. He decided to begin his attack with shorter, shallower, narrower, and cheaper canals in the Kissimmee and Caloosahatchee basins—where they could start draining millions of acres of wetlands he had purchased in those basins, and could create a steamboat channel from Kissimmee to the lake and out to the Gulf, instantly connecting his company town to the world. Disston wanted to recoup his investment as soon as possible, and he believed that less ambitious canals would provide faster results that would help him claim more land. He figured that if the drainage wasn't perfect at first, he could always dig more canals later with the proceeds from his land grants.

So Disston sent one dredge south along the chain of lakes to Lake Okeechobee, slicing off a few of the Kissimmee River's hairpin turns. A second dredge headed east along the Caloosahatchee to its headwaters, then slashed through three miles of sawgrass to connect the river to Lake O. The planned canals east to the St. Lucie and south through the Everglades were put on hold. "The groundwork is laid for a reclamation of land that will astonish the country by its fertility," the company boasted.

"The arrival of the dredge," one journalist has written, "was probably the single most important thing that ever happened to South Florida." Dredges would reshape the entire landscape. Dredges would make it possible to grow crops and build houses in wetlands, and most of south Florida was wetlands.

The dredge itself was a lumbering hunk of Industrial Age machinery, and the engineers, boilermen, and firemen who operated them were a rough and resilient bunch, often spending months in the marshes without seeing dry land or other human beings. Dredging technology had come a long way in the half century since men with oxen had dug the Erie Canal, but cutting trenches through the wilderness was still a dirty, labor-intensive job. The steam-powered, smoke-belching dredges were rickety floating factories that looked like giant Erector sets, supporting rotating chains of buckets that scooped up muck and squirted it to the side. They were so unwieldy they could only be towed on windless nights, and Disston's men often had to build jetties and dams to keep them afloat. *Harper's* described one dredge at work in the Kissimmee valley: "The huge crane swings; the timbers groan; steel and iron rattle and clang; the cough of the engine is broken by shouts of the men up to their waists in water; the anvil clinks; the sharp word of command cracks like a cow-whip; the constant stream of black ooze pours over the sluices; and as the huge iron and steel megatherium toils deep in the marsh, behind it is the clean-cut edge and levees of the new canal."

The dredgemen had to endure the same south Florida badlands that had tortured soldiers during the Seminole Wars—the same mosquitoes, cottonmouths, thunderstorms, and heat—at a time when department stores, streetcars, and electric fans were making civilization increasingly comfortable. The dredges cut less than the length of a football field a day, and the men often had to trek for miles through the marsh to run surveys or find timber they could use as fuel. A Disston engineer named Conrad Menge had to abandon his dredge for a harrowing seventeen-day canoe trip into the Everglades to take soundings for the planned canal south from Lake Okeechobee. "We had to drag our boats practically all the way," he wrote. "Our food supply gave out completely two days and a night before we reached the dredge, and I ate sawgrass buds to stave off hunger."

Ever the stoic, Menge added: "They tasted pretty good."

Menge's near-death experience did not achieve much. His crew dredged thirteen miles south from the lake through the River of Grass, but had to abandon the work after running into an underground rock ledge. So Disston would have no outlet south of the lake through the Everglades, just as he had no outlet east of the lake to the St. Lucie.

STILL, MUCK WAS flying north and west of the lake. In June 1883, less than two years after Disston started carving up the peninsula, an engineer named James Dancy reported on his progress to the fund's trustees. Despite soggy weather, Dancy marveled, the Kissimmee valley's wetlands were drying out: "To the astonishment of all, though it had rained . . . for 24 days in succession, all travelers said the streams were lower than they had ever been. Large tracts within the drainage district heretofore considered undesirable are today improved and susceptible of cultivation." Dancy reported that Lake

Tohopekaliga was four and a half feet lower than ever, and that nearby cattle were grazing in desiccated marshes that once held two feet of water. It was as if the laws of nature no longer applied: "I noticed that small lakes and water ponds, though it rained on me every day but one, heavily, did not rise as they usually do during the rainy season."

A year later, state engineer H. S. Duval filed an even rosier report, declaring a "radical" transformation of the entire region. In the Kissimmee basin, he observed, lakes without canals rose four feet during a month of nonstop rain, while lakes with canals rose less than one foot. The littoral marshes around Lake Toho were replaced by a sandy beach, while the wetlands around the Kissimmee River and Lake Okeechobee were now "a vast pasturage of dry land." In the Caloosahatchee basin west of the lake, settlers were so confident in the drainage company's work that they no longer propped up new homes on stilts: "They look to the curling clouds of the smoking dredge wafted on high as a bow of promise, pledged to exempt them in future from floods."

Dancy and Duval both passed along effusive letters from a Caloosahatchee cattleman named Mr. Frazier, who seemed to materialize whenever the company needed a salt-of-the-earth settler to vouch for its work. "During an experience of 12 years in this vicinity, I have not witnessed a heavier rainy season," Frazier wrote. "Although this is the case . . . the entire country is actually reclaimed—the cypress ponds are dry—and the beds of all the streams . . . are dry or nearly so." After touring the region with Disston's engineers, Duval and Dancy certified that the company had reclaimed nearly three million acres of swamp and overflowed land.

The sawgrass Everglades below Lake Okeechobee was not yet reclaimed, but it seemed only a matter of time. Disston had spent a mere $250,000 on drainage, less than 7 percent the cost of the new Brooklyn Bridge, and the state had already credited him with draining three Rhode Islands worth of wetlands. Disston's other Florida ventures showed great promise as well. The maiden Fort Myers–to–Kissimmee voyage took seven weeks, but soon Disston's steamers were completing the trip in thirty-six hours. His sugar plantation produced U.S.-record yields, and he expanded it to exploit a 2-cents-a-pound federal tariff; it turned out that sugar really did regenerate in south Florida without replanting, a process called "ratooning." After visiting Disston's farm at St. Cloud, the sugar king of Hawaii declared its muck soil "as rich as any that I have ever seen." The federal government's chief chemist agreed they were "superior to any other soil."

Disston was also raking in cash by selling those mucklands. The boosters from the *Times-Democrat* could not decide whether to cheer or cry after seeing Lake Toho's islands, previously worthless because of flooding, commanding exorbitant prices: "There is not one of our little party that does not envy the possessors of the beautiful islands. We are aware that there is not a pocketbook in the crowd that does not contain cash enough to have paid the original purchase price of that, which if we owned now, would be a competence for the remainder of our days, the magical work of turning cents into dollars having been but the labor of months."

This was the original Everglades vision, the Henry Perrine dream of maximum money for minimal work. Disston was making it a reality—not just for himself, but for anyone with the gumption to move his family or his money to Florida. Bloxham's gamble had paid off, and the *Floridian* crowed that the skeptics had been proven wrong:

> None but those who are fault-finders through lack of information, mental perversion or deliberate malice now deny that the Disston sale was an act of the wisest statesmanship. It is useless to ask the latter class of persons to acknowledge its wisdom, and that it opened the doors of Florida to the march of progress which is now making her great among the States. They have said white is black, and they will stick to it with childish pertinacity. They will continue their bald assertions and puerile arguments in the face of splendid facts that almost everyone acknowledges.

Death of a Dream

THEN THE CALOOSAHATCHEE BASIN started to flood again, and those facts no longer seemed so splendid. The area's settlers angrily blamed Disston, and began clamoring for him to dam his new channel to the Caloosahatchee. In 1887, after a review of Disston's drainage work, a state commission suddenly declared most of it a flop. White was now officially black. The state of Florida had joined the skeptics.

The commission did acknowledge that Disston had reclaimed some of the upper Kissimmee valley, where he had dug his widest and deepest canals, and where the steeper slope of the landscape gave those canals their greatest velocity. It also conceded that in the lower Kissimmee valley and the marshes above Lake Okeechobee, "nearly the whole of the vast prairie, extending as far as the eye could reach . . . was dry and apparently fit for cultivation." But the commission credited Mother Nature, not Disston. It claimed that Dancy and Duval had mistaken a temporary drought for permanent drainage, persuading the fund to award Disston nearly 1.2 million acres he hadn't earned. The commission said that, in reality, Disston had only reclaimed about 80,000 acres.

This revisionism has shaped Disston's image as a shady operator. But the commission—appointed amid a flurry of Disston-bashing—did not even hear the company's side of the story. "That style of tribunal, where the injured is allowed no representative, brought on the Revolutionary War," Duval complained. Dancy and Duval did rely too heavily on Disston's version of events, but they were well aware of Florida's wet and dry cycles, and their rain-soaked reports certainly did not sound like they were written during droughts. It's clear from the dueling reports and other accounts that while Dancy and Duval were too kind, the commission was too harsh.

Disston's problem wasn't shadiness; he simply failed to execute his original plan. He successfully drained the upper Kissimmee valley, conveying even more water *into* Lake Okeechobee at an even faster rate. But he barely even tried to convey water *out of* Lake Okeechobee, which was supposed to be his key point of attack. Big canals were expensive, so Disston focused on local drainage in the upper basin, quickly reclaiming land he could sell at a profit. He put off his plan for a more ambitious assault on the lake and the Everglades.

Disston had intended to lower the lake with canals to the east, west, and south, but he only finished the three-mile ditch to the west—the one Chief Engineer Kreamer had warned would "undoubtedly inundate" the Caloosahatchee valley if it were dredged first. That's exactly what happened. With much less capacity and much less slope than the Kissimmee valley canals, the Caloosahatchee canal quickly filled up with silt and sand, sometimes even flowing backward into the lake. The lake, like a trillion-gallon tub with a tiny clogged drain, continued to rise and overflow into the Everglades in summer storms. And lake water that did squeeze west through the twenty-five-foot-wide canal overwhelmed the Caloosahatchee River and sloshed into its floodplain, just as Kreamer had predicted. It was as if the huge tub was being emptied through a leaky straw.

Disston had a lot of work to do if he wanted the lake's outflow to exceed its inflow and stop its overflows. He needed to increase the capacity of his existing canals—by widening them, deepening them, and removing the silt accumulating in them—and dig the other canals in his original plan of attack. But even the hostile commission believed the plan was a good one: "While the company has not progressed as rapidly as may have been desired and expected, the progress made has been sufficient to establish beyond any reasonable doubt the practicability of the drainage scheme." Disston and the fund's trustees eventually reached an amicable compromise, which let him keep the land he had already received, and seek more once he spent an additional $200,000 on drainage.

Disston ultimately dug more than eighty miles of canals, and received 1.6 million acres from the fund. The trustees credited him with "the reclamation of vast areas of rich land and the general improvement of the drainage of the entire country," which does not sound like failure at all. But his investments were buffeted by the nationwide financial panic of 1893, the cancellation of the sugar tariff in 1894, and a pair of freezes that devastated the peninsula's groves and farms in 1895. He mortgaged his Florida holdings for $2 million, but he never did dig canals south and east of the lake, so he never did drain the Everglades.

THE STORY USUALLY ENDS like this: Ruined by the panic and his futile Florida adventures, as creditors prepared to foreclose on his mortgage, Disston blew his brains out in his bathtub in 1896. It's a dramatic story, but there's little evidence to support it.

The panic did force Disston to cut wages at the saw works, but he later rescinded the cuts, announcing that business was recovering beautifully. He did have some setbacks in Florida, but his partners shared in his financial reverses, and their creditors did not foreclose on his mortgage until four years after his death. Disston's estate was valued at $100,000, the equivalent of more than $2 million today, and he carried more than $1 million in life insurance, the nation's second-richest policy. All but one of Disston's obituaries reported that he died of heart disease in bed. So did the coroner's report, and there is no reason to doubt it. Disston exhibited no signs of depression the night he died, attending the theater with his wife, dining with the mayor of Philadelphia. He did complain of fatigue, which had never been a problem in the past.

Buckingham Smith had predicted that the first man to reclaim the Everglades would "be a hero to posterity." It never happened for Disston. The Everglades was still as wet as a waterfall when he died at fifty-one. His family had no interest in pursuing his drainage dream, so it died with him, and his Florida empire crumbled. Some of his land was sold at auction for a pittance. Some eventually ended up in the hands of his distant relatives in the D'Isney family—or, as they were known in America, the Disneys.

But Disston was not a hapless loser chasing a hopeless fantasy. He saved Florida from financial limbo, launching the development of the state. And he began the transformation of the Everglades ecosystem, reclaiming the upper Kissimmee valley. By the time of Disston's death, the area's sawgrass was overrun with prairie grass, its desiccated sloughs were littered with the decayed roots of dead aquatic plants, and its once submerged wetlands were invaded by dry-land red bugs. "These are nature's silent witnesses of a change that has been wrought in the status of the country, showing a new order of things unknown in ages past," wrote state engineer Duval.

This new order was Disston's greatest success. And his failures only reinforced the lesson of the Seminole Wars: The Everglades was a formidable enemy.

SEVEN

The Father of South Florida

*Think of pouring all that money out on a whim! But then
Henry was always bold.*

—Standard Oil president John D. Rockefeller Sr.

"OH, LORD! OH, GOD!" South Florida Railroad president James Ingraham wanted to sleep, but he couldn't stop moaning in agony. His
feet were blistered raw. His legs were jelly. And it was only the third night
of his trek across the Everglades. He still had three weeks to go.

It was the spring of 1892, and the owner of Ingraham's railroad, Henry
Plant, had asked him to survey a line from Fort Myers on the Gulf coast to
the fledgling community of Miami on the Atlantic coast. The twenty-one
men on his team were probably the first whites to cross the Everglades from
west to east since the Seminole Wars, and they were learning all too well
why even grizzled pioneers had avoided its thigh-deep muck and head-
high sawgrass. "Locomotion is extremely difficult and slow," the expedition's log noted. "The bog is fearful and it sometimes seems as though it
would be easier to stay in it than to go on."

The surveyors expected to travel five miles a day, but they averaged less
than three, and frequently got lost in twisted streams and morasses. They ate
through all their food except hominy, which had to be rationed by the
spoonful. Two men got so exhausted they had to be carried in their canoe,
and Ingraham canceled the survey to focus on reaching Miami without any
loss of life. "I was so tired I had lost interest in everything," the expedition's

compass man recalled. "I thought that we were great idiots to come into such a place when we had no wings with which to fly out."

Ingraham decided his sore feet were telling him something: The Everglades was no place to run a railroad. The idea seemed even sillier once the men arrived in Miami, which consisted of two properties on the Miami River: the widow Julia Tuttle's on the north bank and the storekeeper William Brickell's on the south. Ingraham admired Miami's setting on the coastal ridge, overlooking the turquoise waters of Biscayne Bay, and he believed its Everglades backcountry could be drained and reclaimed into "a great tract of land of almost unprecedented fertility." But who would extend a railroad to a settlement without settlers? And who would settle in a settlement without railroads?

This was the catch-22 that had kept south Florida so empty for so long. But it was about to be solved by Plant's rival, Henry Flagler, the Standard Oil baron who had launched a new career as a resort and railroad builder along Florida's east coast. Flagler believed that if he laid tracks and built settlements, people would come. And he was wealthy enough to lay tracks and build settlements wherever he wanted.

Soon Ingraham would be working for Flagler instead.

The Making of a Mogul

HENRY MORRISON FLAGLER was born poor in 1830, the son of an itinerant minister in upstate New York. Henry had to take summer jobs as a farmhand and a stable boy, and while he embraced his father's Presbyterianism, he hated his father's poverty. So he dropped out of school at fourteen and set out for Republic, Ohio, paying his passage on an Erie Canal steamer by working as a deckhand. He had only six coins in his pocket when he arrived, and kept one of them all his life as a memento of his humble origins. It made sense that his childhood talisman was money; six decades later, as one of the richest men on earth, Flagler would still struggle to remove an unused stamp from an envelope. "If I can get this stamp the rest of the way off, we shall save two cents," he told a secretary.

In Ohio, Henry began working as a store clerk for $5 a month, sleeping in an unheated room in the back of the shop, using wrapping paper as a blanket on frigid nights. He routinely declined invitations from friends so that he could work overtime, and he fastidiously put away a few cents from every paycheck. He eventually became a partner in his boss's grain business, and married his boss's daughter, Mary Harkness. And even though Flagler was a strict teetotaler who taught Sunday school and abhorred the use of liquor, he took an interest in Mary's family's whiskey distillery as well. "I had scruples about the business and gave it up," he recalled, "but not before I made $50,000." He was a man of strict compulsions, insisting on punctuality at meals, keeping a meticulous diary of his expenditures, signing his love letters "H. M. Flagler." He looked like a bit of a dandy—tall and slender, with a high forehead, a luxuriant mustache, and a taste for top hats— but he worked like a man obsessed.

Flagler got his start as an entrepreneur during the Civil War, investing in a salt-making venture in Saginaw, Michigan. It flourished during the wartime salt boom, but the industry was soon glutted by competition. Flagler went broke, and had to borrow from Mary's family to pay his workers. A failure at thirty-five, he returned to Ohio in shame. As his frustrations mounted—he invented a horseshoe, but couldn't find a manufacturer—he told a friend that if he could ever pay his debts and get $10,000 ahead, he would retire from business forever. He went back to buying and selling grain, skipping lunch to save money, dreaming of the day he could again afford a comfortable overcoat. "I trained myself in the school of self-control and self-denial," he later recalled.

Flagler soon worked off his debts and relocated to Cleveland, where he rented office space from a business acquaintance, a former grain broker named John Rockefeller. Flagler then persuaded Mary's cousin to invest in Rockefeller's new oil firm, and Rockefeller agreed to make Flagler a partner as part of the deal. Flagler never had to worry about overcoats again.

FLAGLER AND ROCKEFELLER became almost inseparable, walking to work together every morning, then home for lunch, back to work, and home again at night, constantly discussing their moneymaking plans. In the office, they passed letters back and forth until they agreed on every word. Flagler liked to say that a friendship founded on business was better than a business founded on friendship, and theirs was one of the most successful in capitalist history. Rockefeller once remarked that in thirty-five years working together, they never exchanged an unkind word.

They were not so kind to their competitors. Flagler kept a quotation on his desk that summarized the Standard Oil philosophy: "Do unto others as they would do unto you—and do it first." They began with a single refinery; a decade later, they controlled the U.S. oil industry. The Standard "octopus" became the ultimate symbol of corporate monopoly in the Gilded Age, extending its tentacles across the country.

Rockefeller was the top man, but Flagler was his indispensable right hand, creative about the big picture and obsessive about details. Despite his limited schooling, Flagler ran numbers better than most accountants, and knew contract law better than most contract lawyers. It was Flagler— who had learned the risks of unbridled competition in salt—who recognized that cooperation would be the key to oil, and he masterminded the notorious railroad rebates that helped Standard crush its competition by shipping large volumes on the cheap. Flagler also oversaw the company's brutal negotiations with refiners, threatening them with ruin if they refused to merge into the Standard fold. "If you think the perspiration don't roll off freely enough, pile the blankets on him," he once instructed an underling. And while Rockefeller became the symbol of Standard's might, incorporating the business was his partner's idea. "No, sir, I wish I'd had the brains to think of it," Rockefeller told an interviewer. "It was Henry M. Flagler."

If Flagler had a weakness, Rockefeller remarked privately, it was that he could be *too* aggressive: "He was a man of great force and determination, though perhaps he needed a restraining influence at times when his enthusiasm was roused."

"Now I Am Pleasing Myself"

IN THE 1880s, FLAGLER LOST his enthusiasm for the oil business, and scaled back his role with Standard. He did not share Rockefeller's goal of becoming the richest man on earth. And he had no patience for the muckraking journalists who made Standard a national pariah, or the politicians who hauled its executives before investigative committees. During one hearing, a Senate lawyer advised Flagler to stop evading questions. "It suits me to go elsewhere for advice, particularly as I am not paying you for it," Flagler shot back. The lawyer got the last word: "I am not paying you to rob the community. I am trying to expose your robbery!" Flagler grew weary of such abuse.

His private life also entered a new phase. Mary was an invalid, and Flagler spent all but two nights of their marriage by her side. But she died in 1881, and Flagler soon married her young nurse, Ida Alice Shourds, a former actress with flaming red hair and a temper to match. She started dragging Flagler to high-society parties and spending so much of his money on low-cut dresses and garish jewelry that he had to liquidate some of his Standard Oil stock. She wasn't educated, cultured, or mentally stable, but Flagler was smitten.

After their honeymoon in St. Augustine, Flagler became equally smitten with Florida, and began pouring his money and energy into his vision of its future. He discovered that building fancy hotels and creating new communities was a lot more fun than browbeating oil refiners, and he began to shelve some of the Presbyterian thrift that had guided his work for Rockefeller. He once compared himself to the apocryphal drunken church elder who declared that he had previously given all his days to the Lord, and was now taking one for himself.

Flagler considered his Florida projects part hobby, part philanthropy. Like Disston, he loved the idea of making an indelible mark on a virgin wilderness, and transforming a worthless wasteland into a vibrant civilization. He wanted to step out of Rockefeller's shadow, and develop a wild territory that Spaniards, Frenchmen, and Englishmen had been unable to tame. "I can make more money [in New York] in a month than I can in St. Augustine in a lifetime," he wrote. "The improvement of the place has been, and will be, to me a source of great gratification." He gave the city a hospital, a jail, a school for blacks, a city hall, and a grand Presbyterian church. Flagler also built Methodist, Catholic, and Baptist churches, although not quite so grand. His friends called his Florida investments The Hole, but he hadn't entirely forgotten the value of a dollar. "I see that you are wheeling the muck into the church lot," he chastised a contractor working on the Methodist church. "Country sand is good enough for them."

Flagler's work in Florida was part business, too, and Flagler was still a hands-on, hard-nosed businessman. If he was going to build hotels, he was going to micromanage the details down to the height of the floor joists, the shape of the fire escapes, and the designer of the stained-glass windows. (He chose a then-obscure artist named Louis Tiffany.) If he was going to run a railroad, he was going to centralize authority until he owned 9,996 of its 10,000 shares of stock. Flagler was still the kind of boss who noticed the costly blend of cement his contractors were using at the Ponce de Leon.

But the bottom line was no longer his main consideration, and he approved the cement: "I comfort myself with the reflection that 100 years hence it will be all the same to me, and the building better because of my extravagance." When an economy-minded hotel manager suggested that Flagler should fire an overpaid French chef and a top-flight orchestra, Flagler wired back: "Hire another cook and two more of the best orchestras." In Florida, Flagler wanted to create, not just accumulate. "Permanence appeals to him more strongly than to any other man I ever met," one of his engineers observed. His enthusiasm was roused, and he no longer had Rockefeller to restrain him. In fact, Rockefeller was so appalled by Flagler's gauche new wife and his ritzy new lifestyle that he never visited his old friend in Florida, even though he built a mansion near Flagler's railroad in Ormond Beach.

As he began to transform the Sunshine State, Flagler's personal life fell apart. One of his daughters had died as a toddler; now his other daughter died at thirty-three. He became estranged from his only son, a Princeton dropout who resented Flagler's efforts to spark his interest in business. Meanwhile, Ida Alice descended into madness, communing with ghosts through her Ouija board, bragging about her imaginary romance with the czar of Russia, threatening to kill Flagler over his extramarital affairs. And while Ida Alice was delusional, she was right about her husband's cheating. Flagler gave one mistress $400,000 and a Manhattan town house. He then took up with socialite Mary Lily Kenan, thirty-seven years his junior. He also became depressed, which only disgusted him. "Not a day passes but that I call myself to account for what I fear my friends may think is unmanly weakness," Flagler wrote to a friend. "I realize that mine is no exceptional case, but it is no use. . . . This is something immeasurably harder to bear than death."

Flagler tried to push aside his pain by focusing on Florida, gradually expanding his financial and emotional investment in the state. At first, he had limited his interests to St. Augustine and Jacksonville. Then he had intended to stop at Daytona Beach. He had already spent ten times more than he had planned, and south Florida was still a blank space on the map. Flagler figured he would concentrate on north Florida. But after several chilly winters, Flagler realized that north Florida's supposedly frost-free climate was not much warmer than the rest of the temperate South. When he took a trip to the real subtropics 200 miles south of Daytona, he became enthralled by a white-sand barrier island called Palm Beach: "I have found a veritable Paradise!" Flagler also noticed a tangle of scrublands on the mainland, directly across Lake Worth from his new enchanted isle, and West Palm Beach began taking shape in his mind's eye: "In a few years, there will be a town over there as big as Jacksonville."

He could see it already. Palm Beach would be his new American Riviera. West Palm would be a bustling commercial hub. Americans would come to the area to play, and move to the area to stay.

"The Wizardry of the Dollar"

FLAGLER WAS NOW a silver-haired man in his sixties, more dignified than dashing, more driven than ever. He snapped up huge tracts of land on both sides of Lake Worth. He also secured a charter from the legislature promising 8,000 acres for every mile of track he laid south of Daytona, more than twice the amount specified by the internal improvement law. His steel ribbon soon unspooled down the Atlantic coast, to New Smyrna, then Eau Gallie, then Fort Pierce in January 1894. At the height of the nationwide financial panic, Flagler had 1,500 men working on the line.

That February, another 1,000 laborers completed Flagler's most extravagant resort yet, the colonial-style Royal Poinciana on Palm Beach, with twice as many rooms as the Ponce de Leon. It was not only the world's largest hotel but the world's largest wooden structure, requiring 2,400 gallons of paint, most of it in a lemony color known as Flagler Yellow. Flagler's men filled in wetlands along the coast, and landscaped the resort with Australian pines, fast-growing shade trees that controlled the native scrub by blocking its sunlight. In April, the railroad reached West Palm Beach, and land values in the area skyrocketed. "Yesterday a swamp was here," one visitor marveled. "Today you see the wizardry of the dollar."

The Royal Poinciana soon became the Gay Nineties winter hub for the Social Register's exclusive "Four Hundred," attracting Vanderbilts, Carnegies, Morgans, Astors, Fricks, and the rest of America's industrial royalty. The guests enjoyed golf, fishing, yachting, and sunbathing—Flagler employed beach censors to make sure women covered their legs—along with haute cuisine, orchestras, and vaudeville. The guests were served by 1,400 staffers so attentive the resort was known as the Royal Pounce-on-them. Black employees whisked them around in bicycle-powered carriages known as Afromobiles, and entertained them with "cakewalks," minstrel-style dance competitions whose winners got to "take the cake." Suites cost $100 a night, about three months' wages for a typical laborer.

Flagler soon added the Breakers, another swank resort that still operates on the ocean side of Palm Beach. And he commissioned the architects who designed the New York Public Library to build him a $2.5 million Beaux Arts mansion called Whitehall, a fifty-five-room white marble palace stuffed with Spanish tapestries, Renaissance art, the largest pipe organ ever installed in a private home, and period furniture from sixteen epochs in history. The *New York Herald* described Whitehall (which now houses the Flagler Museum) as North America's Taj Mahal, "more wonderful than any palace in Europe, grander and more magnificent than any other private dwelling in the world."

West Palm Beach flourished, too, attracting 1,000 residents in its first year, most of them Flagler employees. Flagler again built the city's churches and civic buildings, paved its streets, and donated land for its cemetery. His aide James Ingraham laid out the town site, and set up a volunteer fire department called the Flagler Alerts. "I feel that these people are wards of mine and have a special claim upon me," Flagler wrote.

FLAGLER THOUGHT HE HAD REACHED the end of his line. But Mother Nature changed his mind that winter, after Florida endured two of its worst freezes in a century, the double whammy that helped doom Hamilton Disston's ventures. In late December, temperatures dipped to fourteen degrees in Jacksonville and thirty as far south as West Palm; in February, snow fell in Fort Myers. Florida's yearly citrus production dropped from more than 5.5 million boxes to 150,000. Flagler dispatched Ingraham to assess the damage, and to hand out seeds and cash to growers willing to give Florida another chance.

Ingraham made his most interesting damage assessment in Miami, sixty-five miles south of West Palm Beach: no damage. Even before the cold snaps, the widow Julia Tuttle had been pestering Flagler to extend his rail-road to Miami. "It is the dream of my life," she once wrote, "to see this wilderness turned into a prosperous country." Flagler had demurred, seeing

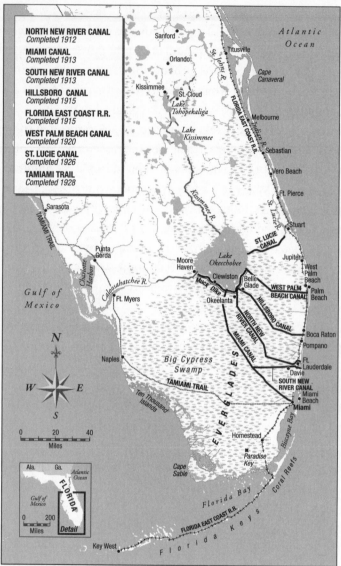

NORTH NEW RIVER CANAL
Completed 1912

MIAMI CANAL
Completed 1913

SOUTH NEW RIVER CANAL
Completed 1913

HILLSBORO CANAL
Completed 1915

FLORIDA EAST COAST R.R.
Completed 1915

WEST PALM BEACH CANAL
Completed 1920

ST. LUCIE CANAL
Completed 1926

TAMIAMI TRAIL
Completed 1928

BY GENE THORP—CARTOGRAPHIC CONCEPTS INC.

Henry Flagler's railroad sparked south Florida's first boom, as tourists and pioneers rode his rails down the Atlantic coastal ridge to new cities like West Palm
Beach, Fort Lauderdale, and Miami. Governor Napoleon Broward then drew
up plans to drain the Everglades with six canals from Lake Okeechobee—one
east, one west, and four southeast through the ridge. The dream of an Empire
of the Everglades drew thousands of settlers to the region's wetlands.

no need for another railroad to nowhere. But now Ingraham returned from a meeting with the ever-persistent Tuttle with unscathed orange blossoms, proof that Miami was below the frost line. Flagler sat silently for a minute, then asked: "How soon can you arrange for me to go to Miami?" Tuttle, her neighbor William Brickell, and local real estate speculators offered Flagler tens of thousands of acres of additional land to bring his iron horse south, and the railway soon chugged down to Fort Lauderdale on the New River, then on to Miami.

Once again, its arrival sparked a mini-boom. Five hundred voters incorporated Miami in 1896; they wanted to call it Flagler, but he declined the honor. The city was still Flagler-dominated; he provided the electric plant, the water works, and the sewage system, as well as churches and public buildings. He built the Royal Palm Hotel with south Florida's first golf course on an old Indian mound along the Miami River. He set up a steamboat terminal with service to Cuba and the Bahamas and began dredging Biscayne Bay for a deep-water port; he had one of his Standard Oil lobbyists secure $300,000 in federal aid. Flagler even donated land for a U.S. Weather Bureau station, hoping to advertise Miami's climate to the nation.

Thousands of soldiers billeted in Miami during the Spanish-American War, and many of their reactions were reminiscent of Seminole War diarists. "If I owned both Miami and Hell, I'd rent out Miami and live in Hell," one complained. But others decided to stay after the war, and the city's population increased tenfold in five years. Miami became the Dade County seat, and Flagler offered his rival Henry Plant teasing instructions on how to get there: Go to Jacksonville and follow the crowd.

FLAGLER PUT INGRAHAM in charge of his real estate operations, ordering him to focus on maximum growth and immigration instead of maximum profit. Flagler was losing several hundred thousands of dollars a year in Florida, but he took the long view. He calculated that every permanent resident would be worth $300 to his railroad alone, ensuring a steady cash flow after the winter tourists went home: "What we want for some little time to come is more settlers, more cultivation and more freights."

Ingraham sent land agents around the world to promote America's winter breadbasket, and sponsored a "Florida on Wheels" railroad car to remind shivering midwesterners what they were missing. He also published the *Florida East Coast Homeseeker,* trumpeting the potential of south Florida agriculture. "Most Productive Soil in Existence," read one typical ad. "Cool Summers, Mild Winters, Pure Water, Perfectly Healthy, No Swamps, Few Insects."

Soon, settlers were launching farming communities all the way down the Atlantic ridge, the so-called "Gold Coast." Below West Palm Beach, Michigan transplants founded Boynton Beach and Delray Beach. A Flagler engineer laid out Boca Raton; southern farmers founded Deerfield Beach and Pompano Beach. The *Miami Metropolis* was impressed by the Japanese immigrants who started a now-defunct farm colony called Yamato, praising "these interesting little people" as industrious tomato and pineapple growers. Danes started Dania, and Ingraham's brother-in-law, Luther Halland, led a group of Swedes to Hallandale. Just north of Miami, Ojus was named for the Seminole word for "plenty," and Flagler helped Henry Perrine's heirs to develop the town of Perrine. He eventually extended his line all the way down to the last dry land at the edge of the Keys, which became Homestead. Dade County's property value increased eightfold in the 1890s, and south Florida was just getting started.

Years later, a reporter for *Everybody's Magazine* asked Flagler if he had really known that settlers would follow his iron into the subtropics. "Did you close your eyes and see the men in the field working?" the reporter asked. "Did you really vision the thing as clearly as that?"

In a matter-of-fact tone, Flagler replied: "Yes."

The interviewer wanted to make sure he wasn't planting ideas in an old man's head. "Please understand me," he said. "Don't let me *suggest* it to you. Did you actually vision to yourself the whole thing? Did you really close your eyes and see the tracks? And the trains running? And hear the whistles blowing? Did you go as far as that?"

"Yes."

"How clearly?"

"Very clearly."

FLAGLER TREATED the Gold Coast's new residents with an unfailing sense of noblesse oblige, spending millions of dollars to promote their welfare. But there was never any doubt who was the noble and who were the serfs. When Flagler invited President William McKinley to Florida in 1898, he sounded like a medieval baron inviting the king to tour his fiefdom. "My domain begins in Jacksonville," Flagler wrote. Even "if the East Coast of

Florida belonged to anyone else," he told the president, "I should venture to say that it possesses very great attractions."

The east coast of Florida really was Flagler's domain. He controlled its transportation and most of its land; he also controlled a host of its newspapers, from Jacksonville's *Florida Times-Union* to the *Miami Metropolis*. Behind the scenes, he also controlled its politics. In West Palm, for example, Flagler personally blocked proposals for a road, a wharf, and a fish house along Lake Worth, and demanded that the local council shut down a brothel.

Flagler's power throughout the state was most notoriously on display after Ida Alice was banished to a sanatorium. Flagler threw his money and influence around Tallahassee to ram House Bill 135 through the state legislature, making "incurable insanity" a legal grounds for divorce. He remarried a few days after the bill passed, and gave Whitehall to Mary Lily as a wedding gift. The so-called Flagler divorce law was one of the most reviled legislative acts in Florida history, widely denounced as a sellout to the state's richest resident. But Flagler felt entitled to a bit of consideration. He had visualized an American Eden, and he had carved it out of the wilderness with his own money. He thought it only fair that he should be able to choose his own wife in his own domain.

"It Would Be a Glorious Undertaking"

FLAGLER'S DOMAIN TRANSFORMED the eastern rim of the Everglades, as railroad men and settlers began cutting and burning the pine forests and hardwood hammocks that grew along the coastal ridge. The fire-resistant pinelands—ideal habitat for red-cockaded woodpeckers and five-lined skinks with electric-blue tails—were converted into turpentine and termite-resistant homes. The hardwood hammocks—shelters for swallowtail butterflies and multicolored tree snails—were "being rapidly destroyed and will soon be a thing of the past," wrote the pioneer naturalist Charles Torrey Simpson. "The charred ruin glares in the sun as a silent and pathetic protest against useless waste and folly."

But while Flagler's domain stretched 350 miles down the peninsula, it was only a couple of miles wide in south Florida. It was devouring the pines, palms, gumbo-limbos, ironwoods, poisonwoods, and mahoganies that had flourished in the higher and drier coastal ridge, but it had yet to penetrate the low-lying wetlands of the Everglades. "There was a most magnificent and gorgeously appointed hotel right in the midst of a perfect paradise of tropical trees and bushes," one soldier recalled after his stay in Miami. "But one had to walk scarce a quarter of a mile until one came to such a waste wilderness as can be conceived of only in rare nightmares."

This "waste wilderness" was still so obscure that in 1897, Flagler's friend Hugh Willoughby, a former naval officer, embarked on a Lewis-and-Clark-style journey of discovery across the Everglades in a dugout canoe. "It may seem strange, in our days of Arctic and African exploration, for the general public to learn that in our very midst, as it were, in one of our Atlantic coast states, we have a tract of land 130 miles long and 70 miles wide that is as much unknown to the white man as the heart of Africa," he wrote. After Disston's troubles, many Floridians doubted the swamp nightmare would ever end.

"Some men believe the Everglades should be drained," one paper scoffed, "while others urge the annexation of the moon."

JAMES INGRAHAM WAS ONE of the believers. He had seen and felt the Everglades with his own eyes and feet, and he remained convinced its waters could be removed by opening vents through its limestone rim, and that its mucklands could become farms of "almost unprecedented fertility." He also thought it made perfect sense for his boss to take on the job. Draining the Everglades could draw new settlers to Flagler's domain, to ship crops on Flagler's trains. Drainage could also expand the domain from a thin strip along the coast to a huge swath of the peninsula. The drainage booster William Bloxham was back in the governor's office, and was again offering Everglades lands as a reward for Everglades reclamation. Flagler had the cash to succeed where Disston had failed.

Ingraham also suspected that Flagler would embrace the challenge of draining the Everglades, the opportunity to achieve something important that others considered impossible. Ingraham still recalled the stirring words of a companion on his own slog through the swamp, an engineer named John Newman:

With the money spent on hotels in St. Augustine to gratify the luxurious tastes of our millionaires, I believe this land could be drained, and the promoter of such a scheme would have the right to be considered the greatest philanthropist of his age. It would be a glorious undertaking, for charity could ask no nobler enterprise, ambition no higher glory and capital no greater increase than would result from the redemption of this land.

Except for the swipe about St. Augustine hotels, that was just the kind of outsized pitch that appealed to Flagler. Sure enough, the *New York Times* soon reported that as his railroad was snaking down the coast, Flagler was investigating a "land development scheme of monumental proportion" in the interior. The *Times* noted that "very few people believe that the Everglades will ever be other than the rich game land and unhealthy swamps that they now combine to make." But Flagler reportedly believed those swamps could be "transformed into a Garden of Eden," and a sugar bowl for the nation.

In 1898, Ingraham and Rufus Rose, Disston's former drainage superintendent, launched the Florida East Coast Drainage and Sugar Company. Flagler's name did not appear in its corporate papers, but its drainage plan was clearly tailored to his interests.

DISSTON HAD DRAINED the upper Kissimmee basin with local canals, but his grand plan to cut off the Everglades at its source by eliminating overflows from Lake Okeechobee had fizzled. For Flagler, Captain Rose designed a purely local plan that ignored Lake Okeechobee. Instead, he proposed to drain the easternmost Everglades wetlands near Flagler's developments by digging a dozen short canals through the coastal ridge. As Buckingham Smith's report had said, Providence had already begun the job; Rose simply proposed to expand natural Everglades outlets to the Atlantic, "transverse glades" like the Miami, New, and Hillsboro Rivers and Snake, Cypress, and Snapper Creeks. He then planned to wall off the drained tracts from the rest of the Everglades with dikes. About 800,000 acres of seasonally flooded wetlands at the edges of the Everglades would be sucked dry and added to Flagler's domain as sugar plantations, but the permanently flooded sloughs and marshes in the heart of the Everglades would remain as wet as ever.

The company signed a lucrative contract with the Internal Improvement board, and proclaimed that drainage would be "a simple process" in its prospectus: "Where drained by natural means and where reclaimed by artificial means, these soils have produced phenomenally heavy crops." Governor Bloxham proclaimed that the deal would produce hefty profits for investors, rapid growth for south Florida, and "incalculable benefits" to the state, while freeing the nation from its dependence on foreign sugar. The engineering seemed easy, and money clearly wouldn't be a problem. "It may be taken as assured fact that this section of the Everglades will be drained," one paper said. By expanding the transverse glades and helping them flow out to sea, the company would simply accelerate nature's work: "As the bottom of this basin is above tide water, drainage is rendered a certain and simple process."

The politics of drainage, however, was increasingly uncertain and complex.

"I Haven't the Money or the Inclination"

IN THE LATE NINETEENTH and early twentieth centuries, the progressive movement emerged to try to rein in corporate America. The United States was now the richest country on earth, producing half the world's oil and one-third of its iron and steel. Its citizens were consuming Campbell soup, Borden cheese, Post Grape-Nuts, and Hershey chocolate, while enjoying lightbulbs, telephones, automobiles, and airplanes. It was the dawn of the American century, a time of puffed-up national pride and confidence. But there was a growing feeling that average Americans were not sharing in the progress, that business interests controlled the government, and that the balance of power ought to be reversed.

Progressivism was a gospel of science and reason; progressives believed the same pragmatic thinking that was solving great technological and engineering problems could be applied to social problems. Muckraking journalists and social reformers exposed the abuses of obscenely wealthy robber barons and their anticompetitive trusts, as well as the victimization of ordinary Americans who lived in slums, labored in sweatshops, and ate rotten meat. The public began to clamor for action to rein in the abuses—railroad regulations, meat inspections, worker protections, and prosecutions of monopolies like Standard Oil. The "Great Commoner," the Democratic presidential candidate William Jennings Bryan, captivated audiences with his attacks on corporate greed. But President Theodore Roosevelt, who disdained left-wing agitators as much as "malefactors of great wealth," came to embody the spirit of the Progressive Era—not just his trust-busting, but his insistence on energetic government action in general, his rejection of the Republican dogma that elites should be left alone to manage the economy and the country. "I have no command of the English language that enables me to express my feelings regarding Mr. Roosevelt," the usually even-keeled Flagler seethed to a friend. "He is shit."

The progressive spirit spread slowly in Florida, but it did spread, and the Standard Oil magnate turned railroad kingpin inevitably became its prime target. Reformers such as Napoleon Bonaparte Broward—the riverboat captain who was now sheriff in the Jacksonville area—accused Flagler and his ilk of gouging farmers and consumers by manipulating freight rates. The legislature approved a railroad commission by such an overwhelming margin that even the corporate-friendly Governor Bloxham had to approve it, and the commission promptly lowered shipping rates on cabbages and oranges.

In 1900, Bloxham was replaced by an antirailroad, anticorporation progressive named William Sherman Jennings, an Illinois native who happened to be the Great Commoner's cousin. Governor Jennings pushed for an array of progressive reforms, including a pure food and drug law, free textbooks, and a ban on cruelty to children. He also decided to stop Florida's lavish land giveaways to railroads and other corporations.

BY THE TURN OF THE CENTURY, the Internal Improvement Fund that Disston had bailed out in 1881 was in shambles again, its records so chaotic that Jennings had to launch an investigation to figure out whether it owned any land. He found that Florida had given away about 17 million of the 20 million acres of swamp and overflowed lands it had received from the federal government. That left about three million acres of the Everglades in state possession. But the fund had already pledged an additional six million acres to corporations, so it was oversubscribed by about three million. And the Everglades was still drenched.

Jennings decided to shut down the candy store. He declared that state lands belonged to the people, not to railroads that rarely bothered to improve them. He refused to grant a single acre to any corporation during his term, contending that the giveaways violated the spirit of the federal Swamplands Act of 1850, which was specifically intended to promote reclamation. When Flagler's railroad tried to claim 156,000 acres near Cape

Sable—a tract it had been promised by the Bloxham administration—Jennings declared that the promise was not worth the paper it was written on. His message to corporations was simple: If you want state lands, buy them. They responded by filing lawsuits seeking the grants the state had promised. Flagler's railroad claimed more than two million acres.

In this climate, there was not much incentive for Flagler to try to drain the Everglades, not when he wouldn't even get to keep the land he drained. He also worried that an influx of reclaimed lands could glut south Florida's real estate market, reducing the price of his own lands. He did spend more than $100,000 on private drainage canals that reclaimed several thousand acres at the edge of the Everglades, but he never gave Ingraham and Rose the funding they needed to execute their comprehensive plan. Soon Jennings and his fellow trustees canceled their dredging contract for non-performance. The governor had just traveled to California, and had seen how massive irrigation projects were reclaiming the deserts of the West. "Never before had I appreciated the full value of water," he recalled. "As the train went on, mile after mile, day after day through waterless plains, my eyes were opened to the possibilities of the Everglades." Jennings figured it would be much easier to move water off wetlands than it was to move water onto drylands, and he was tired of waiting for the private sector to take up the challenge.

If any businessman could have finished what Disston started, it was Flagler. He would eventually spend $50 million in Florida, and it would barely dent his bottom line. But he clearly developed a bad feeling about the Everglades. At one point, he suggested to Ingraham that a wise investor wouldn't touch the swamp with a ten-foot pole.

"So far as I am personally concerned," Flagler wrote, "I haven't the money or the inclination to take up as big a matter as the drainage of the Everglades."

THAT WAS AN INTERESTING WAY to put it, because there were few Americans with more money or more inclination for big matters than Henry Flagler. In fact, as he entered his eighth decade, Flagler was about to take up his biggest matter: extending his railroad all the way from Homestead to Key West. Flagler believed that America's coral-rock Gibraltar could become a megaport, the nation's access point for a newly liberated Cuba and the soon-to-be-built Panama Canal. Unfortunately, it was isolated from

Flagler's new Florida civilization by more than 150 miles of swamp and ocean. Building a railroad there wouldn't be quite as difficult as the annexation of the moon, but it would cost Flagler at least $27 million—the equivalent of more than half a billion dollars in 2005—and would force him to borrow money for the first time since his salt-making fiasco. It would also be hailed as the engineering achievement of his era.

At the time, though, it was ridiculed as Flagler's Folly. One of Flagler's closest friends gasped: "You need a guardian!" There were only two conceivable ways to lay track to Key West: by slashing through the muck of the Everglades to Cape Sable and then across the muddy shallows of Florida Bay, or by island-hopping through the ocean along the arc of the Keys. The railroad's brochures gave a sense of the challenge: "The financiers considered the project and said: Unthinkable. The railway managers studied it and said: Impracticable. The engineers pondered the problems it presented and from all came the one verdict: Impossible." But once the vision popped into Flagler's head, he could not let it go.

Flagler's railroad was already hemorrhaging cash, and his hotels were barely breaking even. But for a man who couldn't bring himself to throw away a 2-cent stamp, this mission had little to do with money. "It was very strange, at first, for me to work for Mr. Flagler," one of his engineers told *Everybody's Magazine*. "With him it is never a case of *How much will it cost?* nor of *Will it pay?*" Flagler just wanted to make sure his Eighth Wonder of the World was completed before he died.

THE ONLY QUESTION was whether Flagler's railroad would go through the ocean or the Everglades. Flagler's Model Land Company already had holdings in Cape Sable, and laying iron there seemed marginally less insane than cutting across the sea. The residents of the frontier outpost of Flamingo began to fantasize about their tiny fishing village becoming the next Chicago, or at least the next West Palm Beach.

In 1902, a decade after Ingraham's trek, Flagler dispatched an engineer named William Krome to lead a new railroad survey of the Everglades. Krome and his crew spent six months in the marsh, lugging forty-pound packs, wrapping themselves in cheesecloth to ward off clouds of mosquitoes as thick as pudding. "I found a most God-forsaken region," Krome reported. "Of keys, bays, rivers and lagoons there is no end, and it is going to take us much longer to get a survey than I had expected."

After his half year of hell, Krome concluded that Ingraham had been right the first time: The Everglades was no place for a railroad. Flagler would have to connect the dots of the Keys, building concrete bridges as long as seven miles across the open sea—unthinkable, impractical, impossible, but easier than slashing through the Everglades.

Krome also agreed with Ingraham that the Everglades could be redeemed. "The muck with proper drainage will eventually become fine farming land, and the mosquitoes will disappear to a great extent as the country opens up," he wrote. But Flagler had no interest in opening it up himself. He had created a civilization at the edge of the Everglades, but it would be up to his progressive enemies to extend it into the Everglades.

EIGHT

Protect the Birds

Florida has been considered in all respects as a prey and a spoil to all comers.

—Florida transplant Harriet Beecher Stowe

DISSTON'S DITCHES TRANSFORMED the headwaters of the Everglades in the upper Kissimmee basin. Flagler's railroad transformed the eastern rim of the Everglades along the Atlantic Coastal Ridge. But at the turn of the century, the Everglades itself was still essentially the Everglades. Disston and Flagler had begun to attract people to south Florida, but the region was still emptier than it had been during the days of the Calusa.

The Calusa, however, never had rifles. South Florida's newcomers shot deer, bear, gators, turkeys, bobcats—and especially skinny-legged Everglades wading birds. They shot wading birds until there were hardly any left to shoot, and an Everglades without wading birds would not have been an Everglades at all.

In the Progressive Era, bird massacres in the Everglades became a national scandal, inspiring one of America's first conservation crusades, illustrating the potential power of environmental advocacy. Conservation became a national priority. But "conservation" did not always mean then what it means now.

"We Could Scarcely Believe Our Eyes"

WADING BIRDS ARE extraordinarily demanding creatures. A pair of seven-pound wood storks, for example, needs to catch about 440 pounds worth of minnow-sized fish every breeding season. And wood storks are grope-feeders, blindly probing for their prey by swishing their long, slightly hooked bills around in the water. So even though their bills automatically snap shut faster than any other reflex in the animal world when they bump into a fish, they can only forage in shallow water where fish are highly concentrated. The black-and-white storks, which look a bit like pterodactyls, will fly more than forty miles to find a fishing hole, gorge themselves, then fly home and regurgitate some of the food for their chicks. But they will rarely fish in water more than two feet deep, because they end up expending more energy swishing than they gain from eating.

The natural Everglades offered a kind of Restaurant Row for wood storks and other wading birds, a shallow-water ecosystem with an extended dry season that could satisfy their picky foraging demands. It was so vast—and encompassed wetlands with such a wide variety of water levels—that they could find drying-but-not-yet-dry marshes throughout their breeding and nesting seasons, starting with the shallowest wetlands at the edges of the marsh in December and progressing inland to the deepest sloughs by March. These shrinking pools supported as many as 600 fish per square meter, attracting spectacular feeding frenzies. Green herons and snowy egrets trolled the edges of the pools; great blue herons and great egrets with longer legs sought slightly deeper water. Each bird had its own foraging strategy, and beaks that had evolved for their favorite foods. Great blues stood still for hours to stab sunfish and gar with their stiletto-shaped beaks, while white ibis grubbed in the mud for crawfish and insect larvae with their downward-curving beaks. Green herons dropped fish food in the water as bait, while reddish egrets spread their wings over the water to cast shadows over their prey.

The Everglades also offered ideal bedrooms and nurseries where wading birds could build nests—usually in cypress clumps or mangrove trees, near alligators that could chase away raccoons and other predators. The density and intensity of these rookeries was astounding, tens and even hundreds of thousands of birds in a bunch. "Here I felt I had reached the high-water mark of spectacular sights in the bird-world," wrote Reverend Herbert Job, a Unitarian minister who was one of the first photographers in the Everglades. "Wherever I may penetrate in future wanderings, I never hope to see anything to surpass, or perhaps to equal, that upon which I then gazed."

As many as 2.5 million wading birds patrolled the Everglades before the late nineteenth century. They reeked of fishy guano, and sounded like a chorus of foghorns, whistles, and screeching babies, but they were amazing to behold. "It was truly a wonderful sight, and I have never seen so many thousands of birds together at any single point," wrote an ornithologist named William Scott. Silvery-black anhingas with S-shaped necks darted underwater to spear fish, then stretched their wings to warm up like avian Karate Kids. Snowy egrets with bright yellow feet seemed to float through the sky, "with the sunlight on their white wings, shining like snow, and then melting from sight like a dream." Brown-and-white limpkins probed the shallows for snails and crayfish by day, then shattered the night with ghastly cries of *kree-ow, kree-ow.* "When do they sleep?" the explorer Hugh Willoughby asked. "Or do they ever sleep?"

John James Audubon, the renowned painter and author, was the first naturalist to fall in love with south Florida's birds. During an 1832 visit, he wrote rapturously of roseate spoonbills that "stalked gracefully beneath the mangroves," great white herons "thrusting their javelin bills into the body of some unfortunate fish," and pelicans sunning themselves on the mud flats: "Should one chance to gape, all, as if by sympathy, in succession open their long and broad mandibles, yawning lazily and ludicrously." In Cape Sable, Audubon rhapsodized about gluttonous ibis, timorous gallinules, and herons crying *wie-wie-wie:* "The flocks of birds that covered the shelly beaches, and those hovering over head, so astonished us that we could for a while scarcely believe our eyes."

Audubon was so astonished that he almost forgot to shoot those flocks out of the sky. "The first volley procured a supply of food sufficient for two days consumption," he recounted. "Our first fire among a crowd of the Great Godwits laid prostrate sixty-five of these birds." Audubon preferred to paint freshly killed specimens, but he didn't need sixty-five fresh godwits for his art. He just liked shooting birds.

So did the winter pleasure-seekers who descended on Florida later in the century. They picked off birds from the decks of steamboats, much like the western travelers who blew away buffalo from the windows of trains. These "sportsmen" reignited the outrage of the abolitionist crusader Harriet Beecher Stowe, who published Florida's first environmental broadside in

1877, "Protect the Birds." She accused Florida's winter visitors of waging a "war of extermination" against its feathered natives, a war she considered as evil as the slave trade: "The decks of boats are crowded with men, whose only feeling amid our magnificent forests, seems to be a wild desire to shoot something and who fire at every living thing on shore."

But these trigger-happy intruders barely dented populations of wading birds and other wildlife, because they rarely strayed from the rivers. Tourists didn't venture into the fastnesses of the Everglades, where most of the bird rookeries were hidden. In the nineteenth century, people still avoided the Everglades unless they had an extremely good reason to be there. The love of hunting was not usually a compelling enough reason.

The love of money was a different story.

The Killing Fields

IN FEBRUARY 1886, a birdwatcher named Frank Chapman conducted an experiment in Manhattan, identifying 160 bird species on two strolls through the Ladies' Mile shopping district. This was no winter ornithological miracle. The birds were all dead, and perched atop the heads of stylish ladies. Of the 700 women's hats spotted by Chapman, 542 were festooned with feathers, the most elegant with "aigrettes," the dainty nuptial plumes of courting wading birds. A few of the most expensive hats served as pedestals for entire birds.

At the height of the fad, an ounce of feathers cost more than an ounce of gold, which provided an excellent incentive to brave the Everglades. Plume hunters sold spoonbill skins for $5, great white herons for $10, flamingos for up to $25. At a time when average per capita income was less than a dollar a day, plumers gladly supplied cheaper birds as well, selling tricolor and great blue herons, reddish and snowy egrets, pelicans and owls for a dime to a half-dollar per skin.

"What do you hunt?" the ornithologist William Scott asked one Florida plumer.

"Almost anything that wears feathers," the man replied.

In 1886, the American Ornithological Union estimated the annual nationwide carnage at five million birds. One Florida agent shipped 130,000 plumes in a year. A plumer named George Cuthbert slaughtered $1,800 worth of herons, egrets, and spoonbills on a single trip to a rookery he found in a mangrove jungle near Cape Sable. Birds were big business, and competition was fierce. Cuthbert used the proceeds of his hu~~nt~~ to buy half of Marco Island, which became some of America's priciest ~~r~~eal estate. Cuthbert's crewman was murdered by a rival plumer for refusing t~~o~~ divulge the rookery's location.

Florida's most notorious plumer, Jean Chevelier—a curmudgeonly Frenchman who shot out the St. Petersburg area's rookeries, then relocated to the Ten Thousand Islands—gathered 11,000 skins in a single season. "There were plume and song birds of every description that the Creator had placed here to beautify and adorn Man's Paradise, but the lawless marauders just about destroyed everything that came in reach of their powder and lead," one critic complained. The logkeeper for one Chevelier expedition into the Everglades catalogued the destruction of 1,397 birds of thirty-six species, recording the daily slaughter with matter-of-fact entries like: "Louie killed eleven birds, Guy killed eight, and I killed nineteen." Over one rookery, the flocks were so thick that Chevelier became disoriented and laid down his rifle. "Mine God, 'tis too much bird in this country, I cannot shoot," he gasped. The logkeeper had no such problems: "I killed two night herons, two [tricolor] herons, and sixteen reddish egrets."

If Stowe was repelled by tourists taking potshots from steamboats, it's good she never witnessed the systematic slaughter of a plume hunt. It was more like a harvest than a hunt. At the height of nesting season, plumers patiently shot out rookeries one bird at a time, leaving rotting carcasses and helpless chicks to be devoured by raccoons, crows, and buzzards. They used quiet weapons like Winchesters or the Flobert—a rifle favored by French aristocrats for after-dinner target shooting inside their chateaux—so their shots sounded like snapping twigs. The birds rarely noticed them, and when they did, the adults rarely left their nests for fear of abandoning their young. The ornithologist Scott described the remains of another rookery in Charlotte Harbor:

> Hundreds of broken eggs strewed the ground everywhere. Fish crows and both kinds of buzzards were present in great numbers and were rapidly destroying the remaining eggs. I found a huge pile of dead, half-decayed birds lying on the ground which had apparently been killed for a day or two. All of them had the plumes taken off with a patch of skin from the back, and some had the wings off. I counted over 200 birds treated this way. . . . I do not know of a more horrible and brutal exhibition of wanton destruction than that which I witnessed here.

ᵗ to
al
o

                        provided income for Seminoles, but they practiced
                  'e exploitation, refusing to wipe out entire rook-
                  . enough of the old birds to feed the young of the
                  . observed. "The white man kills the last plume bird he
            .g the young ones to die in their nests, then returns a few
        .est he might have overlooked a few birds."

     .s kill-them-all strategy took its toll. Roseate spoonbills, snowy egrets,
ₐeat white herons, and short-tailed hawks nearly vanished from Florida.
The wild flamingos that so enchanted Audubon—and inspired the name of
the village at the tip of Cape Sable—did vanish from Florida. The lime-
green-and-carmine Carolina parakeet was hunted to extinction. There was
only one pair of reddish egrets left on the peninsula, and only one rookery
for brown pelicans, a clump of mangroves off Vero Beach called Pelican
Island. "I don't think in my reincarnation, if there is such a thing, that I want
to come back to Florida," sighed the author and outdoorsman Kirk Munroe.
"They are killing off all the plume birds. I remember when the spoonbills
on the beach in front of my house made such a racket it was almost unpleasant. Now they are all gone."

The Roots of Conservation

THIS WAS A REVOLUTIONARY CONCEPT for Florida, the notion that a
human victory over nature might not represent progress, the idea that
spoonbills might have value regardless of their usefulness to human beings.

The early descriptions of the peninsula were relentlessly utilitarian:
Cypress stands were evaluated for timber quality, marshes for agricultural
potential, rivers for navigability. Wildlife was described as "game." In 1837,
for example, the author of *The Territory of Florida* noted that alligators made
"excellent leather," panthers were "particularly destructive to calves," and
flamingos were "excellent food." These were natural resources in the literal
sense, valuable only insofar as they could be exploited by human beings.
Occasionally, a writer like Buckingham Smith lapsed into lyricism, con-
fessing that "the effect of such visit to the Pa-Hay-Okee upon a person of
romantic imagination, and who indulges his fancies on such subjects, it may
be presumed, would be somewhat poetic." But Smith hastened to make it
clear that he was by no means that kind of person; from a practical stand-
point, the Everglades was worthless, and ought to be reclaimed.

The industrialization and deforestation of the latter half of the 1 teenth century prompted a few Americans to think about nature in n ways, laying the groundwork for the environmentalism of the twentieth century. Henry David Thoreau, the bard of Walden Pond, worshipped at the altar of nature, and raged against its exploitation by man. "We need the tonic of wildness," he wrote. "We can never have enough of nature." Nature was his God, his nurse, his balm, and he denounced the destruction of trees and animals the way others denounced the murder of people. What is a country, he asked, without rabbits and partridges? He rejected his era's anthropocentric worldview: "I love Nature partly because she is not a man, but a retreat from him." He loved all of Nature—even the most dismal swamps: "I enter a swamp as a sacred place, a sanctum sanctorum."

Thoreau inspired devout preservationists like John Muir, who founded the Sierra Club in 1892. Muir fought like an avenging angel to protect the wilderness—not for the sake of man, but for the wilderness itself. He wanted to preserve the nastiest rattlesnakes and ugliest alligators as well as rabbits and partridges, and he argued that man was conceited to fancy himself more precious than other beasts. "They . . . are all part of God's family, unfallen, undepraved, and cared for with the same species of tenderness and love as is bestowed on angels in heaven or saints on earth," he wrote.

While Thoreau was meditating at Walden, George Perkins Marsh was developing a less radical but no less alarmist philosophy of conservation, more sympathetic to the use of natural resources but not to their abuse. Marsh—who dabbled in manufacturing, farming, lumber, real estate speculation, and politics as well as writing—was a dedicated capitalist who believed in man's dominion over nature and the pursuit of progress. But he also believed in moderation. He warned that Americans were knocking nature out of balance, risking dire consequences to themselves and their descendants by obliterating forests and extirpating entire species: "All nature is linked together by invisible bonds, and every organic creature, however low, however feeble, however dependent, is necessary to the well-being of some other." While Thoreau raged that civilization was destroying nature, Marsh fretted that the wanton destruction of nature would end up destroying civilization, and the human race. "The earth," he warned, "is fast becoming an unfit home for its noblest inhabitant."

Marsh inspired utilitarian conservationists like the first director of the Forest Service, Gifford Pinchot, a European-trained forester who shared his

ental degradation had led to the collapse of ancient
d championed the "wise use" of America's natural
ation of natural resources is the basis, and the only
ional success." Pinchot, a child of the New York City
a nation's resources to a family trust fund, and while he sup-
spending the interest, he was appalled by timber syndicates that
exhausted the principal, clear-cutting forests into desolate moonscapes. He
did not want to ban logging, grazing, or mining, but he wanted to manage
them rationally and sustainably—to conserve enough trees, grasses, and
ores for future generations to use for logging, grazing, and mining. It was
government's duty to ensure that natural resources were exploited efficiently, so they would produce the most good for the most people over time.

CONSERVATION WAS A CORNERSTONE of the progressive movement, as
vital to the spirit of the age as child labor restrictions or railroad commissions. The progressive Governor Jennings, William Jennings Bryan's cousin,
signed Florida's first forest protection law; his wife, May Mann Jennings,
became one of Florida's leading conservationists. And it was no coincidence
that the central figure of America's conservation movement was the central
figure of America's progressive movement: President Theodore Roosevelt,
a passionate outdoorsman who shared Audubon's fascination with living
beings, as well as his inclination to shoot them.

T. R. began his career as a naturalist at age seven, composing meticulous
zoological treatises about the bugs, reptiles, rodents, and other specimens in
the "Roosevelt Museum of Natural History" that he curated in his bedroom.
"All the insects that I write about in this book inhabbit North America," he
wrote in his *Natural History of Insects*. "Now and then a friend has told me
something about them but mostly I have gained their habits through ofserv-
a-tion." As a teenager, Teddy became obsessed with birds, documenting
every *cheech-ir'r'r* and *fl'p-fl'p-trkeee* in his diary, publishing an ornithological
guide to the Adirondacks when he was eighteen. Roosevelt's enthusiasm
for nature went well beyond "ofserv-a-tion"; he also became an ardent big-
game hunter. But as the great herds of the West dwindled, he began to
denounce the "swinish game-butchers" who wiped them out indiscrimi-
nately. In 1888, he helped found the Boone & Crockett Club, one of
America's first conservation groups to fight for their protection.

After entering politics, Roosevelt maintained sympathy for both strains of conservationism—Thoreau's aesthetic and spiritual revulsion to all attacks on nature, as well as Marsh's more practical and ecological opposition to needless and overzealous attacks. He was still a devoted nature-lover who exploded with enthusiasm while camping with John Muir under the sequoias of Yosemite. "This is bully!" he yelled. "I never felt better in my life!" But he was also an economic expansionist who put his friend Gifford Pinchot in charge of his conservation agenda. As president, Roosevelt protected 230 million acres of public land—including the Grand Canyon, Mount Olympus, and Alaska's Tongass and Chugash forests—but his proudest conservation achievement was the Reclamation Act, which promoted ecologically destructive dams and irrigation schemes throughout the arid West. Roosevelt loved the outdoors and its creatures, but he believed that mankind had not only a right but a duty to exploit nature: "Conservation means development as much as it does protection."

The preservationist and utilitarian strains of American conservationism found common ground in the defense of Florida's birds. It was hard for anyone to defend the systematic extermination of such lovely creatures, just so upscale matrons could wear feathers in their caps. "This is the last pitiful remnant of hosts of innocent exquisite creatures slaughtered for a brutal, senseless, yes, criminal millinery folly," Reverend Job wrote after photographing a shot-out rookery. This became a mainstream view, supported by almost everyone except hatmakers and hunters. By 1900, "Audubon societies" devoted to bird protection laws were forming nationwide, despite the trigger-happy proclivities of their namesake; the Florida Audubon Society's honorary officers included Flagler as well as his political nemeses, Jennings and Roosevelt. The society helped Governor Jennings and a state senator named W. Hunt Harris to usher a plume-hunting ban through Florida's legislature—not known as a hotbed of environmentalism—authorizing jail sentences of ten days and fines of $5 per bird.

In 1903, Frank Chapman, the shopping-district feather-counter who had become the curator of New York's Museum of Natural History, visited Roosevelt to request additional protection for Pelican Island. The president turned to his aides and asked if any law prevented him from declaring the five-acre mangrove key a federal bird sanctuary. They didn't think so. "Very well then," Roosevelt said, "I so declare it!" Pelican Island became the first of America's 535 national wildlife refuges. "Birds should be saved for utilitarian reasons— and moreover, they should be saved for reasons unconnected with dollars and cents," he said later. "To lose the chance to see . . . pelicans winging their way homeward across the crimson afterglow of the sunset . . . is like the loss of a gallery of the masterpieces of the artists of old time."

The butchery of wading birds just didn't feel right. Even a hardened Everglades pioneer named Charles McKinney, who routinely celebrated victories over nature in newspaper columns under the byline "Progress," was overcome by guilt after his one plume hunt. He could not bear to watch the group of crows gnawing on some orphaned egret chicks whose mothers he had just killed. Somehow, that didn't feel like Progress at all. "It looked too hard for me," McKinney wrote. "I decided that I did not think it was doing God's service, and I never went on that kind of hunt anymore."

A Martyr to the Cause

THE STATE HAD NO SHERIFFS canoeing the Everglades beat, so the National Audubon Society hired its own bird warden to enforce the pluming ban, a thirty-two-year-old ship captain from Flamingo named Guy Bradley, a reformed plumer who knew every nook of south Florida's rookeries. Kirk Munroe, who recruited him for the job, described him as "a sturdy, fearless fellow, filled with a righteous indignation against the wretches who . . . are using every effort to kill off the few remaining birds of that section."

Bradley had tagged along on his first plume hunt at age seven. As a teenager, he had been one of the shooters on Jean Chevelier's expedition in the Everglades, the "Guy" in entries like "Guy killed eight." And as a young man working as a surveyor for Flagler's railroad, he had done some plume hunting on the side. But he later renounced bird slaughter, and eagerly accepted the $35-a-month Audubon job. A short but rugged man with thinning hair and a bushy mustache, Bradley wasted no time making his presence felt in the Everglades, patrolling nesting sites, issuing citations, even searching the boats of gator hunters to make sure they were only hunting gators. He also gave tours to visiting ornithologists, who praised his conscientiousness but worried about his safety. Bradley was thrusting himself between armed men and their livelihood in the middle of nowhere, bringing the law to a wild frontier that hadn't exactly been clamoring for it.

The most overt threat was sixty-year-old Walter Smith, an ornery Confederate sharpshooter who had been blinded in his left eye at the Battle of the Wilderness. He had once been on good terms with Bradley, but the relationship had soured over local politics. Then, in the winter of 1905, Bradley ensured Smith's lasting hatred by arresting him for plume hunting, and arresting his teenage son, Tom, twice. "You ever arrest one of my boys again, I'll kill you," Smith warned.

On the morning of July 8, Bradley was home in Flamingo with his wife and two boys when he heard gunfire across Florida Bay. He peered across the glassy water and saw Smith's blue schooner anchored at Oyster Keys, two small mangrove islands where Bradley had once done some poaching with Smith's sons. He could see that the Smiths were again terrorizing the local cormorants, so he rowed his dinghy across the bay and announced that he was arresting Tom again. Smith demanded to see a warrant. "I don't need one," Bradley replied, according to Smith's account. "I saw them shoot into the rookery and I see the dead birds. Put down your gun, Smith." Smith claimed that Bradley cursed at him and wildly fired his nickel-plated pistol, burying a bullet in the schooner's mast. Smith then shot Bradley to death with his .38-caliber rifle.

Smith was arraigned in Key West by the local prosecutor, W. Hunt Harris, the senator responsible for Florida's bird protection law. But Smith knew how justice worked on the frontier, and a week later, Harris showed up to court as *Smith's* attorney. A lawyer hired by Audubon to monitor the trial noted that Smith's self-defense claim was outlandish; for starters, Bradley's pistol still had all six bullets in its chamber. But the stand-in prosecutor never challenged Smith's testimony, and never checked his mast for the purported bullet from the unfired gun. Smith was soon a free man, and the carnage resumed in the Everglades. "There is no community sufficiently law-abiding to leave a bank vault unmolested if it were left unprotected," Frank Chapman sighed. "We have given up. We can't protect it, and the rookery will have to go."

But Chapman spoke too soon. Bradley's martyrdom was chronicled in the press in New York, the center of the fashion industry, and pressure began building for a state ban on feather imports. Milliners claimed it would eliminate 20,000 jobs, but the New York legislature eventually passed the bill with only three dissenting votes. The feather craze faded, except among prostitutes, which only hastened its abandonment by everyone else.

As the demand for plumes relaxed, at least 250,000 wading birds came back to the Everglades. It was still an ideal habitat, with abundant food and seasonal dry-downs; it was still the Everglades. The adventure writer Zane Grey, the president of a south Florida fishing club, described a creek near Cape Sable after the return of the white ibis, known as curlew:

Though we saw birds everywhere, in the air and on the foliage, we were not in the least prepared for what a bend in the stream disclosed. Banks of foliage as white with curlew as if with heavy snow! With tremendous flapping of wings that merged into a roar, thousands of curlew took wing, out over the water. . . . It was a most wonderful experience.

TURN-OF-THE-CENTURY conservationists stopped the annihilation of the birds of the Everglades. But they had no problem whatsoever with the drainage of the Everglades.

In fact, Florida's conservationists led the fight to drain the swamp. They saw reclamation as the essence of conservation, an eminently wise use of natural resources. South Florida's leading conservationist, John Gifford of Coconut Grove—the first American to earn a doctorate in forestry, and the cofounder of a national magazine called *Conservation*—declared that the reclamation of the Everglades would be "the greatest conservation project in the United States."

A pedantic Cornell-educated professor with a booming voice and a withering disdain for those who questioned his theories or interrupted his monologues, Gifford was a bank president, a land speculator, and a home builder, as well as an academic, and, like President Roosevelt, he scoffed at the idea that conservation should be synonymous with preservation: "Do not think that conservation is merely saving and hoarding things. Conservation means sane use. Conservation fights for those things which will benefit the greatest number of the present and coming generations." Conservation meant the opposite of waste, and the Everglades—even if it provided a home to pretty birds—was clearly a wasteland.

Gifford threw down the gauntlet to his fellow Floridians, challenging them to convert the useless swamp into a productive civilization:

In southern California, the hand of man has produced a highly developed and attractive region with no resources except vim and climate. In southern Florida, we have the resources, but the vim has been lacking. We have been reposing since the Seminole War. . . . but it is this grappling with nature which develops the latent forces within the man. The coming age is to be an age of conquest, the conquest of nature, the reclamation of swamplands and the irrigation of deserts.

Gifford backed up his words with deeds, importing a thirsty Australian tree called melaleuca to help drain the swamp. He planted a teaspoonful of melaleuca seeds near Biscayne Bay and Davie, and the fast-growing tree with absorbent roots and white papery bark began spreading through the Everglades, sucking the wet out of wetlands. "It is a natural swamp tree," Gifford exulted. "It has few, if any, equals in the plant world."

Not even Gifford could match the vim of *Conservation's* energetic editor, Thomas Will, a self-made man who had worked for Gifford Pinchot at the Forest Service, and would earn a reputation as the John the Baptist of the Everglades. Born in a log cabin on his father's Illinois farm, Will attended country schools for just a few months every winter; with relentless determination, he worked his way through Harvard, and eventually became the president of Kansas State Agricultural College. "He was capable of extreme exertion, both physical and mental, and apparently never tired," his son later recalled. Will was also a born reformer, invariably championing "the underdog against vested interests," losing his job at Kansas State for advocating populist causes. So he moved to Washington to join Pinchot's push for forestry reform, then took over *Conservation* after his appointment expired. In his late forties, he happened to edit one of Gifford's articles on the potential of the Everglades, and decided to visit to see what all the fuss was about.

Will was so inspired that he quit the magazine, and devoted the rest of his life to Everglades drainage and development—first as the region's leading promoter in Washington, then as the founder of a pioneer settlement called Okeelanta. Will started the Florida Everglades Homebuilders Association, the Everglades Farming Association, and the South Florida Development League; he literally mounted soapboxes to preach the virtues of the swamp on the streets of Fort Lauderdale. "Remember, I'm on the job all the time, seeking our Glades' salvation," he wrote to a friend. Will was convinced that the Everglades would be the ultimate conservation challenge of the twentieth century, and it would consume him for twenty-seven years.

But it would take more than Gifford's trees and Will's passion to reclaim the Everglades. It would take drainage canals and political leadership. The man who led Florida into its age of conquest was Napoleon Bonaparte Broward, another crusader for progress and conservation, a politician as colorful and forceful as his name suggested. No one would ever accuse Broward of lacking vim.

NINE

"Water Will Run Downhill!"

Yes, the Everglades is a swamp; so was Chicago 60 years ago.

— Florida governor Napoleon Bonaparte Broward

NAPOLEON BROWARD declared war on the swamp during his 1904 campaign for governor, unfurling giant multicolored maps of the Everglades at campaign rallies, promising to bust a few holes in the coastal ridge and create an instant Empire of the Everglades. "It would indeed be a sad commentary on the intelligence and energy of the people of Florida to confess that so simple an engineering feat as the drainage of a body 21 feet above the level of the sea was beyond their power," he taunted his audiences. Here was a challenge that Henry Flagler had rejected as too big and expensive, but to Broward, it was simple and necessary. If the turbulent floods of the Nile and the Mississippi could be controlled by man, he demanded, then why not the comparatively gentle overflows of the Everglades? His mantra was: Water Will Run Downhill!

Broward was a force of nature, a rugged man of action who looked and behaved like a rustic Teddy Roosevelt. Awestruck journalists gushed about this "man among men" with "a jaw like that of the intense fighter who does not know the meaning of defeat," possessing "the faith that moves mountains, the determination that brooks no resistance and the energy that knows no weariness," not to mention "the driving force of a dynamo and the unswerving steadiness of a trip-hammer." He seethed with contempt for nitpickers "who quibble now, and stand on the bank and shiver and shake, instead of plunging in and doing something." He bristled that if the critics had their way, the Everglades would be a wilderness dominated by Indians until the next millennium: "It might be said of me, and perhaps of every other man who has a desire to accomplish something for the good of mankind, that he belongs to that class 'who rush in where angels fear to tread.'" He served only four years in statewide office, and dredged just a few miles of canals, but it is no coincidence that Florida's drainage era is still known as the Broward Era. "Had it not been for Broward," said one contemporary, "the talk of draining the Everglades might have run on for several more generations without the reclamation of a single acre."

Broward has been vilified by modern environmentalists for his intense assault on the Everglades, but he was considered a staunch conservationist in his day. He supported strict laws to protect fish, game, birds, and oysters, and his top priority was the reclamation of a swamp for agriculture and development. Broward never stopped to think what draining the Everglades might do to the fish, game, birds, and oysters that lived there, but hardly anyone did. The conservationist John Gifford dedicated his book of Everglades essays to Broward, explaining that "the man who makes two blades of grass grow where only one grew before is the proverbial public benefactor, but the man who inaugurates a movement to render 3,000,000 acres of waste land highly productive deserves endless commendation."

Broward was also a progressive—an antirailroad, anticorporation, anti-Flagler populist. His crusade for Everglades drainage was not just a fight for man against nature; it was a fight for ordinary Floridians against "the seductive and enslaving power of corporate interests" who monopolized state lands without improving them. Flagler and other railroad barons, he complained, were "draining the people instead of the swamps." At a time when the richest one percent of Americans owned half the nation's wealth, when forty-two corporate trusts controlled at least 70 percent of their industries, Broward wanted to turn the Everglades into a place where ordinary people could improve their lot in life through hard work. That's what he had done.

"We Were Not Discouraged."

NAPOLEON BONAPARTE BROWARD was born on a plantation near Jacksonville in 1857, a child of Florida's antebellum elite; his grandfather had served on the state legislative committee that first proposed draining of the Everglades. But when Napoleon was five, the Browards fled during the same Union advance that killed the train passenger sitting next to David Yulee. When they returned home after the war, they had almost nothing left. Their slaves had been freed, their plantation looted and burned. They had no home, no mules, no fences; they had to sell most of their land to pay taxes. The Browards could not afford to send Napoleon to school, so he stayed home and cleared overgrown fields by hand. For his first crop, he planted four sacks of potatoes, which yielded only one sack at harvest time. He also planted peas that died, and sugarcane that was trampled by a neighbor's cattle. "We were not discouraged, but immediately went to work," he later recalled.

It was a hardscrabble childhood, and it only got harder: Napoleon was orphaned shortly after he became a teenager; one of Broward's letters suggests that his mother, a longtime invalid, committed suicide after his father became an alcoholic. His father, a Confederate captain who was never the same after the war, caught pneumonia and died after spending a rainy night by her grave. For a while, Napoleon and his brother tended the farm themselves, two miles from their nearest neighbor, sleeping with their rifles by their beds, their knives within reach, their quilts pulled over their heads to drown out the screech owls. He later took jobs rafting logs, splitting rails, and tending an orange grove. Napoleon finally found his calling as a deckhand on a St. Johns River steamboat. Nature, he liked to say, was a better teacher than school—and a lot more fun. After a stint as a crewman on a schooner off Newfoundland, he returned to the St. Johns as a captain during the tourism rush of the 1880s and married his business partner's daughter. But his wife died in childbirth, and his baby died six weeks later. Tragedy seemed to be stalking him.

Broward again buried his sorrow in work, taking over the fastest steamer on the St. Johns, learning every bend of the river, thriving financially as it became a vibrant liquid highway. By the time he was thirty, according to his biographer, Samuel Proctor, he was "every inch the prosperous riverboat captain"—six-foot-two and over 200 pounds, with a neatly trimmed walrus mustache and a sunburn. Broward was well liked for his country bluster and bonhomie, well respected as an upright businessman who refused to allow liquor on his steamer. He was clearly an up-and-comer, expanding his riverboat service into a sizable salvaging, shipbuilding, and dredging business. He bought a lumberyard and a grist mill, and invested in phosphate mines. When he remarried, a local paper called him "one of Jacksonville's strong and manly young men [whose] high character, joined to force and energy, have already given him influence and friends."

IN 1888, Broward's influence and friends got him appointed sheriff for the Jacksonville area. He made an immediate splash by leading raids on illegal casinos, and exposing efforts by gamblers to bribe him. He cemented his reputation for integrity by crusading to keep a particularly scurrilous form of entertainment out of Jacksonville: a boxing match for the heavyweight championship of the world. A judge let the fight proceed, but after Gentleman Jim Corbett defended his title with a third-round knockout of Charles Mitchell, Sheriff Broward charged both fighters with assault. Broward soon emerged as a charismatic leader of Jacksonville's "straightouts," a wool-hat reform faction of the Democratic Party, accusing the silk-hat ruling faction of handing over the state to corporations. He had learned in the phosphate business how railroads puffed freight costs, and he attacked them with folksy rhetoric on the stump. "He is not one of the high-falutin spread-eagle kind of orators, but he is an entertaining talker," a reporter observed. In 1894, when Florida's corporation-friendly governor ousted Broward for trying to crack down on election fraud, that only burnished his reputation as a reformer.

After rising to prominence as a high-minded guardian of the law, Broward rose to legend as an adventure-minded outlaw. When Spain declared martial law in Cuba, President Grover Cleveland warned Americans not to take sides in the conflict, and the Spaniards offered a $25,000 bounty for any "filibuster" who assisted the island's rebels. But Broward led a series of expeditions to Cuba on his tug *The Three Friends,* smuggling guns, explosives, and guerrillas to the junta led by poet and revolutionary José Martí. At a time when the publishers William Randolph Hearst and Joseph Pulitzer were competing to manufacture a war with yellow journalism, the ex-sheriff's filibustering made great copy, as he eluded Spanish gunboats and U.S. revenue cutters in high-speed chases in the dead of night, then jovially denied his involvement in the light of day. Filibustering was lucrative work, but with war lust rising, Broward was hailed as a humanitarian hero. The citizens of Jacksonville threw him a parade, and the more Cleveland's

administration harassed him—boarding and detaining his boat, then charging him with violating neutrality laws—the higher his star rose. His case
eventually reached the Supreme Court, which ruled that the government
had the right to seize *The Three Friends*. But by that time, Cleveland was out
of office, America was heading for war with Spain, and Broward had
regained his sheriff's job in a landslide. The ruling was never enforced.

TEDDY ROOSEVELT ONCE needled Broward that if it weren't for *The Three
Friends,* he never would have become governor. (Broward shot back that if
it weren't for *The Three Friends,* Roosevelt never would have become president.) But Broward's stint as a revolutionary hero was only part of his
appeal in 1904. He was also a native Floridian, a log-cabin pioneer, a self-
made businessman, a reform sheriff, a father of nine, and a brilliant cam-
paigner. He knew he had little chance in Florida's cities, where Flagler and
other railroad men controlled the press, so he barnstormed the backwoods,
making his case to villagers who didn't read the papers. "I'm going
to . . . talk to the farmers and crackers and show them their top ends were
meant to be used for something better than hat racks," he vowed. "I'm going
to make 'em sit up and think. They won't mind mistakes in grammar if they
find I'm talking horse sense."

Broward made the Everglades and the Internal Improvement Fund the
centerpiece of his campaign, warning that his Democratic primary opponent, Congressman Robert Davis, would overturn the Jennings land freeze
and resume the handouts to Flagler and other railroad barons. Broward
pledged not to give any swampland to any corporation. Instead, he proposed to reclaim the Everglades at a cost of only $1 per acre, then sell the
reclaimed farmland on behalf of taxpayers for $5 to $20 per acre. He said
the profits would pay for state government, plus an extra $500,000 a year
for public schools.

Broward's drainage dreams were ridiculed by the state's conservative
papers, especially Flagler's. They printed rumors that Broward was Catholic,
an Apache Indian, a fop who dyed his mustache daily. They called him a
rube and a rabble-rouser. But Broward kept roaming the hinterlands with
his map of the Everglades, portraying their attacks as proof of their corruption: "If a graveyard has been despoiled, you know a hyena has been
abroad, and if an effort to save land for the people meets with the vituper-
ative, slanderous and lying opposition of certain purchasable newspapers,

you know that behind such bureaus of misinformation . . . will be the land-grant corporations of the State." He bashed Davis as a corporate shill, a former lawyer for railroads who still accepted free passes from railroads. "Is it for legal services, or is it because he is a member of Congress?" Broward demanded. "If only as a Congressman, then what consideration does he give them as a Congressman?" Broward also published a campaign autobiography, with Horatio Alger headings like "Pioneer Days of My Early Boyhood," "Rafting Logs for a Livelihood," and "Climbing Up in Life." He once spotted a man reading it on a train, snickering about his early difficulties planting peas. "Laugh if you like," Broward said, "but that book is a votemaker." He was right. Broward upset Davis by 714 votes in the Democratic primary, the only election that mattered in the South.

Broward roared into Tallahassee with a bold, progressive agenda, calling for higher teacher pay, humane treatment of prisoners, a crackdown on corruption, a repeal of the Flagler Divorce Law, and state-run life insurance modeled on a program in New Zealand. But his top priority was to extract the potential of the Everglades for a new class of yeoman farmers, to "tap the wealth of the fabulous muck."

"The Everglades of Florida should be saved," Broward declared in his inaugural address. It was a heartfelt sentiment, with none of the ecological concern those words would imply today. Broward meant that the Everglades should be saved from railroad executives, and from oblivion: "They should be drained and made fit for cultivation."

The Chief Engineer

BROWARD WAS A MAN on a mission, and a man in a hurry. One of his first acts as governor was to appoint outgoing Governor Jennings, another drainage advocate, to be counsel to the Internal Improvement Fund. The two men immediately left Tallahassee to inspect the Everglades with Captain Rose, the former Disston and Flagler drainage superintendent who was now the state chemist. Broward soon sent the legislature a special message on drainage, proposing to start lowering Lake Okeechobee with a canal east to the St. Lucie, the original strategy of Disston's chief engineer, James Kreamer. Broward predicted that the canal would be complete in eighteen months, basing his calculations on Kreamer's old reports, and on "a personal knowledge of the character and quantity of work to be done."

Broward had run dredges for years, and took his own soundings on his trip with Jennings and Rose, but citing "personal knowledge" as the basis for the world's largest drainage project was essentially winging it. Broward refused to commission feasibility studies or engineering surveys; he said he already knew it was feasible to reclaim the Everglades, because water ran downhill. He grumbled that by the time the studies were done, he would be dead and the fund would be bankrupt. He was sure the skeptics would come around once his dredges started digging and the waters of the Everglades started receding: "If my friends will hold the knockers in check, we can soon make a convincing ocular demonstration." Broward also had a legal incentive to make the dirt fly: The rationale for halting the land giveaways had been that the corporations on the receiving end were doing nothing to improve state swamplands, a rationale that would crumble if the state did nothing as well. "I consider the launching of a dredge absolutely essential to the success of the litigation," Jennings advised him.

Broward quickly convinced the legislature to create the Everglades Drainage District, which levied a 5-cents-per-acre property tax in most of south Florida to support the reclamation project. He also wrote an open letter to Floridians proclaiming the feasibility of Everglades drainage and the fertility of Everglades soils, savaging the project's knockers as corporate conspirators intent on controlling every acre of the state: "Shall the sovereign people of Florida supinely surrender to a few land pirates and purchased newspapers and confess that they cannot knock a hole in a wall of coral and let a body of water obey a natural law and seek the level of the sea? To answer yes to such a question is to prove ourselves unworthy of freedom, happiness or prosperity." Broward saw Henry Flagler as the captain of the land pirates, an "insidious enemy" who symbolized New York corporate greed. "He desires to own all lands that are for sale or rent, by any means," Broward wrote.

BROWARD ALSO DECIDED THAT the state would need dipper dredges to drain the Everglades—as opposed to ladder, grapple, or hydraulic dredges—so he traveled to Chicago to buy them. Governors generally delegate purchases of heavy machinery, but Broward had staked his reputation on this project, and he insisted on controlling every detail—not only the politics and promotion, but the design and construction as well.

One day, the governor wrote to tell a contractor to shorten the specifications for a dredge boom from seventy to sixty-five feet; the next day, "after thinking the matter over carefully," he ordered it cut to sixty feet. He wrote another letter asking why a bulkhead had been thickened by one-sixteenth of an inch. His letters to the project's chief engineer, John Newman—the drainage booster who had accompanied James Ingraham across the Everglades, and had predicted that its redeemer would be the philanthropist of the age—were similar marvels in micromanagement: "The dipper should never be hauled back to the rear of the line immediately below the center of the shaft, but should be perpendicular to the horizon, then pulled by the tackle. . . ." There was no doubt who was the project's real chief engineer. "The governor just naturally sweats dope about the Everglades and drainage," one supporter gasped after a three-hour Broward speech. "He is chock full of it and I believe he could talk all night on the subject without ever being at a loss for a word."

Like Disston, Broward abandoned his initial plan to start lowering Lake Okeechobee with a canal east to the St. Lucie. Instead, he decided to begin with diagonal canals through the Everglades from the New River in Fort Lauderdale—where he enjoyed more political support, and where the ditches could help reclaim state-owned swampland as well as lower the lake. By the end of his first year in office, two state dredges, the *Everglades* and the *Okeechobee,* were under construction in Fort Lauderdale—a hamlet of fifty-two residents before the project came to town—and the governor was making monthly visits to make sure they were built right. In 1906, Broward triumphantly launched the *Everglades* into the north fork of the New River. The christening was a bit premature—the state had not even surveyed a line through the swamp—but the governor was determined to get the dirt moving, and start providing convincing ocular demonstrations.

THE CANALS DID MAKE an instant impression, as tomatoes, potatoes, and cabbage began to sprout in former marshlands at the edge of the Everglades. The developer R. P. Davie began promoting his land just west of Fort Lauderdale—then called Zona, because its first settlers came from the Panama Canal Zone, but later renamed Davie—as "The First Improved Town in the Everglades." The populist rabble-rouser Thomas Watson, the editor of *Watson's Jeffersonian Magazine,* declared that Sir Christopher Wren's work on St. Paul's Cathedral was trivial compared to Broward's work in the eastern

Glades. An engineering magazine predicted that the Everglades would soon relieve America of its addiction to foreign sugar, and that Florida was about to become "one of the richest and most important states in the Union."

"It has been said that man can never improve on nature, but one view of this magnificent place contradicts that," another journalist wrote after visiting a sixty-acre Everglades vegetable farm. "Here the hand of man has fashioned out of a once valueless and despised tract of land not only a thing of beauty, but one of great utility."

Who said that man could never improve on nature? Swampland the state had sold to settlers for 25 cents an acre now produced harvests of $600 an acre for tomatoes, $1,000 for lettuce, $1,500 for celery. At a time when farmers were struggling to survive on 160-acre homesteads out west, the farmer Walter Waldin netted $3,400 on six acres in six months in the Everglades—after building a home and feeding a family of five. The Everglades, he wrote, was "a country where fortunes have been and will be made, with probably less exertion, on a smaller body of land, under more pleasant circumstances, and in less time than in any known place on earth"—the original Henry Perrine dream.

Promoters began spreading word of a balmy paradise where the muck was richer than manure, where summer spent the winter, where prosperity beckoned to the lazy and the poor. For a fraction of the cost of a western irrigation scheme, a drainage project was converting the Everglades into a new Nile Valley. "My prophecy is that this great Everglades district will not only develop into a most beautiful and prosperous country, but will prove itself *the* Eden of America," Waldin wrote.

THE KNOCKERS, however, kept knocking.

They attacked the drainage project as "a wildcat scheme," a "sinful waste of the dear people's money," "the death knell of many of the industries of our State," a plot to enrich Broward and a cabal of land sharks. "Of all the foolish ideas that ever entered the brain of man, the draining of the Everglades is the most nonsensical," one paper sneered. The Flagler-controlled *Times-Union* ran cartoons of the fund's trustees robbing children and teachers, and compared the governor to the czar.

Critics wondered why the governor would try to drain such an inaccessible marsh when Florida still had so much land available near existing roads and markets. And they ridiculed his armchair engineering, predicting that his canals would cause fires from overdrainage as well as floods from underdrainage, speculating that kindergartners must have prepared his dollar-an-acre cost estimates. One writer compared the dredges at work in the Everglades to Mrs. Partington, the legendary British storm victim who tried to dry up the Atlantic Ocean with a mop. Frank Stoneman, the editor of a newspaper that would become the *Miami Herald,* was the most persistent skeptic, arguing that it might make sense to try to reclaim small portions of the Everglades with a gradual approach, but not to drain the entire marsh at once without scientific or economic analysis. "What would be thought of a railroad corporation that would start the construction of a road without first making complete surveys of the entire route . . . [or] computing costs?" he asked.

As usual, Broward dismissed his critics as land-grabbing corporate stooges, and newspapers that aired the criticism as "cuttlefish literature." He called for stricter libel laws, sued the *Times-Union* for slander, and refused to sign papers confirming Stoneman's election to a judgeship. He scoffed that if drained swamps really could burn, "the great bogs of Ireland would have been ash heaps long before St. Patrick drove out the snakes."

Broward stepped up his attacks after the state's land grant corporations—which owned 95 percent of the drainage district—persuaded a judge to grant an injunction blocking the drainage taxes. He launched a statewide campaign for a constitutional amendment to overturn the injunction, accusing Flagler and other real estate moguls of scheming to steal the Everglades. "This rich, fertile land, so admirably situated, so rich in its soil, is your land," Broward told one audience. "These corporations want it, and that is why they belittle it. They want to rock you to sleep and then take it away from you." He mused that if he were to change his mind and hand the Everglades to Flagler, "the *Times-Union* would come out in letters big and black and say 'Broward is the greatest governor that Florida has ever had,' and the little *Metropolis* would say, 'me, too.'"

Flagler was a convenient bogeyman, but he had yet to receive a single acre of land from the state, even though he had built 250 miles of track to state specifications. He had poured capital into south Florida when no one else had dared, financing civic improvements, creating jobs, even spending his own money on drainage projects. "Is there any steal in this?" his aide James Ingraham asked. Florida's voters didn't think so, and rejected Broward's amendment. Broward rammed a new drainage tax through the legislature anyway, but the corporations refused to pay it, and the Internal Improvement Fund began running out of cash again. The Broward Era's momentum was stalling.

Reclaiming the Reclamation

BY 1908, DRAINAGE WAS Broward's claim to fame. He was named president of the National Drainage Congress, and President Roosevelt invited him on an inspection tour of Mississippi River drainage projects, the trip on which they traded quips about *The Three Friends*. Broward was even bandied about as a possible running mate for William Jennings Bryan, "because he is the man who is draining the Everglades of Florida."

He wasn't draining it very quickly, though. His administration touted the project's "splendid" and "marvelous" success, hailing "achievements far beyond the most sanguine or hopeful expectation of those in charge." But the *Everglades* and *Okeechobee* had dug only five miles of canals, reclaiming less than 12,000 acres. At that rate, a critic noted, it would take a century to drain one million acres, a mere fraction of the Everglades. Dredging costs were much higher than Broward had anticipated, and two new dredges, the *Miami* and *Caloosahatchee,* were taking much longer than expected to build. Yet his plan seemed to expand by the day. Broward now wanted at least six dredges to dig at least twelve major canals as well as smaller lateral canals. Basically, he planned to dig more and more ditches to carry more and more water until the Everglades was dry.

But all that digging required cash, and the Internal Improvement Fund was paralyzed. It tried to sell swampland, but buyers were scared off by the lawsuits the corporations had filed after Governor Jennings had revoked 5.3 million acres' worth of promised land grants; few investors wanted to pay for land that might turn out to belong to Henry Flagler. For all of his vituperation about "land pirates," Broward realized he had to settle their lawsuits if he wanted to drain the Everglades. He eventually resolved their claims for 13 percent of the disputed land; Flagler settled for 260,000 acres on his claim of 2.6 million acres. The corporations still refused to pay drainage taxes, but now at least the state could raise money for drainage by selling its land. And much of the Everglades still belonged to the public.

BROWARD NEEDED A Disston–style white knight to bail out the fund; and he found one in Richard Bolles of Colorado Springs, a deep-pocketed developer with a silver tongue and a gambler's heart. Bolles looked like a diminutive Vladimir Lenin, with arched eyebrows and a white mustache and goatee, but his expensive suits and free-spending habits reflected decidedly capitalist attitudes. "Money will assuage almost every other grief," he wrote in his diary, "and the want of it, I really believe, is the only thing that in my case, with my nature, gives me unbearable suffering." He reflected that no modest man had ever made a fortune, and that only three traits were necessary for getting rich: impudence, impudence, and more impudence.

Dicky Bolles had all three. The son of a New York doctor, he skipped college to become one of the youngest men ever to hold a seat on the New York Stock Exchange. But he abandoned Wall Street for the Wild West during the silver boom, making and squandering several fortunes in the Colorado mines, spending his money on wine, women, and polo ponies, berating himself in his diary about the fast-paced lifestyle and weak impulse control that led him into a costly divorce and financial ruin. "Suicide is reasonable for such a fool," he wrote. He urged himself to settle down and stop "running out after amusement and excitement, women especially." But then he struck it rich again with Aspen's famous Mollie Gibson mine, which he liked to call the best girl he ever had. Bolles craved respectability, but he admitted to himself that he was a scoundrel at heart: "You have acquired the habit of considering yourself a high-tone, liberal and honest man. Yet occasionally you have been frightened to sickness to see that you are plainly none of these—but low in taste, mean and tricky if you think it safe."

Bolles shifted from mining into real estate late in life, buying gigantic tracts of arid farmland in Colorado and Oregon on the cheap, then subdividing and selling them on the promise of irrigation. Some of his buyers later discovered that their parcels were rocky, mountainous, unredeemable wastelands, "so high that a person could not even carry a bucket of water on top of it, letting alone irrigat[e] it." But by that time, Bolles had cleared millions of dollars in profits, and had decided to try to replicate his success marketing western drylands with Everglades wetlands. "The people of our country are land-hungry," he explained. And the governor of Florida was investor-hungry.

Bolles met Governor Broward and ex-Governor Jennings at the 1908 Democratic convention in Denver; William Jennings Bryan had decided to choose a midwestern running mate, so Broward had plenty of time for other business. By the convention's end, the Florida men had brokered a deal for Bolles to buy 108,000 acres in the Everglades from a land grant corporation. Bolles then agreed to buy another 500,000 acres from the Internal Improvement Fund for $1 million, the same amount Disston had paid for eight times as much land. Once again, the deal rescued the fund from a financial crisis. But Disston had pledged to drain the Everglades on his own. Bolles got the state to pledge to do the job for him.

Broward and his fellow trustees promised to dredge at least four major diagonal canals from Lake Okeechobee through the Everglades and the coastal ridge, following the natural outlets of the transverse glades. The canals would empty out to sea through the Miami River in Miami, New River in Fort Lauderdale, Hillsboro River in Boca Raton, and Lake Worth in West Palm Beach. They would drain Lake Okeechobee as well as the Everglades marshes, allowing the development of millions of acres. The trustees also agreed to expand Disston's short canal west to the Caloosahatchee and out to the Gulf, which would lower the lake even faster. The long-awaited canal east to the St. Lucie was delayed yet again, but the state finally had an official drainage plan with a designated funding source. Bolles declared the project was bringing "a magnificent development to Florida." Broward boasted that "all doubt of its ultimate success has been removed."

Frank Stoneman continued to sound alarms about the "glittering promises" of Everglades drainage. He had little interest in preserving the Everglades—his daughter, Marjory Stoneman Douglas, would feel differently one day—but he fretted about "a large number of dissatisfied people charging this community with being a set of swindlers." He suggested that "the hysterical enthusiasts should at least leave a few words to express their wonder, amazement, surprise, astonishment, admiration, gratification, pleasure, delight and thankfulness" for the first shred of evidence that the project would work. But that evidence seemed to arrive when the state plan was endorsed by the federal drainage engineer investigating the Everglades—the ultimate rebuttal to the knockers.

The Wright Track

CHARLES ELLIOTT, CHIEF of the U.S. Department of Agriculture's new drainage bureau, had written an ingratiating letter to Governor Broward immediately after the 1904 election, offering the federal government's help in reclaiming the Everglades. "You are directing attention to the value of millions of acres which have from time immemorial been regarded as irredeemable and a menace to the healthfulness of the State," Elliott wrote. "The reclamation of this area is of no little importance, not only to the State but to other sections of the country." Broward had welcomed the bureau's assistance, and President Roosevelt's agriculture secretary, James Wilson, had assured him that federal engineers would be glad to help study how, not whether, the Everglades could be drained.

Broward did not really want an investigation of the Everglades, just a rubber-stamp federal endorsement for his drainage plans. But Elliott, who had written two textbooks on drainage, demonstrated an inconvenient penchant for objectivity; he believed in drainage, but he considered the drainage of the Everglades as difficult as it was desirable. Elliott had the gall to suggest that reclamation could cost $50 an acre, and that Broward was "carrying this matter forward in a very energetic and possibly arbitrary manner." Elliott was a fastidious, apolitical engineer, a by-the-book bureaucrat who kept a private supply of pencils for nonofficial use. He was no rubber stamp.

Fortunately for Broward, the lead investigator on the Everglades study was one of Elliott's underlings, James Wright, a former high school math teacher with no formal engineering training and none of Elliott's ethical qualms. Wright's main talent was speechmaking; the bureau often sent him around the country to deliver lectures about the wonders of drainage. When he did oversee surveys, he routinely accepted gratuities from landowners and drainage companies. He once appeared before North Carolina's legislature as a federal official to lobby for the sale of a lake bed, neglecting to mention that he had lined up $45,000 in kickbacks—twenty times his annual government salary—that were contingent on the sale.

Wright was a smooth political operator in a world of engineering nerds, and he was happy to do Governor Broward's bidding in the Everglades. Shortly after he began working in Florida, he tipped off a Broward aide and drainage lobbyist named J. A. Dapray that railroad men were giving members of Congress tours of other southeastern swamplands. "Mr. Wright is afraid that the federal drainage force which he had hoped to have at work in Florida may be sent to the northern Louisiana and Arkansas section, and he himself may be ordered there in charge," Dapray warned Broward in a letter marked "Important and <u>Confidential</u>." "This in Mr. W's opinion would be fraught with very serious consequences to the drainage plans in Florida, and beside would prevent his carrying out his ambitions to help you in the great work of draining the Everglades."

Dapray urged the governor to do whatever he could to keep Wright in charge, predicting a big political payoff: "With Mr. W. at the head, I can foresee for you a great change in public sentiment in the sections that have opposed your plans, and there would be a great tidal wave of popularity that could . . . sweep you from the Executive Office into the U.S. Senate." Clearly, the rubber stamp had arrived. "I feel sure he can be trusted as a loyal friend of Florida drainage and yourself," Dapray wrote.

With Broward's help, Wright retained control of the Florida study, and made little effort to conceal his support for Broward's project. He gave a booster-style speech in Miami before his survey was even complete, announcing that there were "no engineering difficulties to overcome in the draining of the Everglades." And in March 1909, Wright released some predictably rosy excerpts of his report to Florida officials and land developers—before his bosses in the drainage bureau had even seen it.

The Wright report declared reclamation to be perfectly feasible, endorsed Broward's plans, and seconded the governor's $1-an-acre cost estimate. Wright predicted that Everglades muck would produce America's most prolific crops without a pound of fertilizer. And he suggested that Florida could use its canals for dry-season irrigation as well as wet-season drainage. Everglades land companies quickly began citing the Wright report in advertisements promoting the Everglades as the "Tropical Paradise," the "Land of Destiny," "Nature's Gift to Florida." They promised "the richest land available," thanks to a project "assured to be thorough and complete." They argued that since a federal engineer "cannot afford to give out an exaggerated statement, his valuation of the land should be accepted as conclusive proof that the buyers of Everglades [farms] hold the most valuable assets obtainable." Bolles paid a Kansas City firm $400,000 to help his Florida Fruit Lands Company advertise the "Poor Man's Paradise," the "Land of Opportunity," the "Magnet Whose Climate and Agriculture Will Bring the

Human Flood." His ads even quoted Secretary Wilson warning that "doubting Thomases who were waiting for the Everglades to develop before buying would regret it all their lives."

That Wilson quotation was probably phony, but the Wright report was real, and the federal seal of approval was like an engineering endorsement from God. Even the skeptical Frank Stoneman had to admit that "the tentative experiments that have been made seem to point to the eventual success of drainage." Magazines predicted that the Everglades would soon supply winter vegetables to every American east of the Rockies, that its downpours would "disappear within a few minutes." William Jennings Bryan declared the project "one of the greatest enterprises on record."

By 1909, the state had dug only thirteen miles of canals, less than one-sixth of Disston's output. But John Gifford announced that the time for debate was over: "The drainage of the Everglades is well under way, and almost every unprejudiced person who visits this work becomes an enthusiastic convert. This is no idle dream or wild land scheme, but a feasible, practical piece of good business. . . . It is not a complex engineering problem; it is merely a matter of digging, so that the water in this great Everglades basin can flow into the sea."

AFTER WRIGHT'S ENCOURAGING words leaked into view, major Everglades landowners finally agreed to support the drainage project with 5-cents-an-acre taxes. In exchange, the state agreed to turn the project over to a private contractor, and to hire Wright as Florida's chief engineer at twice his federal salary. "We do not believe a more competent, honest, energetic and thorough man could be found than Mr. Wright," a legislative committee enthused. The new contractor purchased the state's four dredges, added five of its own, and pledged to speed up the work to excavate 200 miles of canals by 1913. After a cursory review, Wright announced that Broward's original plan would proceed without revisions.

The Broward Era was back on track. "There is no 'if' nor 'but' nor 'maybe,'" one booster publication declared. "The reclamation of the Everglades does not depend upon the life or the promise of any one man or any one corporation. IT IS BASED ON THE PLEDGE OF A STATE." The land syndicates dispatched propagandists across the nation to promote that pledge, promising "the greatest opportunity of the century," highlighting "the evident cooperation of the Government." The Florida Everglades Land Sales Company sponsored an Everglades lecture series by the editor-turned-promoter Thomas Will, published Everglades books by the conservationist John Gifford and the farmer Walter Waldin, and distributed *Florida Everglades Review* and *Everglade Magazine* "to report on the progress of America's Latest Empire."

"The Great West of Horace Greeley's day no longer exists," read the company's brochures. "Were Greeley alive he would now say: Go South, Young Man, Go South!"

Even Flagler's *Florida East Coast Homeseeker,* which had attacked the drainage project for years, now printed a special Everglades edition with a dredge on its cover, featuring articles titled "America's Winter Garden," "An Irrigation System Designed by Nature," and "Cities in the Everglades—Why Not?" The *Homeseeker* predicted that within a decade, the population of the new Palm Beach County would skyrocket from 5,000 to 750,000: "It is hard to keep within the bounds of belief in foretelling the future . . . as the possibilities here are greater than in any other part of the Union. The resources are unlimited and the opportunities unsurpassed."

UNFORTUNATELY FOR BROWARD, the Wright report was not done in time to sweep him into the Senate in 1908. He was narrowly defeated by Duncan Fletcher, who had worked as a railroad lawyer and had railroad support but managed to portray Broward as the railroad candidate, accusing him of settling Flagler's land claims on Flagler's terms. Fletcher also distributed brochures warning that "Broward's Drainage Scheme Will Bankrupt the State," lambasting the governor for plowing ahead without planning or funding after his constituents rejected his drainage amendment.

But that was just campaign talk. Senator Fletcher became a loyal supporter of Broward's drainage scheme after the election, pulling strings to get Congress to publish the full Wright report in 1911, a promotional masterstroke that implied the entire federal government had endorsed Florida's reclamation project. "United States Official Indorsement" trumpeted ads for the Florida Everglades Land Company. "First time in the history of the Government such a thing's been done!" In a letter to the Everglades promoter Thomas Will, who had helped persuade Senator Fletcher to take on the issue, the company's president exulted: "It is a peach!"

Broward's successor, Albert Gilchrist, a genial moderate who had distanced himself from drainage and the progressive movement during his campaign, also embraced the project after touring the Everglades with Broward during his transition. Gilchrist preached the gospel of the Everglades with a convert's fervor, and the pace of dredging accelerated almost tenfold during his term. A natural booster with a background in Florida real estate, Gilchrist gave exuberant speeches around the country promoting the Everglades, introducing himself as "the governor of a state draining 7,488 square miles."

"Opposition is rapidly disappearing," wrote the editor of the *Miami Metropolis.* "On the Glade land, with a little intelligence and application, a man may get larger returns on less capital than in any business that I know anywhere." Here was a dream world where pioneers could drop peels in the soil and watch potatoes sprout, where "sunstrokes and heat prostrations have never been known," where the drainage question, in John Gifford's scientific opinion, was no longer a question at all: "It is a question only in the minds of doubting Thomases who are prejudiced, who are ignorant or who belittle every project in which they have no hand and out of which they can make no rake-off."

IN FACT, drainage was still a question in the minds of some federal engineers, as a U.S. Department of Agriculture form letter made clear. Distributed to all prospective investors who requested information on the Everglades, the mimeographed circular warned that land values were "still largely problematical." It also cautioned that drained mucklands could be highly flammable, and usually required fertilizers to support crops. "Undoubtedly much time will yet be required before any considerable areas will be habitable and fit for cultivation," the letter said.

In other words, the department was alerting the public not to believe the Wright report—until the letter was abruptly withdrawn from circulation.

"I Call It Graft"

BROWARD AND JENNINGS continued to promote the Everglades after leaving the public sector. Both former governors took jobs with Dicky Bolles—Jennings as a lawyer, Broward as a celebrity spokesman. As the public face of Everglades drainage, Broward was paid $400 a month to provide testimonials for Florida Fruit Lands holdings. "I believe this Company is thoroughly responsible and reliable, and will comply with whatever representations or propositions they may make to you," he wrote one prospect. Never disclosing his own connection to the company, he told a North Carolina man that its Everglades lands were so good—and so certain to be reclaimed within two years—that he was buying some himself. Broward also peddled his name to the rival Florida Everglades Land Sales Company for $1,000, authorizing its officers to use the letter he had been sending on behalf of Bolles with one significant change: "You may strike out the words 'Florida Fruit Lands Co.' and it will imply to the Everglade lands owned by any company."

Broward stopped working for Bolles to run for Senate again in 1910. But the relationship became an issue when one of his Democratic primary opponents, a muckraking editor named Claude L'Engle, revealed that Broward and Jennings had received lucrative payoffs from Bolles while they were still in office. Bolles had given both progressive stalwarts 27,000 acres of land—a parcel approximately the size of Boston for each. "Some call this a bonus," L'Engle said. "I call it graft."

Broward claimed the tracts were "commissions" the politicians had earned by brokering Bolles's private 108,000-acre land deal at the 1908 convention, and insisted the work had nothing to do with their public duties. "I had a right to do it," he said. "There's no legal or moral law against a governor making a real estate transaction." But real estate brokers rarely receive *50 percent* commissions. The more plausible explanation was that Bolles bribed his new friends to ensure favorable terms for his subsequent 500,000-acre purchase from the fund, a transaction directly related to their public duties. Jennings, in fact, drew up the contract for that deal, and excused himself as the fund's counsel so that he could represent Bolles at the hearing where it was approved; his temporary substitute, the state's attorney general, ended up on the Bolles payroll as well.

The people did not care. Despite gallstone attacks that limited his campaign, and a new allegation of a $24,500 cash kickback from Bolles, Broward fought off L'Engle on his left and a procorporation incumbent on his right to capture the primary. Even if Broward had made some rake-off, voters did not believe his passion for Everglades reclamation was motivated by anything but a genuine desire to develop south Florida. They knew Broward had envisioned an American Italy rising out of the swamp, and they had every reason to believe his impossible dream was about to become a reality. Even L'Engle, in the same speech that accused Broward of rank corruption, conceded that his drainage project was a marvelous achievement, that history would revere him long after his mistakes were forgotten, that there were not enough bricks in Florida to build a monument tall enough to do him justice. "If Broward had the good fortune to be located in one of our north-central states," one admirer wrote, "without a question of a doubt he would be elected president."

BROWARD NEVER DID become president, or even senator. He died suddenly after another gallstone attack before he could take office. He was fifty-three. "The stunning and hardly comprehensible announcement . . . fell with the same awesome effect that the sudden death of a near relative would bring," reported the *Metropolis*.

L'Engle was right: Broward's graft was instantly forgotten, and he was mourned across the state as "a great man in every sense of the word." Even his antagonists at the *Times-Union* saluted "his brain, his force, his power," and the Fort Lauderdale area—which had planned to call itself Everglades County—decided on Broward County instead. He was eulogized as a public servant of modest means, although his widow later netted a $167,500 profit from the land he had received from Bolles, the equivalent of $2 million in 2005 dollars.

Governor Broward reorganized Florida's universities, banned child labor in its factories, expanded its roads, and erected more state buildings than all his predecessors combined. But his most important legacy was his Everglades project—not only his four canals, but his no-turning-back attitude. Most Floridians had already accepted that the Everglades ought to be drained, but Broward made them believe that it must be drained, and would be drained. Florida now had a plan, with the money and the will to execute it. A month after Broward's death, the fund sold 50,000 acres in the Everglades for $15 an acre—sixty times what Disston had paid three decades earlier, and seven and a half times what Bolles had paid before Broward sent dredges into the swamp. The deal, Governor Gilchrist said, "placed the fund on Easy Street, with ample funds to complete the work." The Everglades land rush that Broward had always envisioned could now begin.

"The Florida Everglades will be dry in two years—that is the latest Big Fact for farmers," Flagler's *Homeseeker* declared. "There is no longer any doubt about this enterprise." A Florida legislative report officially concluded that the Everglades would soon be the garden spot of the world, and that its farmers would never have to fear frost, drought, or flood: "Your committee is of the opinion that the drainage operation will ultimately be a great success, and is one of the greatest undertakings of the age."

MEANWHILE, FLAGLER SPENT his twilight years on his own great undertaking, his overseas railroad to Key West. He put a dozen dredges and as many as 4,000 men on the job, far more than the Everglades ever had. Three hurricanes ravaged his construction sites, killing 140 men and wiping out huge chunks of track. The Roosevelt administration indicted his railroad for peonage in 1907, charging that workers were being held in the wilderness against their will. But the charges were eventually dropped, and the old man eventually proved his doubters wrong.

Flagler was eighty-two when he rode the first train to Key West. He was nearly deaf and blind, but he was still mentally keen. Long reviled as a rapacious robber baron, he was now hailed as a conquering hero, a living embodiment of progress. Thousands of citizens chanted for Uncle Henry; his workers gave him a plaque; a children's choir sang in his honor. "I can hear the children, but I cannot see them," Flagler whispered, overcome with emotion. Governor Gilchrist compared the railway to the Panama Canal, and one reporter compared Flagler to Moses, raising his rod and parting the sea. "Now I can die happy," Flagler said. A year later, he did.

Flagler was wrong about Key West. It was too small to become a great port, and too isolated to become anything more than a funky tourist attraction in the automotive age. The overseas railway was a financial disaster, and in 1935, it would be destroyed by the most powerful hurricane in U.S. history. It was rebuilt as the Overseas Highway.

But Flagler was right about south Florida. In his mind's eye, he had envisioned a flood of humanity following his tracks down the coast. The flood was on its way.

TEN

Land by the Gallon

The real estate propaganda said: "Take a tent, a bag of beans and a hoe, clear a few rows in the sawgrass, plant the seed, and in six weeks you will have an income." . . . That may have provided an income for the land offices, but the settlers found out differently.

—Everglades settler John Newhouse

The Great Utopia

THOUSANDS OF NORTHERNERS descended on Fort Lauderdale in the spring of 1911, transforming the piney-woods hamlet at the mouth of the former New River—now the North New River Canal—into a swarming tent city. "The Village of Yesterday Today a Seething Mass of Bustling Humanity," one headline blared. The lines stretching out of the post office were so long that one mail-seeker thought he had stumbled into a bank run. One land company set up a booth for befuddled newcomers to find their friends—and perhaps buy some real estate in the Everglades while they waited. That, after all, was why everyone was in town.

The visitors had arrived for the Everglades land lottery, a gimmick concocted by Dicky Bolles, who had used similar methods to sell parched land out west. The contestants all agreed to pay Bolles $240 over two years, in exchange for swampland to be chosen in a random drawing—anywhere from ten to 640 acres in the Everglades, plus a town lot along the coast in a planned community called Progresso. Most of the "swamp boomers" would only receive ten acres, but the Bolles literature had assured them that was just as good as 100 acres of farmland up north. Journalists proclaimed that south Florida's long-awaited explosion of immigration and capital was on its way. "The air of expectancy pervading this place is equal to that possessing a kindergarten the day before Christmas," the *Metropolis* observed.

Few of the lottery players had ever laid eyes on the Everglades, which gave the air of expectancy a bit of a nervous edge. But ex-Governor Jennings, still on the Bolles payroll, reminded the crowd that the state had solemnly promised to drain the swamp, even if it had to spend every cent in the Internal Improvement Fund. One man asked whether the state would cut lateral farm ditches as well as major canals, but Jennings refused to get any more specific. "The State's going to drain the land," he repeated.

He didn't say that just because of his corporate connections, or because of his progressive politics. He said it because he believed it. Almost everyone believed it.

EVEN SKEPTICAL JOURNALISTS BELIEVED IT.

In April 1912, a dry month in a dry year, the Internal Improvement board sponsored a four-day media tour of the Everglades, showing off its canals and two farms to about two dozen newsmen. The visitors heard from all the leading drainage advocates, including Governor Gilchrist, ex-Governor Jennings, Captain Rose, Dicky Bolles, Thomas Will, and James Wright, who assured them that by year's end the Everglades would never overflow again. The reporters received the same hard-sell treatment that Bolles gave his prospective buyers, including a trip up the Caloosahatchee River on his steamboat, *Queen of the Everglades,* followed by a stay at his Bolles Hotel on Lake Okeechobee's south shore, the first two-story building in the Everglades. A *Grand Rapids News* correspondent bought 100 acres on the spot.

One might have expected the reporters to exercise caution before concluding that drainage was a sure thing; the "unsinkable" *Titanic* had sunk a week earlier, puncturing the era's aura of invincibility and infinite progress. But they all left rhapsodizing about the Everglades as "the Garden of Eden without a serpent," "the Great Utopia," a cornucopia where "you simply tickle the soil and the bounteous crops respond to feed hungry humanity." Several reporters seriously suggested that Everglades knockers might be traitors. Their glowing testimonials reveal how compelling the project seemed to unbiased eyes:

"The most superlative adjectives are required to tell the bare truth and to give the actual facts. . . . It was not my intention to write what must sound like a land agent's advertisement, but one cannot begin to do justice to this embryonic paradise without seeming unduly enthusiastic and visionary." —*Chicago Telegram*

"Seeing is believing. Hence, to see the Everglades in their present state, with the drainage system only partially completed, is to believe—to be certain, for that matter—that the absolute and complete reclamation of the Everglades is not a possibility, but an assured fact." —*Cleveland Plain Dealer*

"I had read with the proverbial grain of salt the stories that have come north about the results that would follow the completion of the drainage project. Now that I have seen what has been accomplished by the engineers, I have to admit that the truth is more wonderful than any of the promises and predictions I had read before visiting Florida. A new name must be found for this land, for within a few months The Everglades will exist no more." —*Canada Monthly*

"I can think of no sufficient expressive adjectives and will therefore say simply that 'Them's my sentiments.' Since my return north I have been preaching the gospel of the Everglades, and I believe every other man of the party I accompanied is today an Everglades evangelist." —*Sioux City Tribune*

Evangelist was the right word. The imminent redemption of the swamp was hailed as a victory of faith as well as progress, an expression of divine will. "God made the Everglades and He made them for a good purpose," the *Chicago Record-Herald* reporter explained. "He also made man to discover His purposes in Nature, and it is evidenced that man is doing so in carrying out the drainage of this great tract of land." It was inconceivable that a benevolent God, after ordering mankind to take dominion over the earth, would expect wastelands to lie fallow when they could be transformed to feed the hungry: "Now comes man, driven by necessity to complete God's plan . . . Presto! North America's tropical winter garden is made ready for the sower." Reclamation was God's work, and as surely as water would flow downhill, Americans would improve the Everglades into a Promised Land.

Seeing was believing, but it wasn't knowing. "The absolute and complete reclamation" of this "embryonic paradise" was about as sure a thing as the *Titanic*.

Water, Water Everywhere

THE INITIAL EVERGLADES BOOM was a faith-based process, fueled by the breezy drainage assurances of land syndicates, state officials, and the oracular James Wright. The market concluded that the reclamation of the Everglades was inevitable, and Bolles and his fellow swamp barons sold 20,000 parcels in a matter of months. But as additional months passed without the miraculous redemption that had been so energetically promised, settlers and speculators began to experience a crisis of faith. One Iowa man, after inspecting the swampland he had purchased in the Bolles lottery, uttered the memorable line that still haunts Florida's real estate industry: "I have bought land by the acre, and I have bought land by the foot; but, by God, I have never before bought land by the gallon." As one Illinois schoolteacher discovered while wading through the marsh in a fruitless search for her new farm, the Everglades still "looked like a wheat field in June until you got close. Then you would see water, water everywhere."

The swollen canals from Lake Okeechobee to the coast were making floods even worse, spilling excess water from the lake into the lower Glades during storms. When a Chicago man named R. H. Little arrived in south Florida in 1912 to check out the Everglades, he heard a local on the platform muttering about "another trainload of land suckers." Little bought anyway, but soon concluded that the local had been right. Another pioneer called the "lurid literature" of the land syndicates "a mass of glossy deception," and offered his own description of the swamp: "The mosquitoes in the Everglades are fearful, the gnats are blinding and the morning fog looks like a sea. There are all kinds of snakes, every kind of bug that lives, and my folks cannot keep the worms out of their clothes and food." Swamp boomers joked that they would have to attach their plows to boats, or grow fish instead of crops; one quipped that he couldn't visit his land because he didn't have webbed feet. The Everglades became so synonymous with unfulfilled drainage promises that the unwieldy phrase "rave on, Everglades, we'll get you drained someday" became the local equivalent of "yeah, right" or "when pigs fly." One Miami resident summed up the plight of the swamp boomers in a letter urging the U.S. government to ban Everglades land sales. Every day, she wrote,

> some poor, deluded victim arrives here in Miami to find the acres which he has bought, and which have been described to him as a very gold mine for productiveness, as much as eight feet under water, and with no present prospect of that water disappearing. These land companies are flooding the country, especially the Middle West, with the most fabulous misrepresentations. We who live here know how absolutely cruel are the sufferings of these misguided creatures. The Everglades may be drained some day, but that day has not arrived.

The misguided creatures began suing Bolles, who stood to clear more than $2 million in profits from his lottery; one newspaper decried his scheme as "one of the biggest land swindles in history." Federal prosecutors in Kansas City indicted Bolles and other Florida land promoters for mail fraud, accusing them of falsely assuring midwesterners that drainage would be complete by 1913, and that Everglades soils were so fertile they could be sold as fertilizer. Even the fund's trustees began to distance themselves from the land rush, claiming they had "nothing whatever to do with these companies, know nothing of their plans, methods of selling or contracting to sell their holdings" and "cannot in any way endorse or recommend any private enterprise."

Still, a parade of current and former Florida officials traveled to Missouri to testify for the defendants and defend the honor of the Everglades. They insisted the state was still committed to reclaiming it, and complained that the main stumbling block was the "agitations and misrepresentations" generated by the trials, which were slowing down the land sales they needed to finance the drainage work. "I read something not so long ago about there being six feet of water on some of your streets here in Kansas City when you had a flood," ex-Governor Gilchrist sniped on the stand. If public officials were still promoting the Everglades as an imminent paradise, the promoters asked, how could it have been a crime for them to say the same thing? Bolles only bought his land after the state promised to drain it, and only sold it after the Wright report declared that the drainage plan would work. One judge tossed out his indictment, agreeing that he had simply relied on official drainage assurances, declaring that "the action of Mr. Bolles throughout the entire transaction is that of an honest man."

The Bolles indictment was later reinstated, and several of his fellow promoters were convicted by midwestern juries. But most of them had only echoed the assurances of the government. Bolles was an incorrigible wheeler-dealer, and his initial intentions were far from pure, but he ended up believing in the Everglades as strongly as any of the buyers he had supposedly swindled. He died before his case could go to trial, but to the end, he was still buying land in the Everglades, and still predicting a spectacular boom. It turned out that the most flagrant Everglades swindle was the Wright report.

What Wright Got Wrong

THE U.S. HOUSE COMMITTEE on Expenditures in the Department of Agriculture was not known as a marquee congressional panel, but in 1912, its investigation of the department's role in promoting the Everglades land rush attracted breathless newspaper coverage in forty-three states. "If the people want to be humbugged, I am perfectly willing to let them be humbugged," said Congressman Frank Clark of Florida, who requested the hearings, "but not by a partnership between a great government and land speculators." The Everglades hearings produced 1,759 pages of testimony, with scandalous revelations almost daily. It came out that Secretary Wilson had scuttled the form letter warning about "problematic" land values under pressure from Everglades promoters like Thomas Will and ex-Governor Broward. Wright confessed that on at least four occasions, he had accepted under-the-table "commissions" from interested parties while conducting government surveys in North Carolina, lamely explaining that he hadn't read the department's ethics rules. The press also chronicled the entertaining feud between Governor Gilchrist, who showed up uninvited to denounce the hearings as a conspiracy to discredit the Everglades, and Congressman Clark, who responded that the governor was a pinhead.

But the central drama of the hearings was the story of Wright's wildly enthusiastic report about Everglades drainage, and how it got foisted upon the public. The hearings revealed that by the time Elliott got to see the report, Wright had already circulated the giddiest excerpts among Everglades land companies, allowing them to claim the imprimatur of the department. "With only a cursory examination in the field and no critical review in the office, engineering plans for this vast reclamation work—the largest drainage project in the world—were favorably recommended to the public, bearing the approval of the Department of Agriculture," the committee noted.

That would not have been a problem if the report had been solid. But Elliott quickly realized that it sounded more like a real estate promotion than a technical evaluation. And a bright young engineer in the drainage bureau, Arthur Morgan, recognized that Wright's problems extended far beyond irrational exuberance; he was also "completely incompetent as an engineer." So Elliott rewrote the report, softening its cheerleading tone and removing some of the more glaring errors. But Secretary Wilson then refused to publish the corrected version, saying he was fed up with the Everglades. "I don't want you to say anything more about the Everglades to anybody—not a thing," he told Elliott. "The state and those people down there are engaged in a promotion scheme, and we don't want to have anything to do with it."

It was too late for that. Thanks to Wright, the department was already on record as endorsing the Everglades drainage project. The department's involvement only deepened after Thomas Will slipped a copy of Wright's original report to Senator Fletcher, who got it published in its entirety as a congressional document—even though the leaders of the drainage bureau had already rejected and rewritten it. And when Elliott had the temerity to complain, Wright pulled strings at the department to get the squeaky-clean Elliott and his top lieutenant fired for trumped-up financial irregularities. "The Everglade interest is all-powerful," Elliott later wrote. "I see now that a nice good report would have been worth a lot of money to me."

THE SIX MONTHS of hearings produced only circumstantial evidence that Wright had been rewarded financially for his "nice good report." But they provided damning evidence that the report was a mess of bad data, bad analysis, and bad recommendations, and that reclaiming the Everglades would be much harder than Florida's leaders thought.

Wright's most obvious mistake was radically underestimating the canal capacity needed to drain the Everglades. Wright's proposed canals were much too shallow, too narrow, and too few. That's because he designed them to remove a maximum of ten millimeters of rain in a day—even though gauges in Fort Myers had once recorded 297 millimeters in a day. Wright made his calculations by assuming a constant daily rainfall, ignoring the region's dramatic fluctuations between flood and drought. So he underestimated how much rain would fall and how much would need to be removed, and overestimated how much would evaporate. Morgan noted that if his outlandish evaporation predictions had been correct, Lake Okeechobee would have dried up without human intervention.

Wright's second error was expecting the same gravity canals to provide irrigation as well as drainage, as if water would flow wherever it was needed whenever it was needed without prompting. Without powerful pumps capable of moving water in a hurry, Morgan warned, south Florida would remain at the mercy of the weather. If Lake Okeechobee were kept low for drainage purposes, an unexpected drought could parch Everglades farmland and ignite muck fires. And if the lake were kept high for irrigation purposes, a sudden storm could drown Everglades farmland and create floods in the new communities downstream.

Wright also dramatically underestimated the cost of drainage, not only by omitting pumps and lowballing the necessary canal capacity, but by predicting ludicrously low dredging costs that turned out to be less than one-fifth the actual cost. Wright also ignored the high cost of maintaining canals, a problem exacerbated by the gentle gradient of the Everglades, which produced currents too slow for the canals to scour themselves out, and by the explosive proliferation of the water hyacinth, an attractive but invasive weed that had clogged almost all the state's waterways since a well-intentioned gardener named Mrs. Fuller imported it to Florida in 1884.

Morgan found other serious errors as well. Wright assumed that five-foot-deep canals would carry five feet of water; in fact, there was a two-foot difference between depth of cut and depth of flow, which meant the actual capacity of Wright's proposed canals would be much less than his estimated capacity. Wright also discounted the risk of "subsidence," the possibility that drained Everglades soils would burn up or blow away, which would further diminish the capacity of his canals. And while Wright did admit that additional lateral canals and farm ditches would be needed to drain individual properties, the state had no plans to dig them or pay for them. How were thousands of Illinois schoolteachers supposed to coordinate the drainage of their ten-acre farms?

* * *

IN AUGUST, THE COMMITTEE released its report on the Everglades fiasco, concluding that Wright's conduct "cannot be too severely condemned." Elliott got his job back with a pay increase, and Wright had to quit his state job overseeing the drainage work. But Elliott never recovered his health after his ordeal, and Wright made a soft landing, getting hired by the contractor whose work he had been overseeing.

Wright blamed his newfound notoriety on "a bitter and acrimonious political campaign" to slander the Everglades. "This tirade of abuse and misrepresentation has placed the entire project and everyone connected with it in a false light; it has created dissatisfaction among the many purchasers of Everglades lands; it has destroyed the confidence in the project to such an extent that the future of the work is jeopardized," he wrote. It never occurred to Wright that the lost confidence of the purchasers might have anything to do with the submerged condition of their land. He insisted he was "more firmly convinced than ever that the plan of reclamation adopted by the State and now being carried out is the best and most economical one that could have been selected."

Not even the land syndicates believed that anymore. The Florida Everglades Land Sales Company—which had aggressively championed the Wright report, and had helped get it published by the Senate—now commissioned its own independent review by three hydraulic engineers. Their secret assessment was even harsher than the House committee's, declaring that Wright's original version was "totally inadequate to accomplish the drainage of the Everglades," and that even Elliott's revised version was "still too optimistic." The consultants agreed that reclamation was feasible, but concluded that the capacity of Wright's proposed canals would have to be expanded at least 800 to 1,200 percent—not including the extra capacity needed to provide separate canals for irrigation as well as drainage, and to compensate for the expected "shrinkage" of as much as 40 percent of the muck in the Everglades. They warned that private landowners would still have to build additional dikes, ditches, and pumps in order to make agriculture profitable in the Everglades, and rejected Governor Broward's Jeffersonian dream of small farms as economically unrealistic for the Everglades. Echoing Captain Rose's suggestions to Henry Flagler, the consultants also recommended piece-by-piece reclamation rather than an all-at-once approach, urging the company to wall off and drain its own land immediately, instead of waiting for the state to reclaim the entire swamp.

Finally, the consultants recognized that controlling Lake Okeechobee and draining the Everglades were two separate problems. They urged the state to rely exclusively on direct east–west canals to the St. Lucie and Caloosahatchee to reduce lake levels and prevent overflows to the Everglades, so that the long diagonal canals through the marsh could be used solely to carry away local rainfall, and would no longer worsen floods downstream. Water does run downhill, but in the Everglades, there wasn't enough downhill to carry huge volumes of water all the way from the lake to the sea.

The consultants' report was a remarkably gloomy document, considering that their client had been one of the loudest voices in the Everglades choir just a few months earlier. But the scandals had brought the land rush to a screeching halt, and the syndicates were as unsure as anyone about the future of their holdings. *Everglade Magazine* warned that "under no circumstances should any purchaser make any arrangements to migrate to the Glades until we advise him that his land is ready for occupancy." After his run-ins with prosecutors, even Bolles revised his brochures, blacking out suggestions that fertilizer would be unnecessary in the Everglades. Florida swampland was now a national punch line, lampooned in cartoons of befuddled suckers touring their underwater farms, and helpless widows handing their life savings to land sharks. National magazines published tales like "The Plunderer: A Story of the Florida Everglades," and booster publications complained that "in many localities in the North, the mere mention of the term 'Florida Everglades' suggests a swindle."

As the real estate market fizzled, swamp boomers stopped making land payments, syndicates stopped paying drainage taxes, and the Internal Improvement Fund nearly went broke again, dipping below $24,000 in reserves. It was becoming clear that draining the Everglades would not be quick, cheap, or easy after all.

Back to Broward

TO RESTORE CONFIDENCE in the Everglades, Florida needed a credible new drainage plan. In 1913, it outsourced the job to sixty-five-year-old Isham Randolph, one of America's best-respected hydraulic engineers. Randolph had served on the Panama Canal board and had helped recommend its design. He had also overseen the Chicago Drainage Canal, a gargantuan project best remembered for reversing the flow of the Chicago River.

In October, Randolph's Everglades Engineering Commission published an impressive-looking report, stuffed with hundreds of pages of maps and meteorological charts. Its strongest recommendation was a familiar one: a massive canal to send water east from Lake Okeechobee into the St. Lucie River and out to sea. Both Hamilton Disston's chief engineer and Governor Broward had initially proposed this as a first step, to end overflows from the lake into the Everglades, and the land company's consultants had recommended it as a next step, to relieve pressure on the diagonal canals. The commission agreed that Lake Okeechobee was the key point of attack, that controlling lake levels was a separate problem from draining Everglades rainfall, and that the St. Lucie Canal was the way to solve it. "Without that canal," Randolph wrote to ex-Governor Jennings, "the efforts now being expended are a sheer waste of money." The commission noted that the St. Lucie Canal would also fulfill the Menéndez dream of a shipping shortcut across the peninsula, which was reasonable, and even suggested that it could power a hydroelectric plant, which was not.

The commission also agreed that much more canal capacity would be needed to drain the torrential rainfall in the Everglades. The Miami and North New River Canals were complete; the Hillsboro and West Palm Beach Canals were under way, along with several smaller ditches to the Atlantic. But the Randolph commission proposed an extensive latticework of additional canals to suck the peninsula dry—some parallel to the main southeastern ones, some draining southwest to the Gulf along Shark Slough and other outlets. The commission also explicitly warned that even these canals would not complete the job; private landowners would have to dig and maintain ditches on their own properties as well, or else the Everglades would soon "revert to the swamp conditions which now prevail."

The Randolph commission estimated that the project would ultimately cost $24.6 million, twelve times what the state had spent so far. Still, its report was relatively upbeat, describing Everglades drainage as "entirely practicable" and the cost per acre as "very small." It predicted that muck soils would subside less than eight inches, which was wrong, and it asserted that "in the Everglades violent floods are inconceivable," which was tragically wrong. And even though Randolph privately harbored doubts about the

Broward Era canals, his report claimed they were "worth . . . every dollar that they have cost," and would "serve a useful purpose in the great scheme of reclamation upon which the State has embarked; a scheme which has only to be carried to completion to make fertile fields of a watery waste and a populous land where now no man dwells."

It was a delicate moment for the project, and Randolph didn't want to sink it. In fact, Randolph formed an Everglades engineering company, and became one of the project's leading boosters.

THE RANDOLPH COMMISSION did provide a modest reality check for the Broward Era. But in the wake of the land scam indictments and the Wright report, its barely measured optimism and ambitious reclamation plan allowed the Broward Era to continue. It gave the land syndicates enough confidence to pay drainage taxes again, which gave the state enough confidence to float bonds to try to finance the project.

Bond buyers had been scared away by the Everglades scandals, so the dredging work still lagged far behind schedule. In 1915, impatient drainage boosters organized the Back to Broward League, clamoring for the state to hurry up and fulfill the late governor's dream. In a pamphlet with Broward's photograph on the cover, over the caption "Florida's Favorite Son," the league argued that "we are actually losing thousands of settlers each year through the odium attached to the Everglades operations." One cartoon portrayed masses of potential pioneers turning their backs on the swamp: "They Came to Boost, but Went Away to Knock When They Found the Glades Not Drained as Promised." Another depicted baby egrets asking their mother when they would have to move out of the Everglades. "Not during the present administration," the bird replied.

But the Back to Broward movement never gained much political traction, because the establishment was already committed to Broward's dream. Florida officials were eager to drain the swamp and replace its egrets with people. The scandals bogged down their work and ended their guarantees of overnight success, but they gradually proceeded with their new plan, led by a new engineer who personified Florida's slightly chastened but generally undaunted attitude toward drainage.

WRIGHT'S REPLACEMENT WAS a thirty-four-year-old Tallahassee native named Fred Cotten Elliot, a trim, clean-cut engineer with a military bear-

ing, a Wall Street haircut, and a formidable aura of confidence and competence. He was the kind of bureaucrat who always seemed indispensable,
even when things went wrong on his watch, which they often did.

Elliot finished twelfth in his class at the Virginia Military Institute, one
spot ahead of a cadet named George Marshall, then toiled as a civil engineer in the subways of New York City and the mines of Mexico and the
American West. But he had always dreamed of returning home to Tallahassee to help develop Florida, and the Everglades job was a perfect fit. He
would serve fourteen governors, and would still be working to drain the
Everglades when his old VMI classmate was launching the Marshall Plan to
revive Europe after World War II. Elliot led the first comprehensive survey
of the Everglades, designed an early version of the "swamp buggies" that
are still driven in the Everglades, and engineered many of the canals that
still crisscross the Everglades.

Elliot was a conservationist who loved to go hunting, fishing, and
boating, but the kind of conservationist who believed in developing natural landscapes, especially those with the economic potential of the
Everglades. Like Wright, he was sure that the reclamation of the Everglades was not only feasible but inevitable, and that the swamp was destined to become as valuable as America's richest gold mines. "The
wonderful lands which you are now rescuing from inundation will
become, when drained, a national asset of great value, inexhaustible and
perpetual," Elliot said in a 1913 speech in West Palm Beach. "This great
but as yet undeveloped resource . . . will continue to develop and
advance with ever-increasing fruitfulness. The limit of its possibilities
can scarcely be measured."

But unlike Wright, Elliot cautioned that the Everglades would not be safe
for settlement until the drainage project was complete. The marsh might
look dry in the winter dry season, he warned, but it was foolish to expect
it to stay dry in the summer rainy season. He wished he could keep people
out of the Everglades until it was truly prepared for cultivation, fretting that
"notwithstanding the catastrophes which are liable to occur . . . many settlers are swarming to that section.

"Those who rush in, regardless of these uncertain conditions," Elliott
said, "should bear in mind that the blame lies with their own folly and
impatience, and not with the drainage scheme. I wish you would bear that
point in mind."

Hope City

SETTLERS WEREN'T REALLY SWARMING into the Everglades. But they were starting to trickle.

In the second decade of the twentieth century, as America's population passed the 100 million mark, a few thousand pioneers braved the Everglades frontier, founding muck towns with sunny names like Chosen and Utopia and Hope City. As a settler named John Newhouse wrote in a memoir of Okeelanta, the first community in the upper Glades, most of the newcomers expected "a life of ease, plenty and independence" in a frost-free, flood-free agricultural paradise. "Then the hard work and sweating began," Newhouse recalled. So did the frost and floods.

The editor-turned-promoter Thomas Will was the driving force behind Okeelanta, buying the land four miles south of Lake Okeechobee, devising the plan for a farmer's utopia in the middle of nowhere. His son Lawrence, later a folksy "Cracker historian," was one of the five original pioneers who traveled up the North New River Canal to Okeelanta in 1913. There was nothing around the canal but seven-foot sawgrass, and the lakeshore was "blamed near vacant of human life," he wrote in his *Cracker History of Okeechobee.* Aside from a few catfish camps and tar paper shacks, "it was still just as the good Lord had fashioned it." Will cursed the brochures—some written by his father—that had described the Everglades as almost mosquito-free: "They came in swarms, they zinged and bored, they even brought droves of fireflies to light up the massacre." His father soon moved down from Washington, and they lived together in a twelve-by-sixteen-foot shack, toiling in the squishy black soil, battling "the muck and the muck itch." Even the elder Will, who had abandoned a comfortable life in academia, government, and journalism to pursue his vision for the swamp, soon realized that "farming here is not the Cock-sure thing we may have thought."

The pioneers who had harbored the old Everglades dream of maximum profit for minimum work were in for a rude awakening. They had no phones or refrigerators, and little access to credit or labor. A trip to Fort Lauderdale for supplies could take two days; deliveries were sporadic, and usually came FOB—Flung Over Board. Realtors handed out snapshots of a quaint sign one family had posted in its yard, advertising "A Happy Home in the Everglades," but the Chicago transplant R. H. Little recalled the sign as a symbol of shattered expectations. "Apparently their home was not so happy," he wrote in his memoir, "as they left a few months later."

The settlers had to brave a wilderness squirming with snakes, gators, and lizards. One of Okeelanta's first women was particularly unnerved by the blue-tailed skinks that scurried around her palmetto-thatched shack. "Jim says don't kill any, they are wonderful bug killers," she wrote in her diary. "So far they have not depleted the supply of roaches." Insects were the central fact of pioneer life, even more than loneliness or shortages. "If'n a man was to put his mind to it, I reckon that a plumb book could be written about the insects that used to infest these Everglades," Lawrence Will recalled.

There were redbugs in the custard apple woods, while hard-shelled, thousand-legged worms covered the sawgrass ground. We had flying ants with red hot feet which came in swarms on windless, muggy after-noons, when you felt plumb beat up anyway. There were yellow gnats that didn't bite, but filled your eyes, your nose, your ears, buzzing like a swarm of bees, deerflies which picked out a shady place under your hat brim or your chin, and bored in until you swatted him; and in the spring time, horseflies, good golly, how they could bite; and naturally, all summer long, mosquitoes by blue millions.

The glowing descriptions of the Everglades as a miracle garden also turned out to be a stretch. Farming the swamp was a bitter struggle, and most early settlers in the upper Glades were northerners with little agricul-tural experience; the southerners who hunted and fished around the lake called them Dumb City Dudes. Their mules sank into the soft muck, and a sixteen-foot-long amphibious tractor developed for the Everglades—known as the Juggernaut—also turned out to be useless. In the marshes, the settlers had to hack down the sawgrass with scythes or machetes, then yank out the roots with hoes or potato hooks, then try to burn it away without igniting the precious soil. In the richer custard apple (or pond apple) belt that lined the lake's southern shore, they had to clear and burn an even more forbidding jungle of gnarled trees and thorny shrubs.

It could take a farmer two months to prepare an acre of the Everglades for planting, and preparation was no guarantee of success. Three weeks after

Newhouse arrived, for instance, a cold snap swept through the Everglades. Settlers tried to protect their crops by burning trash, hanging lanterns in their fields, and covering their plants with muck, but the frosts recurred all winter, and the harvest was ruined. The pioneers also learned that the soot-black organic muck that had lured them to the Everglades in the first place was not as perfect as it looked. Beans and tomatoes sprouted quickly, but then wilted, yellowed, and died from a mysterious "reclaiming disease." Marsh rabbits devoured cabbage, blight destroyed celery, and cattle died of malnutrition. Only potatoes flourished. In 1915, the U.S. Department of Agriculture published a report suggesting that most Everglades soils were ill-suited to any agriculture, although settlers made sure to burn every available copy at a public bonfire in Fort Lauderdale.

Many of the frontiersmen gave up and returned north. Utopia and Hope City withered away. So did Gladesview, Gladescrest, Fruitcrest, and Gardena. The usually upbeat Thomas Will began to worry that any more bad publicity about the Everglades "simply kills the whole thing—and it's dead enough now, Heaven knows, with a lot of the old buyers."

STILL, FOR PIONEERS WHO KNEW BETTER than to believe the promises of land syndicates, and understood that success in America's last frontier would require hard work and good luck, all was not lost. They learned to outfit their mules with steel snowshoes that kept the beasts from sinking. They discovered that the copper-based fertilizers they had applied to their potatoes helped control reclaiming disease. Thomas Will developed an Everglades plow, and several manufacturers rolled out more effective Everglades tractors with oversized wheels, rotating knives, and other swamp accessories. And poor black laborers from the South and the Caribbean began migrating to the Everglades to work as field hands, allowing the white settlers to expand their operations.

For several relatively dry years, the settlers enjoyed ideal weather for winter vegetables, with just enough summer rain to keep the soil moist through the growing season. Lake Okeechobee retreated several miles, exposing its rich bottomlands for cultivation, allowing farmers to "raise cabbages on land which a few years ago was the home of the turtle and the catfish." And soaring demand for food during World War I pumped up prices for Everglades produce, prompting celebratory headlines in the *Palm Beach Post:* "Cane, Potatoes and Corn Factors in Great Wealth of Everglades," "New

Day Dawning in Everglades Development," "The Florida Everglades an Empire of Wealth and Potency Unequaled by Any Like Area in the Country."

The problem was getting crops from farm to market. The first roads in the Everglades were muck piles; they were impassable in the rainy season, and shrouded drivers in black dust in the dry season. But soon railroads arrived to connect the growers to civilization and consumers. James Ingraham, whose faith in the Everglades had never dimmed, linked Henry Flagler's east coast line to the tiny outpost of Tantie on Lake Okeechobee's northeast shore, a boost for farms as well as the local catfish, cattle, timber, and turpentine industries. A few years later, Henry Plant's line pulled into Moore Haven on the lake's southwest shore. And prosperity followed the tracks. Within a year, Okeechobee—the new name for Tantie—had 1,100 residents, electric lights, and an ice plant for the catfish industry. Boosters hyped it as the future Chicago of the South, and tried to make it the state capital. Moore Haven became the largest town in the Everglades, with two theaters, another ice plant, a bank with half a million dollars in assets, and an amazingly relentless real estate operation. "I don't know which was advertised more, Moore Haven or Coca Cola," Lawrence Will wrote.

Soon Everglades land began to sell again—not like it had sold before the scandals, but enough to spur development. Just before his death, Bolles laid out the town of South Bay south of Lake Okeechobee. Moore Haven's Marian O'Brien, America's first woman mayor, helped found nearby Clewiston. William Jennings Bryan bought marshland below the lake, as did Alton Parker, the 1904 Democratic presidential nominee. The steel magnate Henry Phipps bought several thousand acres east of the lake, and hired Chicago's city planners to design Port Mayaca. They proposed to dig up the area's wetlands to create artificial lakes and inlets with names like Sapphire, Emerald, Crystal, and Opal, then use the displaced muck to create artificial high ground for waterfront homes and golf courses, an early vision of the dredge-and-fill alchemy that would guide south Florida's future development. "I have watched the development of the Everglades for a long time," Phipps said. "I have seen that once vast wilderness gradually molded into a place of human habitation . . . I'd be just a trifle ashamed of my judgment if I let an opportunity like this get away."

No investor was more enthusiastic about the Everglades than William "Fingy" Conners, a foul-mouthed Buffalo shipping tycoon who shared power over New York's Democratic machine with Tammany Hall. With a linebacker's build, a prizefighter's nose, and a mouth "as round and menacing as a cannon," Conners had been renowned for his brawling as a young longshoreman, and had built a multimillion-dollar freight empire through similarly bare-knuckled tactics. "There are no rules in his fighting, any more than there used to be in his slugging days on the Buffalo docks, unless it be the bull-rule—rush and gore and never go back," one reporter wrote. But for all his brutality, Conners was a dreamer at heart; he was the inspiration for Jiggs, the gruff but lovable Irish bricklayer-turned-millionaire in the popular comic strip *Bringing Up Father.*

Conners began dreaming about the Everglades after a trip up the West Palm Beach Canal when he was sixty years old. He knew nothing about agriculture, but he immediately launched the state's largest vegetable farm near Canal Point east of Lake Okeechobee. When that failed, he started an even larger cattle ranch on a nearby tract he named Connersville. When that failed, he bought another 12,000 acres of marshland. Conners explained his faith in the swamp in a state agriculture bulletin, offering a concise version of the Everglades creed: "Balmy sunshine, wonderful climate . . . unlimited opportunities for development, wonderful productivity . . . nothing like it on earth."

THE MOST WONDERFULLY PRODUCTIVE land in the Everglades was in the pond apple belt, a strip of dark, swampy, almost impenetrable jungle along the lake's southern rim, about fifty miles across but only two miles wide. Before the settlers arrived, the area was dominated by a thick forest of scraggly pond apple (or custard apple) trees with cream-colored blossoms and yellow fruits, blanketed by a dense mat of green moonvines that blocked out the sun. There were also pop ash, cypress, and elders, amid lacy ferns and gourds found nowhere else on earth. Lawrence Will loved wandering through these moody woods, brushing away spiderwebs, "the silence broken only by a hawk's lonely scream." In 1913, the noted botanist John Kunkel Small, curator of the New York Botanical Garden, found the area "picturesque beyond description . . . one of the most beautiful spots I had seen."

But as Will noted, the custard apples made the mistake of growing in the most fertile soils of the Everglades, protected from frost by the lake, "and when farmers found this out, it was goodbye custard apples." When Small returned a few years later, he found that "the whole region is a waste." He snapped photos of a desolate landscape littered with twisted trunks and branches; pond apple wood was too soft to use as timber, so it was burned to expose the spongy peat below. "The natural features of that region are duplicated nowhere else, and unfortunately they are fast being destroyed," wrote Small, a Renaissance man who identified fifty varieties of ferns in south Florida, played the flute for the Metropolitan Opera, and counted Thomas Edison and Henry Ford among his friends.

Small recognized that the fires were not only obliterating the forest, but ravaging the soils underneath, accelerating the subsidence that James Wright and Isham Randolph had cavalierly dismissed. Fires were natural phenomena in the Everglades, but they were normally caused by lightning during rainstorms, and were usually limited by water tables near or above ground level. Now they were being set by people in dry conditions, and the combination of drought and drainage canals was lowering water tables so drastically that the fires were spreading underground, smoldering for months and burning away the desiccated soils. One member of Small's party fell into a crater where a subterranean fire had consumed the muck; such fires, he said, "were so numerous that the region might well be designated 'The Land of A Thousand Smokes.'"

The naturalist Charles Torrey Simpson, who accompanied Small on the expedition, was similarly disgusted by the destruction of such a unique place. With a shaggy John Muir beard and a let-it-alone John Muir worldview, the "sage of Biscayne Bay" had protected a tropical hammock on his own Miami property as a wilderness reserve, and he wished the government had done the same for the pond apple swamp:

> All the glamour and mystery which once surrounded the great lake, all the wildness and loneliness . . . peace and holiness are fast disappearing before the advance of the white man's civilization. . . . It should have been preserved as a state or government reservation where its rare flora and rich fauna, its mystery and beauty, could have been kept forever.

Most of the pioneer towns that survived in the upper Glades sprouted from the pond apple swamp, including Moore Haven, Belle Glade, Chosen, and Pahokee. But even some development-minded settlers echoed the nature-minded scientists, lamenting the rapid destruction of a singular wilderness. It had thrived for millennia, and they had erased it almost overnight. "I was grieved at the loss of much of the natural beauty of the original growth of trees and shrubs that were here . . . two years ago," R. H. Little wrote. "I thought it a great mistake not to keep at least 100 acres of this virgin vegetation reserved by the state, so the present and future generation should have the opportunity of admiring the original beauty of the land and lake shore prior to the advent of the white man."

The pond apple swamp was gone forever, sacrificed on the altar of progress. But south Florida's conservationists did manage to preserve one slice of the Everglades.

Saving Paradise

SIMPSON HAD TRAVELED throughout the tropics, from the islands of the Caribbean to the coast of West Africa, but his favorite place was a hammock island in the middle of the sawgrass Everglades. "My eyes," he once wrote, "have never rested on any spot on earth as beautiful as Paradise Key."

The island felt like a lush rain forest stranded in the marsh, packed with gumbo-limbos, mastics, and other hardwoods, shrouded in Spanish moss and thick vines. It had drooping clumps of shoestring ferns that clung to its cabbage palms, and resurrection ferns that seemed to spring to life after storms. Simpson once hauled away a forty-pound sack of iridescent orchids; they were so abundant that their removal felt no more destructive than snapping a few blades of grass off a lawn. But Paradise Key's most memorable feature was America's largest stand of royal palms, presiding over the Everglades like ancient monarchs. "Their great smooth white stems appeared everywhere, and one could look up, up, up away into the intensely blue sky where their glorious crowns were tossing the sea breeze," Simpson marveled.

Paradise Key could easily have gone the way of the pond apple swamp. Flagler's railroad had wiped out many of the region's finest hammocks, and the settlers who followed the railroad down the coastal ridge had cleared many more. Paradise Key also faced growing threats from developers, especially after James Ingraham built a road from Homestead to Flamingo right through it. And the Everglades drainage project was carrying away much of the water that had surrounded and protected Paradise Key, exposing its virgin forest to wildfires for the first time in centuries.

Simpson, Small, John Gifford, and the nature writer Kirk Munroe all wrote passionate defenses of Paradise Key, but the conservationists who ultimately saved it were women. Gifford's wife, Edith, and Munroe's wife, Mary, persuaded the Florida Federation of Women's Clubs to launch the first battle for its preservation. Henry Flagler's widow, Mary Kenan Flagler, agreed to donate his company's land around Paradise Key for a state park. And the wife of ex-Governor Jennings, the persistent May Mann Jennings, pushed the park through the state legislature after becoming president of the women's federation.

Mrs. Jennings had an elegant bearing and a finishing-school education, with training in piano, needlepoint, and French, but she was also a politician's daughter and a politician's wife who understood the levers of power. Women did not yet enjoy the right to vote, but conservation was one issue where they could exert influence—thanks in part to their garden clubs, and in part to a widespread belief that defending beauty was women's work. In 1914, Mrs. Jennings began to throw her weight around Tallahassee, staying as a guest in the governor's mansion, persuading the internal improvement trustees to donate additional land in Paradise Key. Her husband then drafted a bill to create Florida's first state park, and she mobilized her federation—including many other wives of politicians—to lobby the legislature to act before its two-year recess.

This was harder work. Many legislators dismissed Paradise Key as a wasteland, and ridiculed the idea of a park. "If the park tract is so dense and useless," Mrs. Jennings groused, "I do not see why the men are so anxious to keep it if we are so anxious to have it." Time was of the essence—vandals were already digging up royal palms—so she literally lobbied until she dropped, besieging legislators with letters and visits before falling ill with exhaustion. Her husband picked up the slack, and the bill passed a few minutes before the session's midnight deadline.

In November 1916, Mrs. Jennings formally dedicated the 4,000 acres of Royal Palm State Park "to the people of Florida and their children forever." Royal Palm saved only one-tenth of one percent of the Everglades ecosystem, but it would one day become the nucleus of Everglades National Park, introducing millions of visitors to the Everglades through Anhinga Trail, Gumbo-Limbo Trail, and the Royal Palm Visitors Center. The new park also set a vital precedent. Florida had given away land for decades, but this was the first time it had done so for a purpose other than development and exploitation.

*   *   *

IF NEWCOMERS WERE TRICKLING into the Everglades, they were streaming into the Gold Coast. In the 1910s, Miami's population swelled nearly
sixfold to about 30,000, and southeast Florida's population quadrupled to
about 70,000. The nation was entering a prosperous new age of leisure,
mobility, and easy credit, and middle-class "Tin Can Tourists" began driving their Model Ts down to Florida for vacations, attracted by new mass
marketing campaigns featuring palm trees, sun-kissed beaches, and
"bathing beauties" whose knees were scandalously uncovered.

But the early stirrings of progress were already starting to take an ecological toll. Conservationists had succeeded in protecting wading birds
and Paradise Key, but Everglades "swamp rats" filled their skiffs with
gator hides, otter pelts, and frog legs, and developers tore up the coastal
ridge for houses, farms, roads, and resorts. Miami had the world's highest per capita consumption of concrete, most of it fashioned from limestone gouged out of the Biscayne Aquifer. Simpson observed how
massive quarries were becoming "the dumping ground for the offal of a
rapidly growing town," jammed with discarded trunks, stoves, tubs,
crockery, and even automobiles. He also catalogued the new phenomenon of "roadkill," from the smashed crabs and grasshoppers that practically covered local highways to sparrows, rabbits, minks, and a dozen
varieties of snakes.

There were now nearly four times as many people living in south Florida
as there had been at the time of European contact, and they made their presence felt on the land. Work crews started building a Tampa-to-Miami highway called the Tamiami Trail across the Everglades and Big Cypress, an
engineering marvel that would block the north-to-south flow of shallow
water down the peninsula as abruptly as any dam. Lumbermen, oystermen,
and fishermen plundered natural resources, while settlers poured sewage
into streams and estuaries. And a dynamic midwestern entrepreneur named
Carl Fisher—the father of the automobile headlight, the Indianapolis 500,
and America's first transcontinental highway—began carving an overgrown
barrier island into the winter playground of Miami Beach, hacking away its
deep-rooted mangroves, and expanding its shoreline with millions of cubic
yards of white sand dredged from the bottom of Biscayne Bay. By 1920,
"Crazy Carl" had remodeled a worthless spit of swampland into a destination resort, but he had also ravaged a formerly pristine habitat for croco-

diles, pelicans, shrimp, crabs, and a rainbow coalition of fish. "The jungle itself seemed to protest in every possible way against this intrusion by man," Fisher's wife recalled. "But Carl had started something, and it was not easy for him to give it up."

The Everglades drainage project was starting to remodel the ecosystem as dramatically as any developer, transforming the crystalline transverse glades into drab canals choked with weeds and silt. For example, the Miami River, once called "as beautiful a stream as ever flowed through an unbroken wilderness," was now the Miami Canal, straightened, deepened, and polluted beyond recognition; its picturesque rapids had been dynamited into oblivion, and its dredge spoil dumped along its banks. Meanwhile, 34,000 acres of the Everglades had been converted into farms, and much of the rest was parched by ditches, drought, and the Tamiami Trail. "The drying up of the Glades, due to the various canals, is playing havoc with the birds here," one surveyor wrote. "The finer ones are fast disappearing. They lack feeding grounds." The water table was dropping fast, drying out springs that once bubbled to the surface on Cape Sable and within Biscayne Bay, reducing the downward hydraulic pressure caused by the weight of fresh water—the "head"—that kept salt water from the region's estuaries from intruding into its aquifers. By 1920, Miami's overpumped well fields at the edge of the Everglades were turning salty.

The declining water table was also fueling the fires that raged in overdrained Everglades wetlands. Broward had ridiculed the idea that a swamp could catch fire, and Randolph had predicted soil shrinkage of no more than eight inches, but some of the Everglades had already lost three to five feet of the black muck that had inspired so many pioneer dreams. This was not only the result of subterranean fires; it was also caused by "oxidation," the exposure of historically flooded organic soils to the open air. The aeration of the muck breathed life into long-dormant microbes in the soil, which consumed organic material that had accumulated underwater over thousands of years. The soils then dried into powder and blew away on windy days, kicking up dust storms so violent that pioneers "could hardly get out of the house without wearing goggles." In a ferocious book called *From Eden to Sahara: Florida's Tragedy,* John Kunkel Small warned that south Florida's unique wetland ecosystem would soon be a barren desert, just like other regions of its latitude:

"Drainage and burning have become such a fad, or even a mania, that the land will soon be unable to support vegetable life, which in turn supports animal life!"

THE BROWARD ERA WAS STILL in full swing, and most conservationists still supported the development of the Everglades. John Gifford issued the first call for the ecologically disastrous Tamiami Trail, arguing that an "Ocean-to-Gulf" highway across the swamp would "open up to settlement a wonderful back country," knitting together the east and west coasts "for the future development of South Florida." May Mann Jennings defended the Everglades drainage project in a letter to a fellow activist, and warned that criticizing it publicly would only give Florida "a black eye." But as the Everglades continued to wither, a few of their colleagues began to wonder if conservation really should mean development more than preservation. These heretics did not believe that God had created man in order to "improve" or "redeem" nature; they found God's grace in nature itself.

Charles Torrey Simpson, the most eloquent of Florida's preservationists, suggested a radical new ethic grounded in Thoreau and Muir, in which Floridians no longer considered themselves superior to nature, and stopped trying to subdue and exploit it at every turn. "Only Florida's climate is safe from vandal man," he wrote "and if it were possible to can and export it, we would until Florida would be as desolate as Labrador." Simpson called for man to start respecting and protecting the wilderness, instead of trying to sell it and replace it with his so-called civilization:

There is something very distressing in the gradual destruction of the wilds, the destruction of the forests, the draining of the swamps, the transforming of the prairies with their wonderful wealth of bloom and beauty—and in its place the coming of civilized man with all his unsightly constructions, his struggles for power, his vulgarity and pretensions. Soon this vast, lonely, beautiful waste will be reclaimed and tamed; soon it will be furrowed by canals and highways and spanned by steel rails. A busy, toiling people will occupy the places that sheltered a wealth of wildlife. . . . In place of the cries of wild birds will be heard the whistle of the locomotive and the honk of the automobile.

We constantly boast of our marvelous national growth. We shall proudly point someday to the Everglade country and say: Only a few years ago this was a worthless swamp; today it is an empire. But I wonder quite seriously if the world is any better off because we have destroyed the wilds and filled the land with countless human beings.

In retrospect, those words would seem prescient. But first, the Empire of the Everglades had to be developed. The wilds had to be destroyed, and the land filled with countless human beings.

ELEVEN

Nature's Revenge

What's the matter with the Everglades?

—*Everglades News* editor Howard Sharp

Boom!

I N THE FIRST HALF of the Roaring Twenties, the stream of newcomers to the Gold Coast suddenly gathered into a tidal wave.

South Florida enjoyed one of history's wildest land booms, with speculation rivaling the Dutch tulip craze and immigration exceeding the California Gold Rush. Ford was cranking out a Model T every ten seconds in 1925, and it felt like most of them were heading straight to the Sunshine State. "Was there ever anything like this migration to Florida?" one paper asked. "From the time the Hebrews went into Egypt, or since the hegira of Mohammed the prophet, what can compare to this?" The pilgrims included celebrities such as the boxers Gene Tunney and Jack Dempsey, the actors Errol Flynn and Rudolph Valentino, the violinist Jascha Heifetz, King George of Greece, and the humorist Will Rogers, as well as brand-name businessmen such as John Hertz, Alfred Dupont, Albert Champion, R. J. Reynolds, and J. C. Penney. Henry Ford and Harvey Firestone tried to grow rubber trees near LaBelle, and the financier E. F. Hutton built a 118-room villa called Mar-A-Lago on Palm Beach for his wife, the cereal heiress Marjorie Merriweather Post. But ordinary Americans also headed to Florida to seek fortunes, enjoy vacations, or retire in comfort; their new Mecca was the beach, not the swamp; their attractions were sun and fun, not coal-black soil. The suntan, once a symbol of labor, became a symbol of leisure.

Most of the human flood followed Flagler's railroad and the new Dixie Highway down the coastal ridge, the relatively high and dry ground that had attracted south Florida's early settlers. Fort Lauderdale's population tripled, West Palm Beach's quadrupled, and Miami's quintupled, while upscale boomtowns sprouted in Boca Raton, Hollywood, and Coral Gables. Florida was all the rage: Baseball teams arrived for "spring training," movie theaters offered "air conditioning," and boosters joked that it would soon be possible to golf the entire length of the state. President Warren Harding played a particularly memorable round on Miami Beach, using one of Carl Fisher's promotional elephants as a caddy. Even some of the Indians cashed in on the boom, moving to roadside villages where tourists paid to watch them wrestle alligators, sew patchwork, and weave baskets.

It was the Jazz Age, a fizzy time of rising hemlines and soaring markets. Life on the south Florida frontier felt especially out of control, like a never-ending party that lacked adult supervision. Liquor flowed freely at the height of Prohibition, with saloons opening across the street from police stations, government officials publicly siding with bootleggers, and rum-runners stashing stills around the Everglades. Madcap drivers routinely flouted traffic laws, and thought nothing of parking in the middle of crowded downtown streets. Crime became so rampant in Miami that the Ku Klux Klan offered to take over the policing of the city.

Nothing was crazier than the real estate market. A veteran who had swapped an overcoat for ten acres of beachfront after World War I found the property worth $25,000 during the boom. A Miami entrepreneur bought and resold a lot for a $10,000 profit during a stroll down Flagler Street. A screaming mob snapped up 400 acres of mangrove shoreline in three hours for $33 million; some of the speculators were so desperate to buy lots in the future Miami Shores that they threw checks at salesmen. "Hardly anybody talks of anything but real estate, and . . . nobody in Florida thinks of anything else in these days when the peninsula is jammed with visitors from end to end and side to side—unless it is a matter of finding a place to sleep," said the *New York Times,* which began devoting an entire real estate section to Florida, as if the state had joined the New York metropolitan area. Farmers on the coastal ridge stopped growing tomatoes and started growing Yankees; the few holdouts who didn't want to unload their fields and groves to developers had to put up Not for Sale signs. Carl Fisher and other developers raced to dredge and fill Florida's bay bottoms and lake bottoms, converting state-owned submerged lands into lucrative privately owned waterfront properties, while wheeler-dealers flacked metropolises-to-be such as Indrio, Idlewyld, Fulford-by-the-Sea, and Picture City, which was supposed to become the new hub of the American film industry.

Most of the grandiose developments existed only on blueprints, or on phantasmagorical advertising posters. "The majority of these depicted an entirely mythical city," one salesman recalled, "with gleaming spires and glistening domes making up an idealized blend of Moscow and Oxford, except that they were invariably rising out of a tropical paradise in which lovely ladies and marvelously dressed gallants disported themselves under the palm trees." Even the booster who wrote *The Truth About Florida,* a 260-page book of propaganda defending the state as a sound long-term investment, had to admit that "greedy realtors" and "get-rich-quick speculators" were driving short-term prices beyond the realm of rationality, that some fly-by-night realtors were selling the same lots to multiple suckers, that "thousands of newly arrived Florida land owners are taking part in one of the greatest gambling spectacles ever witnessed in this country." Land-by-the-gallon cartoons resurfaced nationwide, featuring drowning men in front of signs announcing "CHOICE LOTS." It was only fitting that the infamous swindler Charles Ponzi—already convicted of fraud after the collapse of his "Ponzi scheme" in Massachusetts—resurfaced near Jacksonville with a new name and a new pyramid scheme, promising to triple investments in his land company in two months. Money was pouring into Florida, and greed merchants were following the money.

How insane was Florida during the boom? One land speculator reportedly parlayed two quarts of bathtub gin into $75,000. *Harper's* reported the following monologue from a passenger on a bus heading down the Dixie Highway to Miami:

> Florida? Wonderful! Came with a special party two weeks ago. Bought the third day. Invested everything. They guarantee I'll double by February. Madly absorbing place! My husband died three weeks ago. I nursed him over a year with cancer. Yet I've actually forgotten I ever had a husband. And I loved him, too, at that!

★   ★   ★

MIAMI WAS THE EPICENTER of the insanity. The value of its building permits soared 1,300 percent in those five frenetic years, while the volume of
its real estate transactions skyrocketed 1,700 percent. In 1925, after the
abolition of Florida's income and inheritance taxes accelerated the land
rush, the *Miami Herald* shattered the world's newspaper advertising record;
the *Miami News* printed one 504-page edition that weighed in at more than
seven pounds. "Are you aware of the fact that Real Estate is the best investment for savings as it is the REAL basis of all wealth?" asked a typical ad
for a subdivision. "To speculate in stocks is risky, yes, even dangerous, but
when one buys Real Estate he is buying an inheritance."

In 1920, Miami had one skyscraper; by 1925, it had thirty high-rises
under construction, 974 platted subdivisions, and forty-nine land offices
doing business on a single block. One hotel leased its dining room, coffee
shop, lobby, cigar stand, and phone booth to land outfits. Motor-mouthed
"binder boys" in knickers known as "acreage trousers" mobbed the streets,
pestering pedestrians to buy and sell contracts that often changed hands
three times a day. "Bird dogs" trolled train depots and ship docks for fresh
prospects, while planes and motorboats dragged banners touting beachfront
property five miles from the sea. As the land shark played by Groucho Marx
observed in *Cocoanuts,* a comedy of the Florida boom, "You can even get
stucco! Oh, boy, can you get stucco . . ."

But buyers weren't complaining. That widow on the bus was probably
gullible to believe the value of her land would double by February, but it
probably did. On Miami Beach, for example, property assessments
exploded from $250,000 to $44 million in the town's first decade. Will
Rogers quipped that Carl Fisher had replaced the island's water moccasins
with fancy hotels, jazz orchestras, and New York prices, and "rehearsed the
mosquitoes so they wouldn't bite you until after you bought."

South of Miami, an equally energetic builder named George Merrick
sold $150 million worth of property in his master-planned suburb of
Coral Gables, featuring curving boulevards, gracious esplanades, Venetian
canals, Spanish architecture, and ambitious plans for a University of
Miami. Merrick deployed 3,000 salesmen and a fleet of pink buses around
the country to promote his "City Beautiful," buying billboards in Times
Square and full-page ads in national magazines. He paid $100 a week to
a talented Wellesley-educated reporter named Marjory Stoneman Douglas,

who had moved down from Massachusetts to divorce her alcoholic husband and write for her father at the *Herald;* journalists often supplemented their incomes in those days by doing publicity work, and Miami's papers routinely ran real estate press releases unedited. Merrick also paid $100,000 a year—double Babe Ruth's salary for the Yankees—to William Jennings Bryan, who had just served as secretary of state, to promote Coral Gables real estate. The Great Commoner liked to say that in Florida you could tell a lie at breakfast that would come true by nightfall, and with land values rising faster than its sleaziest hucksters ever predicted, his observation made sense.

North of Miami, the orgy of development was just as wild. George Goethals, the engineer who had overseen the Panama Canal, helped convert a square mile of pine flatwoods and tomato farms into a $40 million "Dream City" called Hollywood. The architect Addison Mizner designed Boca Raton in his signature Mediterranean Revival style of tropical pastels, attracting high-rolling investors from the Vanderbilt and Dupont families, along with the cosmetics queen Elizabeth Arden, the sewing machine heir Paris Singer, and the songwriter Irving Berlin. The shoreline villages above Palm Beach enjoyed such a population surge that their leaders formed Martin County, a name chosen to ensure the support of Governor John Wellborn Martin; it was one of nine new counties created below Kissimmee in those five manic years. Greater Miami's leaders modestly declared the region "the Most Richly Blessed Community of the Bountifully Endowed State of the Most Highly Enterprising People of the Universe."

THE MANIA SPILLED WEST into the eastern Everglades as well. As land on the coastal ridge and along the beach grew scarce, promoters turned their eyes to the edges of the swamp, envisioning a suburban extension of the Gold Coast as well as an agricultural backcountry. "The wealth of south Florida . . . lies in the black muck of the Everglades, and the inevitable development of this country to be the great tropical agriculture center of the world," wrote the *Herald* columnist Marjory Stoneman Douglas, who would come to value the Everglades for quite different reasons.

The largest development west of the ridge was Hialeah, where the renowned aviation pioneer Glenn Curtiss helped convert 14,000 acres of marshes and dairies into a gangster's paradise, featuring illegal casinos, illegal greyhound and thoroughbred tracks, an illegal jai alai fronton, and illegal speakeasies that sold a pungent moonshine called Hialeah rye. On the other side of the Miami Canal, Curtiss helped build Country Club Estates, the future Miami Springs, a suburb as staid as Hialeah was raucous. He also founded the nearby town of Opa-locka, its name inspired by the Seminole word for hammock, its over-the-top Moorish architecture inspired by *The Arabian Nights.*

A few miles further up the canal, the Pennsylvania Sugar Company pursued the enduring dream of an Everglades sugar bowl, developing cane fields and a state-of-the-art mill on a swath of muck known as Pennsuco. Ernest "Cap" Graham, a rough-hewn mining engineer who had prospected for gold in South Dakota before serving as an army captain in France, moved to the marsh to run the operation. Graham was a gruff, stubborn man of principle who could curse a blue streak but never touched alcohol; he later entered politics to take on the gangsters who controlled Hialeah, and ran them out of town. Like his hero, Theodore Roosevelt, Graham was drawn to the challenge of the frontier, and the Everglades was one of the last places in America where a man could tame a real wilderness. Cap is best remembered as the father of Phil Graham, the late publisher of the *Washington Post,* Bill Graham, the lead developer of Miami Lakes, and Bob Graham, a governor and U.S. senator, but he was also the father of the Everglades sugar industry.

The developments were served by two new highways slicing through the Everglades, bringing the Tin Lizzie to America's last frontier. In 1923, a convoy of "Tamiami Trail Blazers" made front-page news around the world by driving from Fort Myers to Miami, even though there was still a forty-mile gap in the road. The journey took three weeks, nearly as long as Ingraham's trek on foot three decades earlier, as the caravan repeatedly bogged down in the soupy marsh. The Trail Blazers might have starved if not for the deer hunting of their Indian guides, and an emergency airlift by the Miami Chamber of Commerce. But the heavy publicity persuaded Barron Gift Collier, a streetcar advertising entrepreneur who had bought one million acres of southwest Florida swampland, to finance the rest of the road. The legislature showed its gratitude by establishing Collier County, and Marjory Stoneman Douglas, an enterprising journalist with a tendency to get carried away—she was once assigned to cover an enlistment ceremony during World War I, and ended up enlisting herself—wrote two poems celebrating the "greatness" of the highway, another position she would reconsider after its ecological impact became clear.

Meanwhile, the irrepressible Fingy Conners had discovered that his latest purchase of Everglades marshland was virtually inaccessible. After unleashing a tirade of profanity, Conners calmed down and spent nearly $2 million building a private toll road through the Everglades alongside the West Palm Beach Canal, linking the booming coast to Lake Okeechobee. When he opened the Conners Highway on July 4, 1924, Florida's governor dubbed him "The Great Developer," and boosters compared him to Flagler, Charlemagne, and even St. Patrick. "The barriers of America's last frontier . . . fell here today," the *Palm Beach Post* reported.

It looked like the boom was about to overrun the Everglades. Ads for a planned Everglades subdivision called Royal Palm Estates claimed that it would be linked to Jacksonville by ten railroad tracks. Ads for Caterpillar bulldozers bragged they were "Conquering the Everglades," grading the Tamiami Trail through "impenetrable swamp." The *Herald* staged a $100 contest for its readers to devise a new name for the Everglades, to help shed its soon-to-be-outdated reputation as a wilderness "inhabited by Indians, rattlesnakes and alligators." The winner suggested "Tropical Glades."

"The Everglades is calling to the . . . farmers of the land: 'Come and use me and make me useful, and I will reward you a hundred, yes, a thousand-fold,'" said a *Herald* editorial titled "Come South, Young Man." "It is calling to the builders: 'Come and build; build for those who will follow in your wake, as follow they must and will.'"

"We All Came to the Glades Too Soon"

BUT THEY DIDN'T FOLLOW. The human tidal wave did overflow into the eastern Glades, but at the height of the land rush, pioneers were abandoning the upper Glades. "The Everglades has lost population while practically every other part of the state has grown," said the *Everglades News,* an unusually candid booster publication based in Canal Point. "Manifestly there is something the matter with the Everglades."

There was no mystery about the problem: The Everglades was not yet reclaimed. In dry years, its canals carried away needed water. In wet years, those canals could not handle the excess water. The resulting floods extinguished muck fires and slowed down soil subsidence, but they also washed out farms in the upper Glades, swamped the Tamiami Trail Blazers and the Pennsylvania Sugar Company, and chased away settlers as intrepid as Thomas Will, who dubbed the period the Dark Ages of the Everglades. "I only hope the old rule, 'no lickin', no larnin',' may hold," Will wrote.

In 1922, the region was almost entirely underwater from September to February; Lake Okeechobee rose five feet, recapturing its bottomlands from the cabbage farmers. "It surely presented a desolate scene," wrote the pioneer R. H. Little. "The water spread way back from the lake into the sawgrass country, which made it all look like a vast sea." That's what the natural Everglades had always looked like in wet cycles, but it was a problem now that people were trying to live and farm there. Little finally got a crop into the ground that March, but it was destroyed by a hailstorm in April, which was followed by heavy downpours all summer. And in 1924, an astounding forty-six inches of rain fell on Moore Haven in just six weeks, flooding the region for nine months. "We began to realize there was not much truth in statements that were made after each flood that it would be the last one," Little said. Most of Moore Haven's settlers fled, and Okeelanta was deserted. At one drainage meeting, an official summed up the plight of the remaining settlers: "The fact is, gentlemen, we all came to the Glades too soon."

TWO DECADES AFTER the Broward Era began, after $13 million of spending and 64 million cubic yards of dredging, Florida's drainage promises were still only promises.

The St. Lucie Canal was poised to become the world's fourth-largest, behind the Panama, the Suez, and the Kiel; "when finished," Fred Elliot wrote, it "will change Lake Okeechobee from the greatest menace to one of the greatest assets of the district." But it wasn't quite finished, so the lake was still an uncontrolled menace. Elliot tried to provide a measure of protection by building an earthen dike along the lake's southern rim, but while the *Palm Beach Post* declared the dike would provide "absolute insurance against any future overflow of Lake Okeechobee," it was basically a squishy pile of muck and sand, only five to nine feet tall and forty feet thick at its base. This mud pile was so vulnerable to erosion that one section in Moore Haven had to be rebuilt five times.

Overflows from the lake were not the only drainage problem in the Everglades. Local rainfall created floods as well, because the district had failed to expand or even maintain its canals aside from the St. Lucie. In fact, the discharge capacity of its diagonal canals had significantly decreased, partly because of the rapid subsidence of the soils around them, partly because of the silt and water hyacinths that were choking them. Howard Sharp, the caustic editor of the *Everglades News,* estimated that less than 2 percent of the land purchased in the Everglades had adequate year-round drainage.

Florida simply wasn't getting the job done. Just as railroads had over-shadowed reclamation in the nineteenth century, roads were a higher prior-ity in the twentieth century; the state was spending more on highways every year than the district had spent on drainage in its history. The state govern-ment, under the perennial control of north Floridians, had yet to contribute a dime to Everglades drainage; taxpayers within the district had shouldered the entire burden.

The federal government was no help, either. The Army Corps had blossomed into America's dominant engineering force, still run by a small cadre of military officers but staffed almost entirely by civilians. It built bridges and hospitals as well as the Washington Monument and Library of Congress, completed the Panama Canal, and oversaw the nation's water transportation network—dredging ports and harbors, while straightening, deepening, armoring, and otherwise manhandling America's unruly rivers into reliable ribbons of commerce. For example, the Corps maintained the navigation channel from the Kissimmee River through Lake Okeechobee and the Caloosahatchee River to the Gulf of Mexico. Water resources were often inconvenient in their natural form; the job of the Corps was to rectify that.

But the men of *essayons* tended to avoid drainage and flood control projects. The Corps had dabbled in flood work under the bombastic Seminole War veteran Andrew Humphreys, who devised a controversial "levees-only" policy for the Mississippi River in the 1870s, dictating the future strategy for controlling America's largest and wildest waterway. But most of the levees were built by local authorities, and General Edgar Jad-win, the white-haired Spanish-American War veteran who led the Corps in the 1920s, still considered drainage and flood control local responsi-bilities. The Corps evaded two congressional directives to study flooding in the Caloosahatchee basin, the first time claiming it had insufficient funds, the second time that the funds had been "misplaced." Jadwin believed his agency's only interest in south Florida was navigation; in fact, the Corps required the drainage district to keep Lake Okeechobee high enough to maintain a steamboat channel, which often exacerbated the region's flooding problems.

The high hopes for reliably dry farmland that had enticed pioneers to the Everglades were fading fast. After one meeting in West Palm Beach, Florida's agriculture commissioner warned the Internal Improvement board that Everglades landowners were losing patience: "I wish to say that gloom seems to be on every hand, and among men who have heretofore stood by the Board loyally and who in the face of everything were optimistic."

EVERYONE WANTED A perfectly calibrated Everglades, with just enough water in dry seasons and not too much in rainy seasons. But the drainage district's unfinished system of gravity canals couldn't ensure that for everyone. South Florida was soon embroiled in its first water wars, and Elliot and his board were attacked by all sides—for allowing fires and allowing floods, for failing to conserve water and failing to get rid of it, for "neglect of duty, inability, incompetence, lack of foresight and every other thing that can be thought of." When a drainage commissioner explained to one farmer that removing his excess water would flood his fellow agriculturalists, the man replied: "To hell with them. I want to gather my corn with a wagon, not a boat."

Farmers in the upper Glades demanded lower lake levels, except during droughts, but fishermen, steamboat operators, and the Corps demanded higher lake levels, and farmers in the lower Glades complained when water released from the lake landed in their fields. Cap Graham protested whenever water stacked up behind the Tamiami Trail and flooded Pennsuco, telling Elliot that if he could not solve the problem, "then I consider the entire Glades proposition hopeless." Elliot finally agreed to dam the Miami Canal, but that only infuriated nearby landowners who absorbed the overflow. "We might make a similar demand that you throw a dam across the South New River Canal a few miles above us and shunt the water elsewhere," they wrote in a sarcastic letter to Elliot. "Where, we do not care, so long as our selfish purposes are conserved." The manipulations did not prevent the Pennsuco sugar plantation from going bust. And they did not stop Graham—who accepted some land from the sugar company as severance pay, and began running cattle in the Everglades—from railing about water management in the Everglades, especially after he became the area's state senator. The people of the Everglades had no more faith in Elliot, said Thomas Will, "than we had in Hindenburg during the last war."

Elliot saw the constant carping as proof of a job well done. He believed he served the public's interest, while the armchair knockers only served their own interests, yelping one day that they were drowning and the next day that they were parched. "A man on one side of the canal wants it raised for his particular use and a man on the other side wants it lowered for his particular use," he wrote. "It seems to be everybody for himself and the devil for all, and everybody knows more about the drainage work than the Board." Elliot exuded authority, and had a gift for making those who questioned his decisions sound like radicals, morons, and money-grubbers. This was the heyday of the American engineer; there would by 226,000 of them by 1930, up from 7,000 in 1880, and their bridges and canals were being hailed as the poetry of the age. Elliot was a model of his profession: booksmart, confident of his control of nature, blissfully untroubled by doubt.

The chief engineer's harshest critic was the fire-eating *Everglades News* editor Howard Sharp, whose paper's motto was "The Truth About the Everglades." Sharp was enraged by Elliot's frequent decisions to hold water in Lake Okeechobee instead of letting it flow southeast through the diagonal canals, accusing him of protecting developments on the booming coastal ridge at the expense of farms in the struggling upper Glades. Sharp constantly sniped at Elliot for managing the Everglades from an office in Tallahassee, and for failing to dredge clogged canals. "The most charitable conclusion for the failure is the lethargy of state drainage officials," he wrote.

In the summer of 1926, when heavy rains raised Lake Okeechobee to the edge of its dike, Sharp demanded releases through the diagonal canals. "The lake is truly at a level so high as to make a perilous situation in the event of a storm," he warned. But Elliot believed that discharges to protect the upper Glades would only flood the coastal communities downstream, and squander water that could be stored in the lake for irrigation that winter. In fact, he believed water shortages were a more serious threat to the Everglades than water surpluses. Elliot told one newspaper the district's lands were "safer from flood or overflow than any other place in the state of Florida I know of."

The Big Blow

BY 1926, THE GOLD COAST BOOM was starting to sputter. The Internal Revenue Service was cracking down on Florida speculators, the Better Business

Bureau was investigating Florida con artists, and national magazines were publishing exposés of Florida scandals. Charles Ponzi was arrested again for fraud and returned to the slammer. "Throughout the country the delusion has developed that any fool, utterly ignorant of intrinsic values, can gamble blindly in Florida real estate and overnight reap a fortune," *Forbes* warned. Ohio passed blue-sky laws to chase away Florida land companies, and the state's banks took out ads warning customers to resist the lure of the Sunshine State. "You are going to Florida to do what?" one ad asked. "To sell lots to the other fellow who is going to Florida to sell lots to you."

Meanwhile, south Florida's infrastructure was beginning to buckle. The Flagler railway was so overloaded with construction materials and other freight that it halted shipments to expand its tracks. Commerce was diverted to the sea, and Miami's harbor was soon as overcrowded as its rail yards, with freighters waiting weeks to unload cargo. Then the 241-foot *Prins Valdemar* capsized at the harbor entrance, blockading the port for a month. The building frenzy stalled, along with the swapping of paper contracts for pie-in-the-sky properties. Land ads vanished from the papers and binder boys vanished from the streets. Gene Tunney, who was selling real estate on the side, had to cancel a local fight because no one could guarantee his purse. "The world's greatest poker game, played with building lots instead of chips, is over," *The Nation* said.

Still, boosters were sure that the slowdown was just a temporary blip, that Florida would keep growing until the sun stopped shining and Ford stopped making cars. Even a mild hurricane that grazed south Florida in July failed to dampen local spirits, especially after U.S. Weather Service meteorologist Richard Gray declared that Miami had little to fear from hurricanes. "There is more risk to life in venturing across a busy street," the *Herald* assured readers. So there was little concern on the Gold Coast on September 17, when a four-inch story noted that a tropical storm was heading for the Bahamas, but was expected to miss Florida. The reaction was similarly muted in Moore Haven that evening, when an engineer received a telegram upgrading the storm to a hurricane, and warning that it might hit Miami overnight. Despite Sharp's tirades about the rising lake, the engineer later recalled, "nobody seemed to be alarmed."

THAT NIGHT, MIAMI was pummeled by the most powerful hurricane in its history. There were gusts up to 140 miles per hour and storm surges as high as fifteen feet, uprooting trees, flinging yachts and grand pianos into the streets, propelling roofs and cottages into the Everglades. "The intensity of the storm and the wreckage that it left cannot be adequately described," Richard Gray wrote. Few Miami residents knew anything about hurricanes, so when the eye passed over the city at 6 A.M., thousands poured outside to survey the damage and thank God for their survival. Gray threw open his door and screamed, redeeming himself a bit for his overconfident forecast: "The storm's not over! We're in the lull! The worst is yet to come!" Many who ignored his shouts were swept away by the "second wind," along with thousands of wood-frame structures built without regard for hurricane-force gales. Miami and Miami Beach were left in ruins. Hollywood, Hallandale, and Hialeah were buried under several feet of debris, and boats were tossed around their streets like bath toys. The only seagoing vessel that managed to ride out the blow, oddly, was the *Prins Valdemar.*

The storm then headed northwest to Lake Okeechobee, where Sharp's warnings of peril came true. Violent winds whipped up the swollen lake and sloshed it south like a 730-square-mile saucer tipped on its side. The lake then ripped through its flimsy muck dike, sending a roaring wall of water through Moore Haven. "Scores of men, women and children were drowned like rats in a trap in the first rush of the flooding waters," a survivor wrote. "Those caught in their beds had not a chance for life as the crazed elements drove the very lake through their windows and doors." One carpenter grabbed his family and outran the surge to higher ground; he lost his home and eleven of his relatives, but saved his wife, his three children, and a single $10 bill. A railroad agent drowned with his wife and five children; the cleanup crew that retrieved his body found a telegram in his pocket warning about the storm, and urging the evacuation of Moore Haven.

In its natural state, Lake Okeechobee had regularly overflowed its southern rim, harmlessly spreading across the Everglades during thunderstorms as well as hurricanes. But the dike, designed to imprison Mother Nature so that people could live in her original path, had only concentrated her fury, gathering the lake's floodwaters until they burst toward their natural destination with explosive force. The people in the flood plain paid the price: The 1926 hurricane killed nearly 400 and left more than 40,000 homeless.

It also burst whatever remained of Florida's real estate bubble. Some lots that had changed hands for $5,000 the year before went back on the market for $100, and just about everyone who held an unpaid contract defaulted. Carl Fisher's fortune vanished. So did Addison Mizner's. George Merrick would lose everything as well, although he did manage to open the University of Miami in Coral Gables a month after the disaster. (Its football team was named the Hurricanes, and its mascot was the ibis, reputedly the first bird to return after storms.) The Model Ts that had snaked down the Dixie Highway for the boom began heading back north. The population of Hollywood, for example, plummeted from 18,000 to 2,000 in a year.

Terrified that hurricane publicity would scare away more visitors and investors, Florida's leaders minimized the damage, denying reports of devastation as rumors and exaggerations, openly discouraging relief efforts. When the *Herald*'s new city editor, a Cincinnati transplant named John Pennekamp, filed a story reporting $100 million in damages, his boss ordered the losses reduced to $10 million. Miami's mayor declined offers of outside aid, and Governor Martin insisted that life was rapidly returning to normal. The chairman of the Red Cross charged that "the poor people who suffered are regarded as of less consequence than the hotel and tourist business in Florida," but official spin continued to portray the storm as a minor inconvenience in paradise. "Miami will be her smiling self again within a short time," the *Herald* said. One booster took out full-page ads pointing out that Florida was still perfectly positioned for growth, that the big blow was nothing compared to floods in the Midwest, "winter diseases" in New England, or earthquakes in California: "Sure, some lives were lost in the hurricane, but hurricanes come only once in a lifetime."

TO HOWARD SHARP and his readers in the upper Glades, the hurricane was no minor accident. It was a negligent homicide.

In a scathing editorial titled "The Dead Accuse," Sharp directly blamed Elliot and the drainage board for the Moore Haven dike failure. He claimed hundreds of lives could have been saved if Elliot had heeded his pleas to release lake water gradually through the diagonal canals, or if the St. Lucie outlet to the east had been completed in time to lower the lake quickly. He demanded Elliot's resignation, and badgered Governor Martin to turn over the administration of the Everglades to the people of the Everglades, who were already paying for the work. In more than two decades, Sharp complained, the board had drained nothing but the district's finances. "The first thing to do is get rid of the men . . . who brought about the death and ruin last September," another critic wrote.

Elliot and his board denied responsibility, insisting that they would have finished the St. Lucie Canal if they had the money, and that releasing water out of the lake before the storm would have inundated the lower Glades without saving the upper Glades. Martin lashed out at the "reckless and foolish" critics who blamed human beings for acts of God. "Of course, the Drainage Commissioners are easier to reach than the Lord is, and they can make this Board more uncomfortable than they can make the Lord," another commissioner sniped. Elliot actually expanded his power during the controversy, taking over the Internal Improvement Fund in addition to the drainage district. He was treated like royalty at an Everglades reclamation conference organized by Florida's business community in 1927. "There is no better drainage engineer than Elliot here, and I believe everybody believes in him," one railroad executive declared.

Elliot soon proposed a new $20 million plan of attack on Mother Nature. The heart of the strategy was seizing permanent control of Lake Okeechobee—by building a taller, wider, and sturdier dike, completing the St. Lucie Canal to the east, and converting the Caloosahatchee into a second lake outlet to the west. Despite his frequent warnings about overly dry conditions, Elliot also proposed to whisk rainfall out of the Everglades for good—by supplementing Broward's four main diagonal canals with at least a dozen shorter and more direct horizontal canals to the coast. Finally, Elliot called for the federal government to help build and finance the new work, a call echoed by grieving residents at Corps hearings in Pahokee, Moore Haven, and Belle Glade. But the old-school General Jadwin was unmoved. He told Congress that "until the resources of local interests and the State of Florida have been exhausted in providing flood control," the Corps should do nothing except promote navigation.

Governor Martin managed to ram a $20 million drainage bond through the legislature, backed by property taxes throughout the district, and Elliott finally completed the St. Lucie Canal, assuring Everglades residents that "floods such as occurred there in 1926 probably will not take place again." But Cap Graham, George Merrick, and other Dade County landowners successfully sued to block the new taxes. Governor Martin accused them of "throw[ing] the brick at Santa Claus," but the drainage district had to shut down its work throughout the Everglades. It had no cash for anything but rudimentary maintenance, and while it did patch up the gelatinous Lake Okeechobee dike, Elliot's plans for a stronger levee were shelved. For the first time since the beginning of the Broward Era, Everglades drainage was on hold.

Elliot wrote a series of grandiloquent reports to his board, detailing his heroic decisions to mothball dredges and lay off staff, highlighting his "clear thinking, straight shooting and careful administration," vowing to save the drainage project by any means necessary. He tried to sound like a general under fire: "There have been hardships, vicissitudes, failures and disappointments in the Everglades, just as in any other frontier country on earth. . . . Untiring effort and intelligent thought, study and work along all lines are needed. It is no country for the weak-hearted." And later: "This is a serious time for the Everglades. There must inevitably be delay, but failure is unthinkable."

Howard Sharp was in no mood for pep talks. He yearned for drainage, too, but he applauded the failure of Martin's plan, calling it "the most dishonest plan of bond-selling to which any crew of pretended honest men ever gave support." His main complaint was that it kept the unrepentant Elliot in charge of the Everglades. The hostility in the upper Glades became so intense that Elliot began to fear for his safety. "We are in far more danger from continued administration from Tallahassee than we are from any outbreak of nature," Sharp wrote.

But the administration continued from Tallahassee. And another outbreak of nature was on the way.

The Monstropolous Beast

FARMERS IN THE EVERGLADES enjoyed the winters of 1927 and 1928. Not only did the weather cooperate, but scientists at a new state-run agricultural experiment station in Belle Glade cured "reclaiming disease." They discovered that the muck and peat soils of the Everglades lacked copper, manganese, and other trace elements, and developed new fertilizers to compensate for the deficiencies. The upper Glades sold $11 million worth of vegetables in those two years, which "lent a halo of romance around the magic word 'Everglades.'" The *Palm Beach Post* announced that "Civilization Is Quickly Taking Backwoods Lands." Blacks in particular flocked to "the muck," as Zora Neale Hurston chronicled in *Their Eyes Were Watching God.* "Folks don't do nothin' down dere but make money and fun and foolishness," says the novel's happy-go-lucky field hand, Tea Cake, before heading to work in the rows of beans below Lake Okeechobee.

But in the summer of 1928, it started raining again, and the lake started rising again. Howard Sharp, now a county commissioner and a candidate for the legislature, went back on the attack, ripping Elliot for refusing again to lower the lake. Elliot was not about to take directions from a politically motivated rabble-rouser with no engineering background—especially since lowering the lake now would look like an implicit admission that he regretted his refusal to lower it in 1926. So he made little effort to empty the lake, even though it rose three feet in a month, even though Sharp kept scorching him in print. "Fred C. Elliot of Tallahassee . . . does not expect overflow conditions in 1928," Sharp wrote. "He did not expect the flood in 1922. . . . He did not expect the flood in 1924. . . . He did not expect the flood in 1926. . . . The chief engineer never expects any overflow conditions." On September 7, Sharp wrote: "Advocates of a high lake level take a terrible responsibility on themselves."

On September 16, another 140-mile-an-hour hurricane smashed into Palm Beach. It was heading northwest, toward Lake Okeechobee.

THE STORM OF 1928 ravaged another 100-mile swath of the coastal ridge, flattening luxury resorts and seaside mansions, burying West Palm Beach's streets under five feet of splintered wood and shattered glass, providing deadly proof that hurricanes do not come only once in a lifetime. "The suffering throughout is beyond words," a coroner wrote. "Individual tales of horror, suffering and loss are numberless." Although the stock market crash was a year away, the Great Depression now began in earnest on the Gold Coast.

But the destruction along the coast paled in comparison to the catastrophe in the Everglades. The meteorologist Richard Gray botched his forecast again, predicting the night before that south Florida would be spared, and there were few radios or phones in the muck lands to catch his last-minute change of heart the next morning. So the upper Glades was taken by surprise when the hurricane steamrolled across the marsh, scattering tractors and barns like tenpins, stripping the sawgrass off the soil, drowning cattle, horses, and even gators. As the storm barreled into Lake Okeechobee, some whites managed to scramble into the region's sturdier homes, packing houses, and hotels, but most blacks had to ride it out in their unprotected shanties in the low-lying fields.

Once again, the lake slammed through the dike like a truck driving through pudding, sending a fifteen-foot-high tsunami through the upper

Glades, drowning the towns of Miami Locks, South Bay, Chosen, Pahokee, and Belle Glade, where the Glades Hotel was the only building left standing after the storm. Survivors clung for life to floating fence posts, tree trunks, rooftops, chimneys, and cows, enduring sheets of rain that felt like they were shot out of cannons, avoiding swarms of cottonmouths that were just as desperate to escape the deluge. "I had thought our storm experiences very trying, but upon hearing what others had to endure ours were trivial," recalled the pioneer R. H. Little, who chopped a hole in his ceiling to hoist his family into a crawl space above the floodwaters, then waited helplessly while their home hurtled half a mile through the darkness. One family rode out the storm in a treetop in South Bay, singing the gospel into the whipping winds: *Hide me, O my Savior, hide, till the storm of life is past.* Wives watched husbands drown; fathers felt children slip out of their arms and into the flood.

The '26 storm had punched a quarter-mile hole in the forty-seven-mile dike; the '28 storm damaged or destroyed twenty-one miles. Hurston provided the most vivid description of the power and the terror in her novel:

> Louder and higher and lower and wider the sound and motion spread, mounting, sinking, darking. It woke up old Okeechobee and the monster began to roll in his bed. . . . Under its multiplied roar could be heard a mighty sound of grinding rock and timber and a wail. They looked back. Saw people trying to run in raging waters and screaming when they found they couldn't. . . . As far as they could see the muttering wall advanced before the braced-up waters like a road crusher on a cosmic scale. The monstropolous beast had left his bed. . . . He seized hold of his dikes and ran forward until he met the quarters; uprooted them like grass and rushed on after his supposed-to-be conquerors, rolling the dikes, rolling the houses, rolling the people in the houses along with other timbers. The sea was walking the earth with a heavy heel.

It was this scene of nature unbound that inspired Hurston's title: "They seemed to be staring at the dark," she wrote, "but their eyes were watching God."

The Okeechobee hurricane killed 2,500 people, mostly poor blacks who drowned in the vegetable fields of the Everglades. It was the second-deadliest natural disaster in American history, exceeded only by the Galveston hurricane of 1900; it was much deadlier than Hurricane Katrina's drowning of New Orleans in 2005, another case of poor blacks in low-lying floodplains betrayed by inadequate dikes. And the death toll would have been much worse if the Everglades hadn't missed out on the Florida land boom. The storm wiped out a third of the residents of the upper Glades, so one can only imagine the carnage if the boosters' predictions of hundreds of thousands of residents had come true.

The actual carnage was ghastly enough. Sixty-three locals took shelter in a farmhouse near Lake Harbor, but only six survived when it was washed away by the surge. Twenty-two others took refuge in a packing house near Chosen, but half were killed when it collapsed. "The complete devastation was simply unbelievable," recalled a cleanup worker named Chester Young. "Ugly death was simply everywhere." With help from vultures circling overhead, Young found bloated bodies hidden under beds, nestled in trees, strewn across fields, and floating in canals; he found one dead man clutching a dead baby. Coffins were reserved for whites, and the soils of the Everglades were far too saturated for burials, so 674 black victims were stacked like cordwood on flatbed trucks and hauled to a mass grave in West Palm Beach. Hundreds of other bodies were tossed into piles, doused in oil, and burned in roadside funeral pyres. "There was so much death in so many gruesome conditions that we became somewhat immune to it," Young wrote.

Once again, Florida's initial reaction was denial. Governor Martin refused to activate the National Guard, claiming the storm had done little damage. But a grisly tour through the Everglades changed his mind; the governor counted 126 corpses along the six-mile road from Belle Glade to Pahokee. After meeting one black farmer who had lost his wife and six of his seven children, Martin got so emotional that he shook the man's hand, an extraordinary gesture for a Florida governor—although the handshake did not dissuade local officials from conscripting blacks into the cleanup at gunpoint. By the time he left the Everglades, Martin was convinced that this time, Florida could not afford to let its pride stand in the way of its needs. "Without exaggeration," he wrote in a telegram appealing for assistance, "the situation in the storm area beggars description."

HOWARD SHARP AND the drainage board's other critics were apoplectic that history had repeated itself. The *Okeechobee News* ran a haunting front-page cartoon of a mother screaming "HELP!" while she struggled to hold her baby above the swirling waters, under the caption: "Two Thousand

Lives Pay the Price of Politics, Indifference and Mismanagement." One man reportedly thrust the bones of a drowned friend at Martin, and blurted: "See what you have done by bringing this disaster!" The legislature finally agreed to give local interests a role in the drainage district, though it hardly mattered now that the district was broke.

Elliot and the drainage commissioners were just as vehement that they should not be held responsible for another act of God. After presiding over his second calamity in two years, Elliot had the gall to claim a measure of vindication, pointing out that he had warned settlers to avoid the Everglades until it was totally drained, and had called for a stronger dike. He accused Dade County critics such as George Merrick and Cap Graham of contributing to the calamity, suggesting that they had scuttled Martin's drainage bond with "inconsequential quibbling over details" and "trivial . . . litigation" over taxes. Elliot and the commissioners also lashed out at the U.S. government, complaining that Army Corps of Engineers navigation requirements had forced them to keep Lake Okeechobee dangerously high, and that their earlier appeals for federal aid had been ignored.

But even Sharp and Elliot were united in their determination that "the Glades will rise again," and survivors vented much of their anger on those who suggested otherwise. The Red Cross, for example, announced at a tent meeting in Belle Glade that it would offer emergency relief, but would not rebuild flood-prone homes and communities in the Everglades. "That tent disgorged as angry an assemblage of ruined farmers as I ever hope to see," Lawrence Will recalled. The resulting backlash spread nationwide, forcing the organization to reverse its sensible policy. Florida's attorney general, Frederick Davis, inspired similar outrage when he clumsily admitted to Congress that the Everglades might be unsuited for human habitation: "I've heard it advocated in certain districts of Florida that what the people ought to do is build a wall down there and keep the military there to keep people from coming in." No matter how many disfigured corpses were floating in the Everglades, it was blasphemy to suggest the abandonment of the swamp.

Still, the disaster of 1928 made it fatally clear that the status quo could not continue in the Everglades. After spending $18 million to dig more than 400 miles of canals, the drainage district was bankrupt and paralyzed, with less than 100,000 of its 4.8 million acres under cultivation—and none of them safe. The unfinished drainage project had only intensified the natural cycle of Everglades fires and floods, while luring pioneers into their horrific path. It had improved some 25-cents-an-acre swampland into $250-an-acre farmland, but not even Elliot could argue that 2,500 dead was acceptable collateral damage. And while Elliot still believed he could finish draining the Everglades and prevent future disasters for another $20 million, the district didn't have 20 cents.

The state government was also strapped for cash, and even in flush times, Florida politicians had never backed up their rhetoric about draining the swamp with financial assistance. The hopelessly impolitic Attorney General Davis told Congress that since most Everglades residents were from outside Florida, "it is mighty hard to get people in other parts of the State interested in whether they perish or not."

FLORIDA NEEDED HELP, and there was only one place to get it.

In 1848, when Senator Westcott first proposed to drain the Everglades, he ridiculed the notion of involving the federal government, scoffing that "thousands of dollars" would be wasted on "steamboats and other expensive apparatus." Back then, reclaiming the swamp seemed like a $500,000 job at most, a simple matter of poking a few holes in the coastal ridge and letting the Everglades pour out to sea. But the Everglades had confounded Buckingham Smith, William Gleason, Hamilton Disston, Napoleon Broward, John Gifford, James Wright, Dicky Bolles, Thomas Will, Fred Elliot, and everyone else who predicted its easy conquest. None of their elaborate drainage plans were ever fully funded or fully executed, which turned out to be a good thing; none of them fully understood the risks of overdrainage, and their ambitious plans could have destroyed the Everglades and consumed its soil.

But the Everglades was already in bad shape, and now it had killed more Americans in one night than the Seminoles had killed in three wars. The Broward Era was over. Florida had failed to conquer the swamp. After the disaster of 1928, the engineers of the United States government took over the war against the Everglades. From that point on, the primary objective of the war would no longer be drainage, but flood control, and the prevention of another disaster of 1928. No lickin', no larnin'.

TWELVE

"Everglades Permanence Now Assured"

There is nothing like it in the world.

— Everglades botanist John Kunkel Small

A Corps Mission

FIVE MONTHS AFTER the 1928 hurricane, Fred Elliot gave a five-hour tour of the lakefront to General Jadwin and another engineer with the U.S. government. As they surveyed the wreckage that still littered the Everglades, and the scattered tents that still housed the storm's survivors, Elliot reminded his guests of the federal government's duty to protect its citizens, its failure to respond to the storm of '26, and its responsibility for Lake Okeechobee as a federal waterway. The lobbying continued that night at a dinner in Clewiston, as a parade of speakers begged the visitors to protect the Everglades, and "those who got the most applause were those who came out for a 100 percent Federal financed program." Jadwin still harbored some misgivings about federally funded flood control, but the second engineer had no such qualms. His name was Herbert Hoover, and he had just been elected president of the United States.

Hoover had witnessed mass starvation during World War I as the U.S. relief administrator, and had just led the response to the cataclysmic Mississippi River flood of 1927 as U.S. commerce secretary. But the president-elect was still overwhelmed by the devastation he saw in the Everglades.

The Okeechobee hurricane had claimed five times as many lives as the Mississippi flood, and tears welled in Hoover's eyes as he accepted a gift of fresh vegetables from the orphans of Pahokee. Hoover was a fervent believer in man's ability to improve nature, and the Everglades clearly required some adjustments. "I'm going to help you with this thing," he vowed.

Today, Hoover is often caricatured as a do-nothing president who fiddled while the Depression burned. But while Hoover was no New Dealer, he was an indefatigable man of action, a can-do engineer infused with the fix-it mentality of his profession. Before his presidency, he was renowned as the Great Engineer for his work in the mines of Australia and China, and the Great Humanitarian for his relief work in Europe and the Mississippi delta; his energetic response to the Mississippi flood had propelled him to the White House. As president, he would spend more on public works than all his predecessors combined—especially flood control projects like the Hoover Dam, which combined his Great Humanitarian desire to ease suffering with his Great Engineer desire to defeat Mother Nature.

Hoover was an avid angler who loved fishing in Florida, and even trolled for bass in Lake Okeechobee during his review of the storm damage. But he was another wise-use conservationist; he once remarked that an engineer could create a waterfall as attractive as any in nature. He also believed that engineers could save the Everglades for humanity. As commerce secretary, he had urged the Army Corps to respond to the 1926 hurricane, to no avail. As president, he ordered the Corps to respond to the 1928 hurricane. The response is still continuing today.

THE CORPS WAS still reeling from the Mississippi flood, a humiliating defeat at the hands of Mother Nature. For decades, the agency had insisted that levees alone would confine the river to its channel. The river had disagreed, staging periodic jailbreaks, but in 1926, General Jadwin had personally declared the problem under control. Most of the levees along the river were locally controlled, but Jadwin had assured Congress they were finally prepared "to prevent the destructive effect of floods." Then the Mississippi had drowned its basin in 1927, leaving 1 million Americans homeless, permanently discrediting the doctrine of "levees-only."

The public demanded a federal response—not just from Hoover, whose relief efforts dominated the newsreels, but from the Corps. Grudgingly, Jadwin submitted a Corps plan to seize control of the Mississippi, featuring

Corps-built levees as well as reservoirs and floodways to give the river room to spread out. The Jadwin Plan was the stingiest proposal submitted to Congress, and the general did not win many friends by testifying that it ought to pass without revisions because he said so. But it still launched the largest domestic expenditure in U.S. history. And it finally accepted federal responsibility for local protection, which made flood control an official Corps mission.

So when Lake Okeechobee reprised its rampage in 1928, President Hoover put the Corps to work, and even Jadwin conceded that "protection must be designed for the extraordinary and unexpected," in the form of a dike "high enough and strong enough to prevent dangerous overflow." The general still tried to dump most of the cost on Florida, but Hoover thwarted his efforts. Over the next decade, the Corps would spend $20 million on the dike—later christened the Herbert Hoover Dike—with local interests footing only 5 percent of the bill. The Corps would never be the same, and neither would the Everglades.

RISING FOUR STORIES above sea level from a concrete base more than a football field wide, the Hoover Dike sent a powerful message that man was in control of the lake. Locals called it the Great Wall of the Everglades—not the wall the attorney general had suggested for keeping settlers out, but a wall encouraging settlers to come in. Even in the depths of the Depression, as Americans huddled in shantytowns dubbed Hoovervilles and slept under newspapers dubbed Hoover blankets, farmers came to the upper Glades to work in the shadow of the Hoover Dike. The combined population of Belle Glade and Pahokee grew from 3,000 after the storm to 9,000 after World War II; farm acreage below the lake doubled, earning the region a prestigious "Army A" award for wartime food production. Vegetables were still the primary crop, but in Clewiston, "America's Sweetest Town," the U.S. Sugar Corporation began building a Machine Age agribusiness empire to pursue the lingering dream of a Florida sugar bowl.

Pedro Menéndez de Avilés imported sugarcane to Florida in 1565, and Indians probably grew it before that around Cape Canaveral, which is Spanish for cane. The Seminoles, David Yulee, Hamilton Disston, Ed "Bloody" Watson, and Ernest "Cap" Graham all tried to grow sugar in Florida as well; Buckingham Smith, General Harney, Henry Perrine, Henry Flagler, Governor Broward, and James Wright all predicted that it would become a lucrative crop in the Everglades. In 1929, when the Southern Sugar Company moved the old Pennsuco mill to the upper Glades, Florida's leading politicians all hailed the new plant as a great step forward for the state. But the firm soon went bust, and was reincorporated as U.S. Sugar under the leadership of General Motors executive Charles Stewart Mott in 1931. With the help of the Hoover Dike, along with its own ditches, levees, and pumps, U.S. Sugar finally succeeded in converting the warm climate and rich muck of the Everglades into money.

By 1945, it was running America's largest sugar operation, with 6,000 employees. Its loquacious president, Clarence Bitting, argued that with more government protection against flood and drought—and a repeal of the government quotas limiting domestic sugar production—the Everglades could provide hundreds of thousands of additional jobs for returning GIs: "It has been demonstrated beyond the peradventure of any doubt that the Everglades has—agriculturally, agro-biologically, agro-industrially and chemurgically—definite possibilities and potentialities for the immediate future far beyond the dreams of the past." Which was a fancy way of saying that the Everglades was ready to make people like Clarence Bitting rich.

The "monstropolous beast," after all, was back in bed for good. The communities of the upper Glades were no longer lakefront towns; they were dike-front towns, forever shielded from their old menace. Lake Okeechobee was no longer the wellspring of the Everglades, overflowing south into the Everglades during downpours; it was a giant reservoir, controlled by men who shunted its water east and west out to sea down man-made canals. The Everglades was cut off from its natural source, just as Disston had envisioned when he launched his attack on the lake. Thanks to the federal government, people could finally live there and farm there without a constant threat of calamity. "Everglades Permanence Now Assured," the *Florida Grower* crowed.

Fire in the Swamp

THE HOOVER DIKE had solved the problem of the murderous lake. But there were still other problems in the Everglades, and the bankrupt drainage district was helpless to do anything about them. The most dire problem was that after centuries of sogginess followed by two vicious floods, the Everglades was becoming a dust bowl.

The Depression years were drought years, and the combination of the new dike, which prevented water from the lake from reaching the Everglades, and the old ditches, which carried water from the sky away from the Everglades, left its wetlands desert dry. Its fresh water table dropped like a boulder, allowing salt water to intrude further into its aquifers every year, contaminating wells and ruining tomato farms along the Gold Coast. Meanwhile, soils that had been accumulating underwater for thousands of years were vanishing with exposure to the air; in Belle Glade—town motto: Her Soil Is Her Fortune—the ground was sinking so quickly that settlers had to add an extra doorstep every few years. Fred Elliot, so myopic about the risks of floods, had been right about the risks of overdrainage—not that he had ever done much about it. Cows were dying of thirst in the Kissimmee Valley, and Miami nearly ran out of potable water. Dried-out wetlands were invaded by opportunistic nonnative trees, especially the melaleucas that John Gifford had imported to drain the swamp, and the Australian pines that Henry Flagler and other developers had planted as ornamentals and windbreaks. "Everglades Drainage Found Too Well Done," one headline declared.

"The saw grass country lies prostrate," Thomas Will despaired. The Everglades was "absolutely dead," his house had burned down, and his tireless effort felt like an awful mistake. "This has cost me a professional career, and every cent of money I had," he wrote. He still believed in the promise of the Glades; if he could just die knowing that it would be ready for human use, he wrote in 1936, "I'll feel amply repaid." But he died a few months later, and the Everglades did not spring back to life. "Citizens of Florida," a New York paper reported in 1939, "came to the sudden realization recently that a vast part of the southern end of their state was on fire." That year, one million acres of the Everglades were incinerated, destroying an estimated $40 million worth of potential farmland, generating so much acrid smoke that schoolchildren in Miami had to cover their faces with wet handkerchiefs. Black clouds hovered over the entire region, driving away tourists, snowbirds, and weak-lunged retirees, frequently forcing highway officials to shut down the Tamiami Trail and Conners Highway for lack of visibility.

BEFORE HAMILTON DISSTON started digging, only 500 people had lived in south Florida. Now there were more than 500,000, and twice as many in the winter. People were the dominant species in the Everglades, and their actions were reverberating all the way down the food chain. Native Americans had tinkered with the ecosystem for centuries, but modern Americans were revamping it in fundamental ways.

Their farms had wiped out the pond apple belt and were displacing the upper Glades. Their subdivisions had wiped out the pinelands and hammocks of the coastal ridge, and were drifting into the eastern Glades. Loggers had cut down 90 percent of the region's virgin timber, ravaging cypress, pine, and mahogany stands for PT boats, houses, furniture, and coffins. Miners were tearing up aquifers and the wetlands above them to quarry limestone for highways and driveways. Fishermen hauled in so many mullet that the price dropped to a penny a pound; their nets left Biscayne Bay almost devoid of fish, and scraped away the submerged grasses that had served as fish food. Swamp rats were gigging 200 tons of frogs every year in the central Everglades alone, while at least a few plumers were back in the business, ravaging rookeries and smuggling feathers to Cuba. Thanks to big-wheeled swamp buggies fashioned out of tractors, and flat-bottomed airboats powered by airplane propellers, even amateur hunters and collectors could penetrate formerly inaccessible fastnesses, slaughtering gators, panthers, and deer, stripping the Everglades of its orchids, palms, and tree snails. "They are going deeper and deeper into the country," said Daniel Beard, a biologist with the National Park Service. "The place is being gutted."

But man's most dramatic alteration of the ecosystem was his disruption of its natural water regime. Lake Okeechobee, the liquid heart of the Everglades, was now cut off from the ecosystem's circulatory system. The transverse glades and other natural outlets that had been the veins of the Everglades had been deepened and widened to drain the water that had been the lifeblood of the Everglades. "Our beautiful streams could not be left alone," Simpson lamented. "Most of them have been dredged, supposedly to facilitate progress, and the debris piled up on the side of the ditch."

As canals lowered the water table, sawgrass invaded drying water-lily sloughs, while switch grass, myrtle, and sweet bay invaded parched sawgrass marshes. Common meadowlarks flourished in the drier conditions, but the wading birds that had symbolized the Everglades dwindled; Beard discovered that one white ibis rookery "which contained more birds than have probably ever been recorded in one place" had simply disappeared.

The stresses of low water wreaked havoc up and down the food chain: as Muhly-grass marshes disappeared, so did the apple snails that laid their eggs in those marshes, along with the Everglades kites that subsisted on those snails. Fewer gators meant more of the Florida gar they liked to eat, which meant fewer of the minnows the gar liked to eat, which meant more of the sandflies the minnows liked to eat.

The fires also brutalized wildlife. Raccoons, possums, and turtles were burned alive; snakes and frogs tried to cool down by burrowing into the muck, which didn't work when the muck itself was on fire. Roseate spoonbills and Audubon's caracaras were almost roasted to extinction. A *National Geographic* writer flew over the burning Everglades and witnessed a "scene of utter devastation," with charred land as far as he could see: "The only living things visible were turkey vultures wheeling low over the blackened ground in search of the carcasses of animals trapped by the fires."

The disaster in the Glades also spelled disaster for more than 600 Indians who still lived there, the descendants of Seminole warriors. The fish and game that had sustained their frontier economy grew scarce. The sloughs that had carried their canoes dried up. They still cherished Pa-Hay-Okee, the adopted homeland that had saved them from extermination. But they had to eat, so many of them moved to small government reservations around the region, or to tourist camps along the Tamiami Trail. Buffalo Tiger was born in the Everglades in 1920, and watched the region's wildlife gradually disappear. "The Breathmaker made the Everglades—water was good, lots of turtles, fish, deer, turkey, all we needed," said Tiger, a tall, weathered man who led the Miccosukee Tribe, an offshoot of the Seminoles. "White people messed it up, so we couldn't live with nature anymore."

THE DRAINAGE EVANGELISTS, convinced that water would run downhill, had never bothered to marshal many facts to justify their plans. As Marjory Stoneman Douglas recognized, they had used a kind of schoolboy logic: "The drainage of the Everglades would be a Great Thing. Americans did Great Things. Therefore Americans would drain the Everglades." But the biblical onslaught of fires and floods suggested that something clearly wasn't working, and inspired the first intense scientific research on the Everglades.

Once again, the federal government played a key role. A brilliant U.S. Geological Survey hydrologist named Garald Parker investigated the region's water problems and calculated that unless south Florida raised its fresh water table 2.5 feet above sea level, salt water encroachment would destroy its water supply. Meanwhile, U.S. Department of Agriculture scientists and state researchers studied the region's soil problems, and demonstrated that only about one-eighth of the Everglades—the upper Glades and the eastern wetlands near the coastal ridge—had soils deep enough for profitable agriculture, shattering the dream of draining the entire swamp. And the National Park Service's Daniel Beard conducted the first comprehensive study of the region's depleted wildlife, demonstrating that many Everglades species were in danger of extinction.

This science reached a popular audience through Douglas, the former *Herald* columnist who had once gushed about farms, highways, and development in the Everglades, but now developed a convert's fervor for the Everglades itself. The editor of the popular Rivers of America series initially asked her to write a book about the Miami River; she replied that it was only about an inch long, so there wasn't much to say about it, but she thought it was somehow connected to the Everglades. Maybe she could write about that instead? She then visited Parker, who introduced her to the basics of the Everglades, explaining that it was not really a swamp, but a subtle flow of water through sawgrass. "Do you think I could get away with calling it a river of grass?" Douglas asked. Parker said he thought she could.

Douglas was a relentless reporter and a fearless crusader; she liked to say that she channeled the energy and emotion that others wasted on sex—which she said she had for the last time in 1915—into her work. At the *Herald,* she had been one of Florida's leading voices for women's suffrage and civil rights, writing passionate columns until she suffered a nervous breakdown in the 1920s. She then turned to fiction, and made a nice living selling poetry and short stories to national magazines. But after accepting the Rivers of America assignment, she poured her energy and emotion into the Everglades. For the next five years, she picked the brains of the region's leading scientists, absorbing Parker's ideas about water and rock, the feather-counter Frank Chapman's ideas about birds, Daniel Beard's ideas about wildlife, and the USDA's ideas about soil. "I was hooked with the idea that would consume me the rest of my life," she wrote in her memoirs.

In 1947, Douglas introduced the Everglades to the world with her best-selling *The Everglades: River of Grass,* a florid description of the region's hydrology, geology, biology, and history—and some of its poetry, too. The book opened with the most famous passage ever written about the Everglades:

> There are no other Everglades in the world. They are, they have always been, one of the unique regions of the earth, remote, never wholly known. Nothing anywhere else is like them: their vast glittering openness, wider than the enormous visible round of the horizon, the racing free saltness and sweetness of their massive winds, under the dazzling blue heights of space.

Douglas was not just describing a place. She was sounding an alarm. In her final chapter, "The Eleventh Hour," Douglas bluntly warned that the Everglades was dying:

> The endless acres of sawgrass, brown as an enormous shadow where rain and lake water once flowed, rustled dry. . . . Garfish, thick in the pools where there had been watercourses, ate all the other fish, and died and stank in their thousands. . . . Deer and raccoons traveled far, losing their fear of houses and people in their increasing thirst. . . . The whole Everglades were burning. What had been a river of grass and sweet water that had given meaning and life and uniqueness to this enormous geography through centuries in which man had no place here was made, in one chaotic gesture of greed and ignorance and folly, a river of fire.

Still, the Everglades wasn't dead yet. It might have been killed if Hamilton Disston had dug his planned canal through Shark River Slough, or if Henry Flagler had run his railroad through Cape Sable, or if Flagler had taken James Ingraham's advice to sink his fortune into drainage, but it had dodged all those bullets. It might not have survived the grandiose reclamation plans of James Wright, Isham Randolph, or Fred Elliot, but they were never completed, because they were never funded. So there was still an Everglades, even if it was degraded.

There was also a movement to save the Everglades. Its leaders did not want to save the Everglades in the Broward sense, by extracting its economic potential for man, but in the literal sense—by preserving a large swath of it as a wild sanctuary for birds, orchids, and alligators, "where all forms of life cease to fear man and he in turn may be an acceptable friend and guest of nature." They dreamed of assuring a different kind of "Everglades Permanence." And their hopes lay with the federal government, too.

"The Appeal Is to Your Heart"

THE NATIONAL PARK SERVICE, created at the height of the Progressive Era in 1916, first proposed a national park in the lower Glades in 1923. The idea was embraced by many of the same Floridians who had fought to save the wading birds and Paradise Key—including Charles Torrey Simpson, John Kunkel Small, May Mann Jennings, and the renowned botanist David Fairchild, head of the USDA's Bureau of Plant Exploration. Fairchild warned that the children of the twenty-first century would revile the greedy and myopic generation that lost the Everglades, that "the newspapers of that day will cartoon us as the most terribly destructive mammals which ever inhabited the earth." Simpson wrote a searing article titled "Everglades Paradise Wrecked by Blunders," while Jennings gave speeches calling for at least some of the Everglades to be "saved for all time for the scientific enlightenment and enjoyment of the peoples of the entire world."

But none of those busy advocates did much to make Everglades National Park a reality before a single-minded, Yale-educated landscape architect named Ernest Coe moved to Miami in 1925. Coe came to Florida to help design the estates of the boom, but arrived just in time for the bust. Sixty years old and unemployed, with no outlet for his boundless energy, he began sloshing around the Everglades in canvas sneakers, often wrapping himself in a blanket and sleeping in the middle of the marsh. Coe fell madly in love with this "great empire of solitude," delighting in its rippling sloughs, billowing clouds, bewitching birds, and "fish of such a diversity of form and color as to make one wonder how Nature could devise such a range." He even loved its venomous snakes, insisting they were "quite disposed to be docile if treated cordially." The Everglades was Coe's escape from the hurly-burly of Miami, and something about its wildness stirred his soul. He called it the Land of the Fountain of Youth, and launched a lifelong crusade to protect it from man's depredations. "It is the spirit of the thing that holds you," he said. "The appeal is to your heart, and arouses in you a deep feeling of wonder and reverence."

Coe's obsession with an Everglades park would outstrip Broward's obsession with Everglades drainage and Thomas Will's obsession with Everglades development. In his first three years in Florida, he founded an advocacy group to lobby for it, persuaded Congress to study it, drew up its tentative boundaries, and took charge of a state commission formed to buy its land. "The blaze that had been lighted in him, the purpose and the power of the idea, would dominate his every moment for the rest of his life," Douglas wrote. Coe lectured about the glories of the Everglades to anyone who would listen—garden clubs, Rotary clubs, strangers he buttonholed on the street. Even the park's supporters learned to avoid the slender, snowy-haired man in the frayed seersucker suit, unless they were in the mood for a soliloquy. In her memoirs, Douglas recounted how her father, the *Herald* editor and drainage critic Frank Stoneman, used to cringe when Coe came by, "because he knew Mr. Coe would read him all the letters he'd gotten and all the letters he'd written." Coe fired off thousands of letters about the park, even when he was bedridden after an accident, even when he was grief-stricken after his wife's death. His commission employed more stenographers than the state attorney general's office. Critics groused that he must be on the payroll of some corporation with a secret interest in the park, since they could not imagine how anyone could be "damn fool enough to spend the time and energy he did unless that was the case."

Coe was not on any corporation's payroll. He really was devoted to the park, and considered it his sacred mission in life. But his moral fervor alienated as often as it persuaded, especially in frontier towns like Chokoloskee and Everglades City. "When a fellow like that gets up before a meeting of honest-to-God crackers and begins to use his high-falutin words, [they] say 'to H—l with that fellow' and they are against anything he is for," observed J. H. Meyer, a surveyor who worked for Coe on the Everglades commission. Coe also tended to veer into hyperbole as absurd as any real estate booster's propaganda. "It is a safe assertion that had this park been in existence for the past 20 years, many of our serious economic problems of today would not be before us," he wrote to President Hoover in 1929. And without bothering to consult any Seminoles, Coe declared that the park would revive their fortunes by providing them jobs as tourist guides, replacing their "pathetic outlook" with "a bright future." When a governor finally did ask some Indians inside the park boundaries what they wanted, they replied: *Pohaan checkish.* Leave us alone.

But no one could question Coe's dedication. In February 1930, for example, he arranged for the federal committee evaluating the park to fly over the Everglades in the Goodyear blimp. Since there was not enough room in the cabin for everyone, he spent the ride dangling below the dirigible in a tiny observer's coop, violently throwing up in a bucket. He still considered the trip a grand success, because his guests got to see great flocks of white ibis and snowy egrets spreading across the sky like confetti, against the backdrop of the watery green carpet that still blanketed most of south Florida. Coe then led the committee on a three-day boat tour, "filled with thrills for every waking hour," including bald eagles, porpoises, spoonbills, fire-orange sunsets, and an army of pelicans, gulls, and herons roosting along the mudflats of Cape Sable. And the committee swiftly recommended his plan to Congress. For all his speech-making and letter-writing, Coe knew that the best lobbyist for the Everglades was the Everglades.

COE'S EXPANSIVE BOUNDARIES encompassed more than two million acres of the southern Everglades, Florida Bay, Ten Thousand Islands, Big Cypress, and the upper Keys, stretching as far north as fifteen miles above the Tamiami Trail highway and as far east as the barrier reefs in the Atlantic. The primary goal was to preserve the ecosystem's vast diversity of habitats in their primitive condition—pinelands and marshlands, estuaries and sloughs, dwarf cypress and elkhorn coral. A secondary goal was half a million annual visitors, but as the botanist David Fairchild explained at a congressional hearing, the Everglades was not Yosemite, and its entertainment value would be only part of its appeal. It would also educate children, provide a unique laboratory for scientists, protect rare flora and fauna from extinction, and "startle Americans out of the ruts which an exclusive association with the human animal produces in the mind of man."

This was a new way of thinking about national parks. "I have been laboring under the impression that the yardstick to use in selecting national parks was that of the showman, that it was the spectacular we were to consider," one congressman told Fairchild, the founder of Miami's Fairchild Tropical Gardens. "Now you were giving us a new thought, and a very interesting one, that a piece of ground which has educational value, scientific value, rises to the height of national park value." Some critics still attacked the park's enabling act as the Alligator and Snake Swamp Bill, especially after a well-intentioned scientist pulled a king snake out of his bag during a congressional hearing. "The Everglades section is almost impassable and is nothing but a snake and mosquito farm, with a climate which no white man enjoys," one opponent wrote. Nevertheless, in 1934, Congress authorized the Interior Department to accept the park, as soon as Florida donated the land.

That was a thornier problem, because the state commission was too busy fighting over Coe's boundaries to buy land. The Izaak Walton League, a sportsmen's group, wanted to cut out two-thirds of the park, especially the game-rich Big Cypress Swamp. Monroe County officials demanded the removal of Key Largo, their most attractive land for development. And Barron Collier and other southwest Florida landowners pushed to exclude the Turner River area, where farms were fetching $200 an acre. The chairman of the commission's boundaries committee, D. Graham Copeland—who was also one of Collier's executives—proposed a new map excluding Big Cypress, Key Largo, and Turner River. Copeland pointed out that the original boundaries were simply "the child of Mr. Coe's brain," and that the revisions would eliminate only one-seventh of the park's land. Copeland had drummed up support for the park among businessmen and landowners who had the power to scuttle the project; he was a reasonable man, and most members of the commission saw his compromise as a reasonable plan.

But Coe believed the Everglades was already compromised enough. He insisted that any reduction would jeopardize the entire park, denouncing moderation in defense of the Everglades as treason to the cause. He argued that Big Cypress was not only a magical landscape in its own right but was vital to ensuring the supply of fresh water to the rest of the park. He fulminated that "to eliminate the Key Largo and marine gardens area from the Park will be on a par with cutting out the geysers from Yellowstone, the bridal falls from Yosemite or the canyon from the Grand Canyon."

These were not widely shared beliefs, even among the park's supporters. May Mann Jennings found Coe's intransigence "absurd," and urged his friends to keep him away from Tallahassee, where he was becoming a major political liability. "He antagonized the [Izaak] Walton League until they hated his very insides," recalled his employee J. H. Meyer. "He rammed Key Largo down the throats of the [Monroe County] boys until they wanted to tar and feather him." In 1937, an annoyed Governor Fred Cone forced Coe out of his state job, and without his leadership, the Everglades commission soon disbanded. Coe kept firing off letters about the park at his own expense, but as one of his admirers acknowledged, his appeals became increasingly "long-winded, obtuse and shrill."

President Franklin Roosevelt's administration also pressed for action, but the drive for the park had stalled. The north Floridians who controlled the state legislature had no interest in saving the south Florida wilderness, especially after oil prospectors began to drop hints about seeking a new kind of black gold in the Everglades. It was almost impossible to raise money for the park during the Depression, and almost impossible to attract attention to the park during World War II. As speculators snapped up Everglades land, and fires raged through the marsh, Jennings began to fear the park dream would be deferred forever: "I am about to die waiting until this thing is ready."

The park no longer needed a crusader on a sacred quest. It needed a practical politician to make the dream a reality. It needed Spessard Holland.

Mr. Florida's Business Proposition

SPESSARD LINDSEY HOLLAND was born in 1892 in Bartow, the rural central Florida town founded by the cattle baron Jacob Summerlin, the "King of the Crackers." Spessard's father, a Confederate war veteran wounded at Kennesaw Mountain, owned a small farm, and he enjoyed a typical cracker childhood there—hunting squirrels and turkeys in the piney woods with a 16-gauge shotgun, fishing bass and trout, milking cows and feeding chickens. He once interrupted a fight with his sister to peel a leech off her foot, and learned to detect a snake by its scent. But the young Spessard Holland stood out like a twenty-five-pound bass in the Bartow pond. He liked to draw, sing operetta, and write poetry, and had such a live pitching arm that the legendary Philadelphia Athletics manager Connie Mack offered him a contract. He was also a straight-A student with a voracious appetite for books, voted most brilliant in his college class, and president of his law school class. He qualified for a Rhodes Scholarship, but World War I intervened, so he volunteered for the Army Air Corps instead and earned the Distinguished Service Cross after he was shot down over the western front.

Holland returned to Bartow after the war as a country lawyer, representing the farmers, citrus growers, and phosphate miners who were fueling the area's prosperity. His practice would later expand into Holland & Knight, now one of America's largest law firms. But he found his own calling in the public sector, serving as a prosecutor, judge, state senator, governor, and four-term U.S. senator, never losing a single election. He was not a scintillating politician, but he was a straight talker, a diligent worker, and a skillful consensus-builder, studying the minutiae of legislation, working the phones at all hours, never breaking his word or losing his temper. He was his state's most popular and powerful public servant for three decades, earning the nickname Mr. Florida; the newsman Howard K. Smith once called him the most respected member of the Senate. Tall, handsome, and solidly built, he even looked like a statesman, especially after his shock of dark hair turned snow-white; with a toga on his shoulders, one colleague said, he could have passed for a senator in ancient Rome.

Ideologically, Holland was a southern conservative who proclaimed his devotion to free enterprise and fought the minimum wage and Medicare, but he was also a master pork-barreler who steered all kinds of government protection and assistance to Florida's agriculture and business interests. He was a committed segregationist, but he authored the constitutional amendment outlawing the poll tax, and twice faced down racist lynch mobs. Zora Neale Hurston once wrote an essay titled "Take for Instance Spessard Holland," citing his personal decency as proof that not all southern lawmakers were "bigoted jumping jacks." Like Governor Bloxham, Governor Broward, Fred Elliot, and other native Floridians who grew up in the backwater era, Holland's foremost concern was his state's development. He was a loyal friend of Florida's sugar, cattle, citrus, and real estate industries, even defending Big Sugar after *Harvest of Shame,* Edward R. Murrow's documentary about the abuse of migrant farmhands. Holland's proudest legislative achievement was the pro-development Tidelands Act, which gave states control of their near-shore waters, allowing the rampant dredging and filling of Florida's coastline. After President Harry Truman went on national television to denounce the law as "daylight robbery," Holland personally defended it before the Supreme Court—and won.

Holland was also a heartfelt conservationist who loved fishing, hunting, and camping in the Florida wilds, and who spent many happy days bird-watching with his wife. He could be transported by the beauty of a red-shouldered hawk in flight or a fawn bent over a watering hole, and he pledged to make conservation one of his top priorities when he became governor in 1940. But he was always a pragmatist first, a wise-use man; he represented people, not hawks or fawns, and he supported conservation for people's sake, not nature's sake. "I do not believe any plan for conservation will get very far unless the average man is considered," he explained. "I mean the man who, after his day's work is over, takes his fishing pole out for an hour or two and comes back with a mess o' fish and can then sit down to a hot supper of bream, hush puppies and black coffee."

EVERGLADES NATIONAL PARK, in Holland's view, was a terrific idea for that average Florida man, recreationally and economically. "I don't think any other project begins to compare in potential to the state as a whole," he said. Now that the government soil studies had made it clear that drainage was no longer a viable option for the entire Everglades, Holland saw the park as a new way to convert swampland into prosperity. He expected visitors to flock to its egrets and crocodiles, spending money and creating jobs at Florida's hotels, restaurants, gas stations, and souvenir shops. "They will see many new things, they will come back, they will stay and invest their money here and help build our state to even greater heights," he said. "To my mind, it is just about the biggest single business proposition now pending." Holland appreciated the mysterious beauty of the Everglades, but he always considered the park a business proposition first. In fact, after oil was discovered beneath the Big Cypress in 1943, Holland warned that if the Everglades could support a petroleum boom, its economics would no longer support a national park.

The wildcat dreams of south Florida becoming a new West Texas did not materialize, so Holland kept pushing for the park. But he didn't push for two million acres. Holland was temperamentally allergic to firebrands like Coe, and the sportsmen, oilmen, farmers, landowners, and Monroe County leaders who were livid about Coe's boundaries were some of his strongest political backers. The senator had represented some of them in private practice, and he saw no point in trying to force them to surrender land they wanted when there was so much land they didn't want. In any case, he knew they would block the park in Tallahassee if their concerns were not addressed, and he believed the park would die in Congress unless it had consensus support at home. Holland was a dealmaker, and while he had no intention of battling his friends for Coe's all-or-nothing plan, he was willing to use every ounce of his influence to make a smaller park happen.

★   ★   ★

THE WORK HOLLAND DID to cut the deals that created Everglades National Park was less interesting than the fact that he did it at a time when the issue had vanished from the political radar screen. When Holland urged *Herald* publisher John Knight to push for the park, Knight had to admit he had never heard of the idea. But as governor and later as senator, Holland devoted countless hours to negotiations that would ultimately shrink the park to 1.3 million acres, slicing out all of the upper Keys, Big Cypress, and everything else north of the Tamiami Trail, the coral reefs, the Turner River area, the marshes of northeast Shark Slough along the park's eastern boundary, and a 22,000-acre tract of farmland inside the park known as "The Hole in the Donut." The revised boundaries cut out three times as much land as Copeland's initial compromise, but they still protected the heart of the Everglades, including Paradise Key, Cape Sable, Florida Bay, most of Shark and Taylor Sloughs, and most of the Ten Thousand Islands. Holland's compromises created America's third-largest national park, behind only Yellowstone and Mount McKinley; it was a deal that every key player except Coe could live with, the kind of deal that modern politicians call a "win-win." Interior Secretary Harold Ickes tried to explain to Coe that something was better than nothing, that the park could always be enlarged in the future. "I want to see the project advanced while there are still park resources left to conserve," Ickes wrote. Even many of the Seminoles, who lost their reservation inside the park, were satisfied with their new reservation on more valuable land in Broward County.

Senator Holland begged and bullied the new plan through Capitol Hill, calling in chits, incessantly reminding his colleagues that the plan had full support in Florida. The only remaining hurdle was getting the Florida legislature to cough up $2 million needed to buy private land inside the boundaries. This task fell to John Pennekamp, the *Herald* editor who had been ordered to soft-pedal Miami's hurricane damage in 1926, and now had been instructed to lead the Everglades push that Holland had requested of his boss.

Pennekamp became legislative chairman of the park commission when it was reinstated after the war, and he decided to try to ingratiate himself with the Pork Chop Gang, the machine of retrograde north Florida Democrats who ran Tallahassee. He wrangled a get-to-know-you meeting with five leading Pork Choppers at a hunting camp in Ocala, and bonded with them over a chicken-and-rice dinner before the group settled into a boozy 10-cent-limit poker game. Pennekamp enjoyed an extraordinary run of luck, taking down one pot by drawing an inside straight, winning another hand with four kings. Soon he had $33 in winnings, and a Palatka senator began razzing him.

"Penny, how much did you say you needed for that park?"

"Two million, senator."

"Why the hell don't you try to get it from the Legislature, instead of our pockets?"

That gave a Gainesville senator an idea: "Why don't we just give him that money when the Legislature meets? Maybe he'll lay off us!" They agreed to do just that.

ON DECEMBER 6, 1947, a month after the publication of *The Everglades: River of Grass,* President Truman formally dedicated Everglades National Park. John Pennekamp served as master of ceremonies, and Senator Holland delivered a gracious speech. Marjory Stoneman Douglas was there, as were May Mann Jennings, D. Graham Copeland, and Daniel Beard, who had accepted the job of park superintendent. Still bitter about the shriveled boundaries, eighty-one-year-old Ernest Coe had vowed to boycott the event, but he changed his mind, and 10,000 spectators erupted when he was introduced as the "daddy" of the park. Over the next few decades, Florida conservationists would fight some of their toughest battles to save areas Coe had originally included within his park boundaries—including Big Cypress National Preserve, Ten Thousand Islands National Wildlife Refuge, Biscayne National Park, and John Pennekamp Coral Reef State Park, as well as the Hole in the Donut and northeast Shark Slough, which would be added to Everglades National Park.

In his speech, President Truman eloquently explained why the federal government was bothering to protect such a forbidding wilderness, hailing the park as a shrine to the diversity and mystery of God's creations, and a source of enjoyment and enlightenment for future generations:

> Here are no lofty peaks seeking the sky, no mighty glaciers or rushing streams wearing away the uplifted land. Here is land, tranquil in its quiet beauty, serving not as the source of water but as the last receiver

of it. . . . For conservation of the human spirit, we need places such as Everglades National Park, where we may be more keenly aware of our Creator's infinitely varied, infinitely beautiful and infinitely bountiful handiwork. Here we may draw strength and peace of mind from our surroundings. Here we can truly understand what the psalmist meant when he sang: "He maketh me to lie down in green pastures; He leadeth me beside the still waters; He restoreth my soul."

But for all his lyricism—environmentalists still quote his line about "the last receiver"—Truman never said that the Everglades should be saved for the sake of nature. He said it should be saved for "the enjoyment of the American people" and the "conservation of the human spirit." He called it a sterling example of "the wise use of natural resources," a phrase favored by people-first conservationists like Spessard Holland, not nature-first conservationists like Ernest Coe. While Truman celebrated the wildness of the southern Everglades, he also called for Americans to "make full use of our resources" in the rest of the Everglades.

In fact, Senator Holland was already orchestrating a plan for the Army Corps to do just that, a plan designed to solve all the problems of the Everglades at once: floods and fires, soil subsidence and salt water intrusion, underdrainage and overdrainage. It was a plan to make south Florida safe for explosive immigration and development, combining century-old drainage dreams with newfangled conservation strategies. For better and for worse, this plan would guide the Everglades into the twenty-first century.

THIRTEEN

Taming the Everglades

Section 21: Water a Common Enemy. It is hereby declared that in said District, surface waters, which shall include rainfall and the overflow of rivers and streams, are a common enemy.

—*Laws of Florida*, Chapter 59-994

"We've Never Had a Water Situation Like This"

AMERICA EMERGED FROM World War II as the richest, most powerful nation in history. It had just manufactured a spectacular arsenal of guns, jets, and bombs; now it redirected its industrial might into big cars, mass-produced homes, and spacious refrigerators for a middle class of GI Joes returning to civilian life and Rosie the Riveters returning to home life. "There never was a country more fabulous than America," a British historian wrote after a 1948 visit. "She sits bestride the world like a Colossus. . . . Half of the wealth of the world, more than half of the productivity, and nearly two-thirds of the machines are concentrated in American hands." That year, Americans broke the sound barrier, and developed the transistor, cortisone, and liquid hydrogen. They also got serious about taming the Everglades.

Before the war, Florida was still a poor, predominantly rural state with fewer residents than Mississippi; it had yet to pass a fence law, so its cattle still roamed free. But two million military men trained there during the war, and many were eager to return now that air conditioners and bug spray

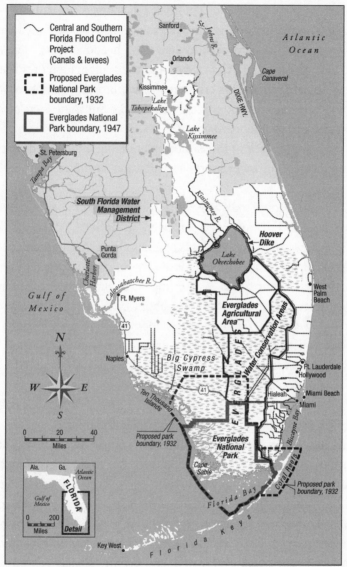

Central and Southern Florida Flood Control Project (Canals & levees)

Proposed Everglades National Park boundary, 1932

Everglades National Park boundary, 1947

Sanford

St. Johns R.

Atlantic Ocean

Orlando

Cape Canaveral

Kissimmee

Lake Tohopekaliga

DIXIE HWY.

Lake Kissimmee

St. Petersburg

Kissimmee R.

Tampa Bay

South Florida Water Management District →

Punta Gorda

Hoover Dike

Lake Okeechobee

West Palm Beach

Gulf of Mexico

Charlotte Harbor

Caloosahatchee R.

Ft. Myers

Everglades Agricultural Area

41

Water Conservation Areas

N

Naples

Big Cypress Swamp

E V E R G L A D E S

Ft. Lauderdale
Hollywood

W — E

41

Hialeah

Miami Beach

Miami

S

Ten Thousand Islands

Buscayne Bay

0   20   40
Miles

Proposed park boundary, 1932

Everglades National Park

Cape Sable

Coral Reefs

Proposed park boundary, 1932

Ala.   Ga.
Atlantic Ocean

FLORIDA

Gulf of Mexico

0   200
Miles   **Detail**

Florida Bay

Florida Keys

Key West

BY GENE THORP—CARTOGRAPHIC CONCEPTS INC.

After the storm of 1928, the Army Corps of Engineers built a dike around Lake Okeechobee, a key step toward taming the Everglades. After more flooding in 1947, Congress authorized the Central and Southern Florida Project, assigning the Corps to finish the job with 2,000 miles of levees and canals. The northern Glades was carved into farmland, the central Glades into reservoirs, and the eastern Glades into suburbs. The southern Glades was preserved as Everglades National Park, although the park was less expansive than Ernest Coe had envisioned fifteen years earlier.

were widely available, and federal mortgage guarantees and veterans benefits offered easy access to housing. Airplanes were also increasingly accessible, ferrying tourists from New York to the Miami headquarters of Pan American and Eastern Air Lines in half a day. And the combination of private pensions and Social Security was creating a newly prosperous class of retirees, who could now spend their golden years on the Sunshine State's beaches, golf courses, and shuffleboard courts.

The only two obstacles to an explosion of immigration and development in lower Florida were water shortages and water surpluses.

IN 1947, AFTER more than a decade of drought, a summer of downpours followed by two fall hurricanes dropped an unheard-of 100 inches on south Florida, washing out hundreds of miles of streets, thousands of homes and businesses, and tens of thousands of acres of citrus groves and vegetable fields. Mother Nature was reasserting her authority, reclaiming the reclaimed Everglades, reflooding just about every wetland that had been drained or paved for agriculture or development—from the pastures of the Kissimmee valley to the farms of the upper Glades to young suburbs like Hialeah, Miami Springs, and Opa-locka in the eastern Glades. She turned most of the region into a shallow lake, reminding its residents they would never be safe as long as she remained on the loose. "Everglades Is Unconquered Despite Man's Great Fight," the Herald declared.

There were no mass casualties, because the St. Lucie Canal and the Hoover Dike kept Lake Okeechobee in bed. But the deluge overwhelmed the undersized Broward Era canals within the Everglades, spreading across five million acres and staying there for months—from Orlando down to Cape Sable, from the Gulf across to the Atlantic ridge. Thousands of cows drowned or starved. Deer were stranded on farm dikes with panthers and rattlesnakes. Septic tanks overflowed, and health officials battled typhoid outbreaks. "We've never had a water situation like this before," said the Red Cross rescue chief.

The floods of 1947 also created a new environmental crisis. To prevent Lake Okeechobee from busting through its dike, water managers had to expel billions of gallons down the St. Lucie, disrupting the delicate balance of fresh and salt water in the estuary at its mouth, the most biologically diverse water body in North America. Those blasts of water also carried silt from the lake toward the fishing centers of Jupiter and Stuart, driving away bluefish, bonefish, and tarpon. "Our St. Lucie River, around which the entire tourist and commercial picture revolves, has been turned into mud soup," one local journalist wrote. "The finest fishing grounds on the East Coast of Florida" was now "a mud hole which no respectable fish would inhabit. . . . Feeling is extremely bitter here."

Feeling was bitter throughout the region, as homeowners who had howled about fires now howled about floods, blaming their plight on water managers who had opened or refused to open various floodgates. Lamar Johnson, who held Fred Elliot's old job at the Everglades Drainage District, now inherited Elliot's mantle as the district scapegoat; one landowner was arrested while waiting to ambush him with a rifle. "I was maligned, threatened, waylaid, investigated by the governor and investigated by a grand jury," recalled Johnson, who packed a .38 revolver on his future visits to the Everglades. But there wasn't much the district could have done to keep more than two million Olympic-sized pools worth of water out of the region's lawns and living rooms. To make this clear to citizens—and to Congress and President Truman—Johnson distributed the "Crying Cow" report, with a cover portrait of a teary heifer up to her belly in water. The report illustrated its preliminary damage calculations—about $59 million in all—with bleak photos of tomato farms, cattle ranches, orange groves, labor camps, suburban cul-de-sacs, military bases, a Seminole reservation, the Hialeah Racetrack, and the Dixie Highway under several feet of water. "Only when the Everglades has an adequate water control and protective system will the agricultural interests and coastal communities feel secure in their investments," the report concluded.

That system would have to prevent extreme drought as well as extreme floods, and it would have to control the Everglades as well as the lake. Designing and building that system, officially known as the Central and Southern Florida Project for Flood Control and Other Purposes, would be another job for the Army Corps of Engineers.

Waters of Destiny

Two decades after the Lake Okeechobee and Mississippi River disasters, the Army Corps was feverishly resculpting the nation's rivers and coastlines, battling to keep water away from people, rearranging nature into more economical arrangements. Its projects helped farmers convert huge swaths of wetlands into croplands, deepened ports for barges and ships, and protected waterfront cities like New Orleans, St. Louis, and Omaha, where a Corps general reacted to a flood by screaming: "I want control of the Missouri River!" Soon he had it, thanks to a series of dams and dikes. And that was a pittance compared to the Corps assault on the Mississippi and its tributaries, where the agency spent billions of dollars on defenses designed to withstand a flood even greater than 1927's, including than 2,000 miles of upgraded levees.

The Corps had its critics, led by Interior Secretary Ickes, who attacked the agency as a "reckless . . . lawless . . . powerful . . . self-serving clique" of concrete addicts, flagrantly disobeying presidential orders, "wantonly wast[ing] money on worthless projects" that kept their employees busy and their congressional paymasters happy. The Corps was supposed to conduct objective cost-benefit analyses of proposed water projects, but critics groused that if a committee chairman wanted, the agency would justify a project to grow bananas on Pikes Peak. "Every little drop of water that falls is a potential flood to the ubiquitous Army Engineers, and they therefore assume it to be their duty to control its destiny from the cradle to the grave," Ickes wrote.

But Corps leaders were unapologetic about their aggressive efforts to develop water resources, and their close relations with powerful congressmen. In their view, America had declared war on nature, and as they made clear in a documentary chronicling their battle to control the Mississippi, they were proud to man the front lines: "This nation has a large and powerful adversary. We are fighting Mother Nature. . . . It's a battle we have to fight day by day, year by year. The health of our economy depends on victory."

The Corps filmed a similar tribute to its war on the Everglades, *Waters of Destiny,* a black-and-white propaganda piece in the style of the Cold War newsreels that warned Americans about the communists under their beds. *Waters of Destiny* was an equally melodramatic account of man's battle to tame "the crazed antics of the elements" in central and southern Florida, his epic struggle against "the maddened forces of nature." It opened with a crash of a gong, followed by a jagged bolt of lightning. A stentorian narrator then introduced the water of the Everglades as the villain of the film:

Hideous . . .
Unrelenting . . .
Shrieking its rage . . .
The vicious scourge of mankind . . .
Burying life and land under its relentless and merciless depths . . .
This is the story of such water—and its mastery by the determined hand of man.

Men had always dreamed about the mastery of nature, but now they were finally achieving it. Their cars and planes were overcoming distance; their air conditioners and insect repellents were overpowering heat and mosquitoes; their bulldozers and concrete were overwhelming wilderness. In this tidy era of U.S. dominance, the Corps was not about to let the Everglades kill thousands of people, or try to dictate where Floridians could or could not live: *"We had to control the water—make it do our bidding."*

The Army Corps plan for the Central and Southern Florida project called for the most elaborate water control system ever built, the largest earth-moving effort since the Panama Canal. It envisioned 2,000 miles of levees and canals, along with hundreds of spillways, floodgates, and pumps so powerful they would be cannibalized from nuclear submarines. The C&SF project was designed to control just about every drop of rain that landed on the region, in order to end the cycle of not-enough-water and too-much-water that had destabilized the frontier and stifled its growth: *Central and Southern Florida just lay there, waiting helplessly to be soaked and dried and burned out again. . . . Something had to be done, and something was.*

This was truly Flood Control and Other Purposes—not only saving lives and property, but reclaiming land, storing water, and promoting economic growth. The C&SF project would subdue the Kissimmee valley into a cattle empire, the upper Glades into an agricultural empire, the central Glades into giant reservoirs, and the eastern Glades into farms and suburbs that would offer the postwar version of the American Dream. "Florida's economists view this soil and water surgery as the most potentially profitable undertaking in the State's history," one reporter wrote.

For centuries, most of south Florida had been considered uninhabitable, but now that people were there and more were on the way, their safety and prosperity had to be assured. Sure, a chunk of the Everglades could remain wild for the national park, but the rest of the ecosystem had to be tamed to serve man. "The easy solution, of course, would have been to leave the Everglades as nature formed them," the *Herald* explained. "Yet to do so would have deprived mankind of thousands of acres of rich land and the chance to enlarge cities in a region with one of the best climates in the world. . . . Too much human effort and treasure have been staked on the usability of the Everglades."

THE C&SF PROJECT incorporated elements of almost every Everglades plan of the last century.

In the Kissimmee basin, the Corps would expand Hamilton Disston's drainage work, deepening and widening his canals in the Chain of Lakes, then channelizing the Kissimmee River to whisk the basin's floodwaters into Lake Okeechobee. The control outlets east and west to the St. Lucie and Caloosahatchee Rivers, originally proposed by Buckingham Smith, would also be expanded to strengthen man's grip on Lake O, and the Hoover Dike would be extended around the entire lake to boost its capacity as an urban and agricultural reservoir. Governor Broward's diagonal canals through the Everglades would be expanded as well, and as the Everglades Land Sales engineers had advised, massive pumps and controls would be installed in the canals to speed the outflow during floods and hoard the runoff during droughts. Gravity just wasn't getting the job done.

The plan's first big innovation was its strict separation of the usable Everglades from the unusable Everglades, a concept that first appeared in Captain Rose's drainage proposal for Henry Flagler. The plan also adopted Rose's call for piece-by-piece as opposed to all-at-once drainage. The work began with a 100-mile-long "perimeter levee" running more or less parallel to the coastal ridge, walling off the Gold Coast and a wide slice of the eastern Glades from the rest of the marsh. Next, the Corps encircled and reclaimed the rich soils of the upper Glades with more levees and drainage canals, creating an Everglades Agricultural Area the size of Rhode Island.

The Corps then built more levees to divide a swath of the central Glades even larger than Rhode Island into three gargantuan "water conservation areas," a more recent plan devised by the Belle Glade research station. The station's scientists had suggested that "rewatering" the central Glades could restore the region's hydraulic head and mimic the natural storage capacity of the Everglades, preventing salt intrusion, soil subsidence, muck fires, and water shortages all at once. The conservation areas would still look like the Everglades, but they would hold onto needed water for farms and cities during droughts, absorb excess water from farms and cities during storms, and recharge the region's aquifers to keep salt out of its groundwater.

The C&SF project, in other words, was about Getting the Water Right. As the *Waters of Destiny* narrator explained in his urgent baritone, the region received more than enough rain to support cows, crops, and people, but nature dumped it in inconvenient spurts, *"turning hard-earned farm profits into devastating losses, ruining homes and businesses, wreaking a devastating havoc that ran into millions upon millions a year."* When the rains passed, the drama continued: *"Arid land, leached and sucked dry, once lush farmland, reduced to dry dust."* Then came the fires, *"leaving a waste and a desolation that was almost absolute."* And then the floodwaters returned, *"doing a damage that could never be repaired, never replaced, never be the same again."*

The simple objective of the C&SF was to smooth out this erratic cycle of deluge and drought. *"Obviously, if the water was in the right place at the right time, if the excess water could be removed in a hurry, then brought in when it was needed, Central and Southern Florida would flower upon the seeds of its own rich resources."* With man in control of the water, the entire Everglades ecosystem would finally be "of use." The vicious scourge of mankind would cater to his needs.

"I Have Not Heard of Any Opposition, Senator"

WATERS OF DESTINY celebrated how the engineers who had helped defeat the forces of Hitler were defeating the forces of water, assaulting the swamp with dredges, draglines, and dynamite, pouring concrete for pump stations that could move two million gallons a minute. *"You've got to have that kind of action to get that kind of rainfall off this kind of land,"* the narrator intoned. With its lingering images of sweaty, muscular men thrusting heavy equipment into squishy, overgrown wetlands, *Waters of Destiny* now feels like pornography of natural destruction. It has become a kitsch classic for environmentalists and even Corps engineers, a monument to human folly. But while the C&SF project and its architects are often blamed for the decline of the Everglades, it's worth recalling that the ecosystem was seriously degraded before the project even began; Marjory Stoneman Douglas was already warning that the "dying" Everglades faced its "eleventh hour." It's also worth recalling that there was overwhelming demand for the project in south Florida; voters in Dade, Broward, and Palm Beach Counties approved the water conservation areas with 96, 86, and 97 percent of the vote.

It's especially worth recalling that while the C&SF was designed to help people, it was expected to help the Everglades as well. The Army Corps claimed that it "would produce substantial benefits from the preservation of fish and wildlife resources," "would not damage or interfere with this great national park," and was "necessary to preserve and restore the unique Everglades region." The National Park Service and the Florida Wildlife Federation supported the project. The U.S. Fish and Wildlife Service said it would "improve conditions for fish and wildlife resources in the Everglades," and agreed to manage its northernmost water conservation area, which became the Loxahatchee National Wildlife Refuge. Douglas hailed the C&SF as "the first scientific, well-thought-out plan the Everglades has ever known." Its congressional godfather was Senator Holland, who had played the same role for Everglades National Park, and saw no conflict whatsoever.

MR. FLORIDA WAS a Democratic backbencher, the newest and lowest-ranking member of the Public Works Committee and its Flood Control Subcommittee. But he again worked his magic to persuade the Republican-controlled Congress to support an ambitious business proposition for his home state. "With near-perfect timing," a reporter wrote, he "brought the measures safely through the legislative maze."

Holland helped shape the project to benefit Florida's special interests, persuading the Corps to expand a $60 million plan into a $208 million plan. He kept in close contact with cattle ranchers, citrus growers, and real estate developers; he put a U.S. Sugar executive in charge of a committee that helped devise the plan; he insisted on one eighteen-mile canal that benefited just fifteen rich landowners. But once the deals were cut and the plan was finalized, Holland made sure every Florida interest publicly supported it without qualification. For example, his friends at U.S. Sugar were upset that it did not include a third outlet from Lake Okeechobee through the Loxahatchee River, but he warned them to keep their dismay to themselves. He once again doubted that Congress would support a huge project for Florida if it sensed any dissension in Florida; in one 1948 speech to the state's cattlemen, he mentioned "unity" seven times in four sentences. Ultimately, the plan was endorsed by almost every newspaper, civic group, and politician in Florida. "Everybody . . . will benefit from this dramatic control of the environment by men," one booster declared, calling the project "one of the greatest examples in America of what man's intelligence and vision can do in converting the erratic forces of nature into solid assets for the vegetable and animal kingdom."

Holland meticulously choreographed a Senate hearing to show off that unity, serving as the project's defense attorney as well as a member of the jury evaluating it. Like Senator Westcott a century earlier, he assured his colleagues that developing the Everglades would benefit America, not just Florida, by producing cheap food, creating homes and jobs for World War II veterans, and protecting bases for Cold War soldiers. He then led a parade of friendly witnesses through a charade of leading questions about the desperate need for water control, often prefacing his softballs with "Isn't it true that . . ." or "Wouldn't you say that . . ." For example, he asked his close friend Irlo Bronson, a leader in the Florida legislature and the head of Florida's cattlemen, whether he knew of any opposition to the C&SF anywhere in the state. "I have not heard of any opposition, Senator," Bronson replied. In fact, the plan did have opponents, including Cap Graham and the Collier family, but Holland made sure none of them got anywhere near the hearing room. After three days of unrelentingly enthusiastic testimony, the chairman of the Public Works Committee declared, "I do not recall that I have ever attended a hearing where such a vast project has been so ably presented in such a short period of time." Congress promptly approved the plan, and ordered the Corps to start moving dirt.

Like Broward's drainage project, this was hailed as a great conservation victory. The Central and Southern Florida Flood Control District—the state agency created to help the Corps manage the project—trumpeted it as "CONSERVATION IN ACTION," preventing the waste of water and soil, improving Mother Nature's plan to serve man. And it wasn't just wise-use conservationists who welcomed the C&SF; Douglas said it would "keep the water of the Everglades in balance just as nature had once maintained it, and in much the same way." White men had almost killed the Everglades, she said, but now its problems were solved: "The ancient southwest course of the grassy river is fully preserved. The water will flow again, as it always did."

*   *   *

IN REALITY, there was no way the water of southern Florida could flow as it always did, not with its route out of Lake Okeechobee blocked by the Hoover Dike, not with its path through the Everglades interrupted and diverted by canals, levees, and the Tamiami Trail. The C&SF plan was about transformation, not preservation; it proposed converting the topsy-turvy Kissimmee River into a ditch, Lake Okeechobee and the central Glades into reservoirs, the northern Glades into farms, and the eastern Glades into suburbs. It did not include U.S. Sugar's destructive plan to channelize the Loxahatchee River, but it did put Everglades National Park's water supply in the hands of the Corps, which even some supporters recognized as a potential ecological disaster. The Park Service, while generally favoring the plan, warned that the park could die of thirst unless the Corps opened enough culverts and spillways to let water through the Tamiami Trail during droughts. The Corps promised to cooperate with the Park Service, but one Audubon Society official, while calling the overall plan "cause for cheering," cautioned that "in the furor over the flood-loss of millions in city and farm properties, the future of the park and its requirements may be ignored."

In fact, an internal report by the new flood control district, run by a former Corps engineer, made it clear that the needs of cities and farms would take precedence over the needs of the park: "This is the wish of the majority of the people. The aesthetic appeal of the Park can never be as strong as the demands of home and livelihood. The manatee and the orchid mean something to people in an abstract way, but the former cannot line their purse, nor the latter fill their empty bellies."

That was a fair reflection of America's postwar politics: Nature was important, but not nearly as important as people, and important only insofar as it benefited people. The environmental movement was still in its infancy, and there was little support for nature in its own right; leading conservation groups such as the Audubon Society, the Sierra Club, and the National Wildlife Federation still focused mostly on preserving birding, hiking, and hunting opportunities for people. Florida's education department published a book promoting conservation, but only wise-use conservation that would promote prosperity. The title was *Florida: Wealth or Waste?*

But the postwar era also produced the early stirrings of ecology, as scientists began to study the interconnectedness of living things and their environments. In 1949, the pioneering ecologist Aldo Leopold, a founder of the Wilderness Society, published *Sand County Almanac,* questioning the notion that nature existed to serve man, calling for a land ethic in which people would be responsible citizens of the earth rather than its conquerors. Leopold would have mourned man's brutalization of the wilderness even if he believed it was economically and ecologically sustainable—he believed that man's ability to mourn this brutalization was what set him apart from the beasts—but he also noticed that the destruction of natural ecosystems often had harmful consequences for people. Leopold defined conservation as "a state of harmony between men and land," and explained how that harmony was being shattered by ditches and dikes that impoverished the soil and reconstructed the national landscape.

Ernest Lyons, the editor of the *Stuart News* and the dean of Florida's outdoors writers, made a similar ecology-based case against the C&SF in an article titled "Flood Control Destroys Last Natural Frontier," attacking the *Waters of Destiny* mentality that sought absolute domination of nature. Lyons warned that the costly "Hollandizing" of south Florida—he was referring to the highly engineered Netherlands, not the senator—would provide land reclamation for the few under the guise of flood protection for the many, drying out wetlands that once helped provide people with "natural flood control." Lyons understood that wetlands served a variety of utilitarian functions—absorbing floodwater, recharging groundwater, and storing and purifying surface water while sheltering fish and wildlife—and he decried the draining of swamps, the straightening of streams, and the damming of rivers as the essence of waste, the opposite of Conservation in Action:

> South Florida started out with a marvelous flood control plan. Nature designed it. It consisted of vast, perpetually inundated marshes and lakes interconnected by sloughs. It was a paradise for wildlife and, more practically, a sensible system of shallow reservoirs in which rainfall was stored to slowly seep into the ground. But being human, we just couldn't leave it alone . . . During dry seasons, private individuals farmed or built areas where old-timers knew inundation was as inevitable as death and taxes. Then when the rains came, we called on Government to take over and operate, with sweeping alterations, the magnificent system God had given us. . . . Now we are calling on Government to be the very God, by the creation of a huge artificial system

of dams, pumps, man-made lakes and controls which must be maintained in perpetuity. . . . Nature's last frontiers of wildlife and last giant units for natural flood control would be destroyed. And Florida would be repeating the folly which conservationists have watched ruin rivers, make droughts and create floods across the nation!

Conservationists know the cure for this evil. Save the swamplands as vast natural reservoirs. Quit being so land-hungry that Nature is left no place to store rainfall. Restore the marshes and little brooks. Cooperate with Nature instead of trying to take all and give nothing.

This was a radical proposal in the tradition of Charles Torrey Simpson, a plea to buy back some of the swamp and overflowed lands that Florida had spent a century trying to unload, and leave them swampy and overflowed instead of transforming them. The early drainage advocates had argued that God wanted man to subdue His wastelands, but critics such as Leopold and Lyons suggested that God's wetlands were not wastelands at all, that man was arrogantly playing God by trying to improve His work. They challenged the Corps and its *essayons* ethic, calling for a cease-fire in the war against nature. "What is a species more or less among engineers?" Leopold asked with poignant sarcasm. "What good is an undrained marsh anyhow?"

"The engineers think only in terms of ditches," Lyons's boss wrote in a letter begging Senator Holland to reconsider the C&SF. "The longer I live here, the more I am impressed with the necessity of stopping this infernal ditch-digging."

THESE WARNINGS OF DOOM were generally ignored. Ditch-digging was still standard practice in postwar America. The only warnings that slowed down the C&SF project at all were accusations that it would reclaim giant parcels of land for millionaires, warnings that happened to be true.

The eastern perimeter levee protected 100,000 acres owned by Alcoa chairman Arthur Vining Davis, south Florida's richest resident. The levees, canals, and pumps below Lake Okeechobee boosted profits for U.S. Sugar, which owned 130,000 acres in the flood control district. A *Harper's* article titled "The Florida Swamp That Swallows Your Money" noted that the planter who chaired the flood control district's board also stood to make millions off the C&SF. As land prices skyrocketed, some congressmen questioned why the feds were paying more than three-fifths of the cost of a

Florida real estate scam. "It appears to me that the federal government is subsidizing the development of Florida here," one congressman said at a 1955 hearing. "I'm just overwhelmed by this."

The House slowed down the project enough that by 1965, five years after its scheduled completion date, it was still less than half done. But it had already produced nearly 1,400 miles of levees and canals, devoured $175 million, and transformed the region like hydrological Miracle-Gro.

The Second Explosion

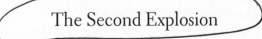

IT IS IMPOSSIBLE TO KNOW exactly how much of south Florida's spectacular postwar growth to attribute to the C&SF, and how much to air-conditioning, DDT, air travel, interstate highways, Social Security, low taxes, national prosperity, or the Florida-worship of veterans who had gotten "sand in their shoes" during the war. Suffice to say that the wheels of this second and more lasting land rush were significantly greased by the promise of water control, the public assurance that the region's natural see-saw between catastrophic floods and drought would be a thing of the past. And this boom boggled the mind as completely as its predecessor in the 1920s. "There is no point quoting statistics. They become outdated almost as soon as they are compiled, but they are almost unbelievable," a *Herald* columnist wrote. "This upsurge is so tremendous it staggers the imagination of men who back their ideas with millions of dollars."

Actually, statistics do give a sense of the initial surge. The entire nation enjoyed unprecedented growth after World War II, but Florida grew at four times the national rate. Before the war, the state had fewer than two million residents, ranking twenty-seventh in the nation and last in the Southeast. By 1965, its population was nearly six million, ninth in the nation and first in the Southeast. Its bank deposits grew 1,250 percent over that period, while its property values jumped 2,000 percent; its soaring new ambition was symbolized by NASA's new space complex at Cape Canaveral. And south Florida, overrun by young GIs as well as retirees—"the newly wed and nearly dead"—grew more than twice as fast as the rest of the state; Hollywood's population skyrocketed from 7,000 after the war to 35,000 in 1960 to 105,000 in 1970. In the 1950s, nearly 1,000 newcomers moved to the Miami area every week, while the West Palm Beach area developed America's highest concentration of golf courses, an honor it later ceded to the Naples area. In the 1960s, Florida platted more new lots than the rest of the nation combined, and south Florida's electric utility experienced America's largest increase in demand.

Tourism also exploded in Florida, from fewer than three million visitors in 1940 to more than 15 million in 1965—when Walt Disney announced even more explosive plans to convert 27,000 acres of marshes around the headwaters of the Everglades into a theme park, and Miami's first cruise ship sailed for the Bahamas. South Florida became America's escape hatch, the vacation getaway for Presidents Truman, Eisenhower, Kennedy, and Nixon, as well as top entertainers, athletes, and Mafia bosses—not to mention the spring breakers who began flocking to the beaches of Fort Lauderdale, an annual rite immortalized by the movie *Where the Boys Are.* Everglades National Park attracted nearly one million visitors a year, although concierges often preferred to steer tourists toward the exotic species at attractions like Monkey Jungle and Parrot Jungle. Flying to Florida was cheap, and passenger traffic through Miami skyrocketed from 600,000 after the war to nearly seven million in 1965. Driving was even cheaper, as northeastern snowbirds headed down the new I-95 to the Gold Coast, while midwesterners pointed their RVs down the new I-75 to the Gulf Coast.

Florida agriculture took off as well, outpacing tourism as the state's top industry until the birth of Disney's Magic Kingdom. Citrus growers doubled their output, cattle ranchers (now required to fence their herds) expanded in the Kissimmee valley, and farmers in the new Everglades Agricultural Area below Lake Okeechobee brought huge swaths of marshland under cultivation. "The River of Grass . . . is retreating before the onslaughts of modern pioneers and yielding its miraculous 'PAY DIRT' for the production of vegetables and other important crops," one brochure said. The most important crop was sugarcane, which was finally living up to its hype in the Everglades. The turning point came in 1959, when a shaggy-bearded leftist named Fidel Castro seized power in Havana, prompting a U.S. embargo on Cuban sugar, along with the repeal of the Sugar Act and its restrictive limits on domestic sugar production. Over the next five years, sugar acreage in the upper Glades quadrupled to 223,000, and eight new mills opened to process the regimented green cane fields.

Dreamers had always envisioned south Florida as a magnet for settlers and tourists and a sugar bowl for the nation. Now their visions were finally coming true. Even factories were testing the region's waters; a manufacturer called Aerojet General Corporation moved to the Homestead area after the state gave it a lucrative option on forty square miles of wetlands near Everglades National Park, and the Corps expanded its new C-111 drainage canal so that Aerojet could barge rocket boosters to Cape Canaveral.

Aerojet never landed a NASA contract, so it never used the canal, although it would later exercise its land option for a $15 million profit before abandoning Florida. In the 1960s, though, the good times were rolling. And as one writer noted, they were all made possible by the C&SF project's "remarkable strides in water control and conservation—a little known but monumental effort of man to control his environment."

ONCE AGAIN, the Miami area was ground zero for the boom, the epicenter of the democratization of leisure. On Miami Beach, developers tore down the mansions of Millionaires Row and built high-rise hotels and condos that stood shoulder to shoulder along the ocean's edge. Harvey Firestone's estate gave way to the Fontainebleau Hotel, where vacationing junior executives from Middle America could catch a glimpse of Jimmy Durante, Marlene Dietrich, or Frank Sinatra's Rat Pack bellying up to the bar; tourists on tighter budgets could choose among a strip of garish neon-inflamed motels a few miles north. The television stars Arthur Godfrey and Jackie Gleason hosted their shows from Miami Beach, exposing millions of ordinary Americans to this land of eternal sunshine, while the state's ballyhoo machine continued to bombard the nation with images of bathing beauties, orange blossoms, and palm fronds. Racial discrimination was still rampant, but "Gentiles Only" signs vanished, and Miami Beach soon supported the world's largest Jewish population outside New York and Israel.

Meanwhile, mainland Miami expanded into a year-round metropolis, Florida's largest and most important city, a bustling center of economic activity and international intrigue that supported the CIA's largest station outside Langley. Miami International Airport in the eastern Glades became the U.S. gateway to Latin America; the Miami Dolphins became south Florida's first major sports franchise; a local eatery called Insta Burger King dropped its "Insta" and took its 37-cent Whoppers nationwide, launching one of America's leading fast-food chains. But Miami's biggest growth industry was growth itself; its skyscrapers were stuffed with mortgage bankers, real estate lawyers, contractors, and the rest of the development industry. Congress had passed new legislation to promote home ownership, and as one observer put it, "the real estate boys read the bill, looked at one another in happy amazement, and the dry rasping noise they made rubbing their hands together could have been heard as far as Tawi Tawi." Concrete-block GI homes quickly covered the coastal ridge "like rows of pole beans," one historian wrote. And that was before Castro's rise to power sent hundreds of thousands of Cubans scurrying to Miami, further swelling the demand for housing.

Many Cuban exiles settled in Little Havana, a Spanish-speaking enclave behind downtown Miami, but others ventured west to less crowded suburbs in the reclaimed Everglades—existing ones such as Hialeah, Miami Springs, and Sweetwater, a town founded by a troupe of circus midgets whose car broke down on the Tamiami Trail, or new ones such as West Miami, West Kendall, Westwood Lakes, and Westchester, now protected from floods by pumps, canals, and the perimeter levee. And Cubans weren't the only newcomers sprawling into the eastern Glades. Cookie-cutter Anglo suburbs also sprouted on cheap wetlands west of the coastal ridge but east of the perimeter levee, including Pembroke Pines, Plantation, Miramar, Margate, Lauderhill, Lauderdale Lakes, Sunrise, Tamarac, Royal Palm Beach, and Coral Springs, where Johnny Carson bought fifty-five acres at an opening day auction. "Live in the Path of Progress!" land ads beckoned.

The Empire of the Everglades was finally coming alive. The new red-roof subdivisions were served by new north–south highways through the Glades, including the Sawgrass and Palmetto Expressways and the Florida Turnpike; Alligator Alley cut east–west across the marsh from Fort Lauderdale to Naples. One dredging firm issued a pamphlet titled *Turning Swamps into Dollars,* proudly touting "The Spectacular Economics of Land Reclamation" that was reshaping the lower peninsula.

In the mid-1950s, south Florida's developed area covered less than 150 square miles. By the late 1960s, satellite photos showed that man's footprint had almost quadrupled. "Conservation and ecology-minded individuals view the disappearance of this last 'frontier' under a cover of concrete and sod with dismay," one professor wrote. "It is most unlikely that their voices will prevail. Demographic and economic pressures are just too great."

ASIDE FROM THOSE FEW "ecology-minded individuals," no one seemed to think that development in the Everglades was anything but a civic improvement. Even Cap Graham's progressive-minded sons built Miami Lakes, a planned community in the eastern Glades that won numerous design awards. The Grahams dredged pastures, low-lying wetlands, and palmetto scrub into artificial lakes that doubled as stormwater catch basins, then piled the spoil into artificial high ground that supported lakefront homes. Much attention was paid to their high-minded decisions to plant trees and dig irregularly shaped lakes, but almost none to the fact that they were tearing up five square miles of the Everglades. "We've been blessed with one of the ugliest pieces of ground anywhere," joked Cap's oldest son, *Washington Post* publisher Philip Graham. "You don't have to worry about the developer screwing it up; we can only make it more attractive."

Graham's quip illustrated how the C&SF flood control project had fueled the perception that all land east of the levee was fair game for developers, that there was now a clearly delineated "wet side" and "dry side" of southeast Florida, that Everglades National Park and the Loxahatchee refuge were the only parts of the ecosystem worth preserving. The original Everglades had included just about everything west of I-95, but few of the new suburban pioneers understood that. They called animal control officers when gators invaded their backyards, never imagining that they might be invading the gators' backyards. They expected their land to remain dry, never suspecting that they were living in former wetlands. They enjoyed their ranch-style homes, well-manicured lawns, and air-conditioned malls, unaware of the wilderness that had been conquered for their comfort. A few hundred homesteaders even built on the wet side of the perimeter levee— forming a sparse community west of Miami known as the Eight-and-a-Half-Square-Mile Area—and soon demanded flood protection as well.

One C&SF brochure reminded south Florida's newcomers that they probably could have rowed across their yard in 1947: " 'Nonsense,' you protest. 'The yard is full of flowers and shrubs. Even the hardest rains don't flood it. The house is bone dry even when it rains day after day.' But there's still a good chance your front porch might make a good boat launching ramp if conditions were the same as they were 20 years ago." But conditions were not the same. That was the beauty of the C&SF. *"Is it worth it? That's an easy one! Look around central and southern Florida today!"* the *Waters of Destiny* narrator urged. Tourists were flocking to its beaches. Settlers were stampeding to its suburbs. Florida was producing sugar, fruit, vegetables, and beef for the nation, and business was booming. The assessed value of land within the flood control district would soar from $1.2 billion in 1950 to $15.8 billion in 1970, stark evidence of the power of human engineering. This was the *Waters of Destiny* vision: *"Flood control must proceed—as fast as humanly possible—so that everyone everywhere can share in the rich results of man's mastery of the elements!"*

Nature's Fool

THE C&SF PROJECT did not extend the glories of flood control to southwest Florida, but that did not stop two Baltimore brothers named Leonard and Julius Rosen from selling nearly half a million acres of swampland there during the boom. The Rosens had gotten rich selling an anti-baldness tonic called Formula Number Nine, featuring the miracle ingredient of lanolin—and the immortal tagline, "Have you ever seen a bald sheep?" The brothers could see that shivering northerners yearned for a piece of Florida the way bald men yearned for hair. Their Gulf American Corporation offered "a rich man's paradise, within the financial reach of everyone," the ultimate miracle elixir.

Gulf American's most ambitious venture was Golden Gate Estates, where the Rosens platted the world's largest subdivision in the middle of Big Cypress Swamp. The Collier family had once dreamed of developing Big Cypress, which was one reason Senator Holland had removed it from Everglades National Park, but the dreams had never amounted to more than a few logging towns. A few Indians lived in the cypress country, and some rough-hewn gator hunters used it as a base for midnight poaching expeditions into the park. But the boom had bypassed the swamp until the Rosens' high-pressure salesmen began depicting it as the new home of the American dream, hawking lots over WATS lines for about twenty times what the Rosens had paid.

The company lured marks around the country to "friendly dinners" with local celebrities, meals inevitably interrupted by frenetic salesmen shouting "Lot number 72 is sold!" and paid ringers yelling "I bought one!" Gulf American operatives applied similar tactics to Florida tourists, trolling for suckers in souvenir shops and the Parrot Jungle, bribing bellboys for leads on gullible guests, offering prospective buyers free weekends at the company's hotel in Cape Coral—where the rooms were bugged to help salesmen customize their pitches.

Golden Gate was marketed as the ultimate in modern living, featuring golf, tennis, and restaurants like "the Country Squire or the elegant Le Petit Gourmet"—all in all, "an entirely new and wonderful way of life," for as little as $10 down and $10 a month. "The wilderness has been pushed aside," crowed one brochure. "With calipers and slide rules, draglines and dynamite rigs, we are literally changing the face of Florida." In fact, Golden Gate was still an inaccessible swamp. It had no golf courses, tennis courts, or restaurants, elegant or otherwise; it did not even have schools, sewers, or phone lines. Its only "improvements" were an elaborate grid of roads and canals, which wreaked havoc with the water table and contributed to a rash of fires. The Rosens sold tens of thousands of lots in Golden Gate, parlaying their $125,000 investment in Florida swampland into a $115 million payout, but only a few dozen homes were built there.

Gulf American's scams made the Broward Era's land-by-the-gallon schemes look like church bake sales. Salesmen told veterans the firm was affiliated with the military. They relocated lots without informing buyers, sold lots fifteen miles from the Gulf as "waterfront properties," and preyed on the senile and feeble. They drove prospects who insisted on checking out their land deep into the swamp, then threatened to make them walk home if they didn't sign a contract. Gulf American eventually pleaded guilty to deceptive sales practices, and might have faced more serious charges if the legislature hadn't required that three of the Florida Land Sales Board's five members come from the real estate industry, and if Governor Haydon Burns—who later landed a lucrative consulting contract with Gulf American—hadn't appointed Leonard Rosen to the board.

THE GOLDEN GATE FIASCO INSPIRED a new round of jokes about ignorant suckers getting stuck with Florida swampland. But some of the buyers knew exactly what they were buying; they just assumed they would be able to resell their swampland for a profit. And that was not such an unreasonable assumption; multibillion-dollar corporations like U.S. Steel, ITT, Westinghouse, and Chrysler were investing in Everglades real estate, too.

Yes, Gulf American told lies, but who could say they wouldn't come true someday? Yes, Big Cypress was a swamp, but why would anyone expect it to stay a swamp? In fact, Gulf American's first development, Cape Coral, marketed just as dishonestly as Golden Gate, is now the largest city in southwest Florida. "As long as the sun shines and warm breezes blow over the Gulf Stream, the Ziegfeld extravaganza of development seems likely to continue," one newsman wrote. Thomas Edison had predicted that 90 million people would discover south Florida someday, and that day had almost arrived.

If some of the cheaper land on the market was still "liable to overflow, and of no use," there was every reason to believe it would eventually be dry. In the Broward Era, reclamation had been an article of faith, but now it was a simple matter of observation. Developers were on the march. Mother Nature was in retreat. The Everglades had been America's final frontier, but the nation's engineers were taming it, remolding it, improving it. The Florida dream of immigration and capital was finally a reality. It was clearly America's destiny to defeat the Everglades, to subdue its wild water, to harness its resources for man's needs and desires:

> *Now it just waits there . . .*
> *Calm . . .*
> *Peaceful . . .*
> *Ready to do the bidding of man and his machines . . .*
> *Central and Southern Florida is no longer nature's fool.*

PART 3

Restoring the Everglades

FOURTEEN

Making Peace with Nature

We must build a peace in south Florida—a peace between the people and their place, between the natural environment and man-made settlement, between the creek and the canal, between the works of man and the life of mankind itself.

—Florida governor Reubin Askew

The Green Revolution

B Y THE LATE 1960s, the Everglades was supposed to be fixed. Thanks to the Central and Southern Florida flood control project, the problems of salt water intrusion, freshwater shortages, soil subsidence, and muck fires were supposed to be solved. Thanks to Everglades National Park, a chunk of the natural ecosystem was supposed to be protected forever.

But central and southern Florida was still nature's fool, yo-yoing between severe floods and even more severe droughts. Salt invaded the wells of every Gold Coast city from Stuart down to Miami. Water managers scrambling to meet skyrocketing demands imposed the region's first lawn-sprinkling restrictions. Scientists concluded that almost all the soil in the new Everglades Agricultural Area would be gone by the end of the century. Gizzard shad and other trash fish crowded out bass and bluegills in the Chain of Lakes, while Lake Okeechobee's catfish were contaminated with DDT, toxaphene, and other persistent pesticides. The Kissimmee basin's wetlands became dry and lifeless pastures. The St. Lucie and Caloosahatchee estuar-

rouded in chocolate-colored scum. The last few dozen panthers
ing on in southwest Florida, but sprawl was crushing their habi-
irs were crushing the cats.

The Everglades itself was no longer the Everglades. The northern Glades
was overrun by sugarcane fields. The eastern Glades was overrun by suburbs. The central Glades was divided into "water conservation areas" that
still looked like the Everglades, but were managed as reservoirs and sewers.
And Everglades National Park was now the National Park Service's most
endangered property, a phenomenon chronicled in articles with headlines
such as "Disaster Threatens the Everglades," "The Imperiled Everglades,"
and "The Killing of the Everglades," and in books such as *The Environmental Destruction of South Florida.* The veteran drainage engineer Lamar Johnson, returning to sawgrass prairies he had surveyed in the 1920s, was
shocked to find them overrun with brushy vegetation: "Marjory Stoneman
Douglas' River of Grass is rapidly becoming, in the vernacular of a native
frogger, a 'hell's nest.' The invasion has become so general that unless it is
controlled the Everglades could become a solid jungle of myrtle, willow,
holly and bay."

In drought years, fires again raged in desiccated Everglades marshes, producing pillars of smoke so huge they grounded the new traffic helicopters
that were monitoring the region's snarls. "I found no Eden, but a waterless
hell under a blazing sun," one park visitor wrote. In flood years, thousands
of gaunt white-tailed deer were stranded in the Everglades to drown, starve,
or succumb to diseases caused by the stress of high water. "This beautiful
part of the world has been pushed to the brink of ecological death by men
who believe that nature has an infinite capacity to give and forgive," said
National Geographic. The park remained vulnerable to forces outside its borders; it was, as President Truman had said, "the last receiver," a final resting
place for urban and agricultural effluvia. "Time is running out for the Everglades . . . and no one knows how to turn back the clock," one author wrote.

Marjory Stoneman Douglas had expected the C&SF project to save the
Everglades, but it turned out to be an ecological menace. It did a terrific job
of draining wetlands and promoting growth, but its expanded canals carried more water out of the Everglades at a time when south Florida's
expanding cities and farms were increasingly dependent on water in the
Everglades. Its flood protection prompted additional development in the
Everglades floodplain, which prompted demands for additional flood pro-

tection. And the Corps and its like-minded partners in the flood control district—often run by former Corps engineers—refused to release water to the park, except when it was already inundated. They manipulated water levels to accommodate irrigation schedules and development schemes, discombobulating the natural water regime to which flora and fauna had adapted over the millennia. "What a liar I turned out to be!" Douglas cried.

Charles Torrey Simpson's dire prophecies were coming true with a vengeance. The Everglades was being tamed and reclaimed into an empire of houses and highways; the honk of the automobile was drowning out the cries of wild birds. And as Simpson had also predicted, many Americans were beginning to regret what they were losing.

AMERICA EXPERIENCED an extraordinary awakening in the late 1960s, a national embrace of the notion that human beings should stop fouling their own nests.

In many ways, the environmental movement reflected the anti-establishment fervor of the Vietnam era, paralleling the antiwar, civil rights, and feminist movements. An array of new green groups—including the Environmental Defense Fund, the Natural Resources Defense Council, Friends of the Earth, and Greenpeace—adopted more confrontational approaches to the defense of nature, spurring old-line groups such as the National Audubon Society and National Wildlife Federation to sharpen their messages as well. But environmentalism's basic goals—clean water, clean air, protecting human health, saving beautiful landscapes—had a broader appeal than most activist causes of the day, an appeal symbolized by the iconic antipollution ad starring an Indian with a tear running down his cheek. Hippies weren't the only ones who wanted to save the whales, the redwoods, and the Grand Canyon, or protect their families from DDT and PCBs. In 1969, a secret poll conducted for President Richard Nixon found that Americans were more concerned about environmental degradation than any other issue except Vietnam.

The rapid change in the American public produced a rapid change in American politics. Pundits joked that every congressman now claimed to be an ecologist—even though few of them had heard of the word a year earlier. Nixon did not care much about environmental issues, but he knew his so-called Silent Majority did. In his State of the Union address in 1970, three months before the first celebration of Earth Day, he called on Amer-

"make our peace with nature," and pledged "reparations for the ....age we have done to our air, our land and our water." Nixon still demonized environmental activists as radical leftists when it suited his purposes, but he created the Environmental Protection Agency and signed an array of environmental laws, including the National Environmental Policy Act, the Clean Air Act, and the Endangered Species Act. "We were worried about losing the garden clubs," recalled John Whitaker, a top environmental aide in the Nixon White House. "We could see the tidal wave coming. These issues were going mainstream."

That's because the environment was in awful shape. The postwar economic expansion had proceeded with almost no environmental safeguards, and progress was revealing its price. The Cuyahoga River caught fire in Cleveland. The bald eagle, the soaring emblem of U.S. power, nearly became extinct. Communities were buried in raw sewage and toxic pesticides, while the proliferation of cars and coal-fired power plants created the phenomenon of "smog." The resulting backlash included some overwrought alarmism—misguided neo-Malthusians predicted that overpopulation was about to overwhelm the world's food supply, while gloom-and-doomers forecast a swift nine-degree increase in the earth's temperature—but there was also cause for real alarm.

The eloquent prophet of this petrochemical age was Rachel Carson, a Fish and Wildlife Service marine biologist who introduced millions of Americans—and their congressmen—to the basic concepts of ecology. In her best-seller *Silent Spring,* Carson warned that DDT and other man-made pollutants were creating "rivers of death," and threatening the entire cycle of life on earth. Carson echoed the advocacy of John Muir, Charles Torrey Simpson, and Aldo Leopold, attacking unbridled capitalism that ignored its impact on the planet, challenging the arbitrary distinction between human and natural environments, warning that mankind's destruction of nature could lead to the destruction of mankind. She repeated their pleas for men to live in harmony with God's creations, instead of trying to dominate the earth with their own creations:

> We still talk in terms of conquest. We still haven't become mature enough to think of ourselves as only a tiny part of a vast and incredible universe. Man's attitude toward nature today is critically important, simply because we have acquired a fateful power to alter and destroy

nature. But man is part of nature, and his war against nature is inevitably a war against himself. We in this generation must come to terms with nature, and I think we're challenged as mankind has never been challenged before to prove our maturity and our mastery—not of nature, but of ourselves.

THE BACKLASH AGAINST man's assault on nature inevitably created a backlash against the Army Corps, which now had a billion-dollar budget and 30,000 employees for its battle against the wilderness. "The rather sudden general awareness of the science of ecology has brought projects which disturb the environment, as Corps projects do, under unprecedented attack," *The Atlantic Monthly* noted. Supreme Court Justice William O. Douglas wrote an article declaring the Corps "public enemy number one"—in *Playboy*.

The Corps was almost a caricature of the pre-ecological mentality, perennially eager to replace Mother Nature's work with its own. Senator Gaylord Nelson of Wisconsin, the founder of Earth Day, called it an agency of subsidized beavers. It was pushing a gigantic dam that would have wiped out Alaska's Yukon Flats, a web of marshes that sheltered more ducks than the entire continental United States, and a gigantic pump that would have dewatered Mississippi's Yazoo Basin, destroying one of America's most magnificent swaths of bottomland hardwoods. It was digging the Mississippi River Gulf Outlet, an environmentally disastrous navigation channel that was instantly denounced as a "hurricane highway" into New Orleans, and the similarly destructive Cross-Florida Barge Canal, an effort to achieve the Menéndez dream of a shipping lane across the peninsula's midriff by plowing through the Ocklawaha River. In fact, the Corps was remodeling dozens of wild rivers into placid barge canals, justifying each pork barrel project by predicting miraculous increases in traffic that never seemed to materialize. One cartoonist routinely depicted Corpsmen as maniacal nature-destroyers who wore pith helmets with the slogan "Keep Busy" and said things like "Look alive, men! Our alert map department has located a free-flowing stream that does NOT have a dam on it!" or "God would have done it if he had the money!" Marjory Stoneman Douglas suggested a Freudian explanation: "Their mommies obviously never let them play with mud pies, so now they take it out on us by playing with cement."

For some critics, the C&SF project epitomized the folly of the Corps. It was the ultimate effort to replace nature's plumbing with man's, a bewildering gridiron of levees, canals, and floodgates with names like L-29, C-44, and G-251, "To anyone who has ever so much as heard the word 'ecology,' the [C&SF] is a horror," one author wrote. "It is an uncaring and terrifying symbol of the triumph of the Engineers and the rape of America." The C&SF brought civilization to the Everglades—but not everyone agreed that civilization belonged in the Everglades. It was supposed to be a triumph in man's war against nature—but it was already causing problems for man, when even President Nixon was proposing a truce with nature.

The notion that man was not the master of the universe was still a revolutionary idea in some circles, as blasphemous as Galileo's insistence that earth was not the center of the universe. But as it choked on exhaust fumes, toxic sludge, and radioactive waste, as its drinking water became dirty and scarce, America was ready for a revolution. Florida was ready, too.

Revolutionaries in Paradise

FLORIDA STILL portrayed itself as a natural idyll, an image bolstered by TV shows starring a dolphin named Flipper and a bear named Gentle Ben, by national advertising campaigns featuring sun-kissed beaches and bays, even by the spread of subdivisions with names like Panther Creek, Eagle Creek, Gator Creek, and other endangered-species Creeks, Woods, Bays, and Lakes. But in reality, Florida was becoming "a leading contender for first place in the nation's chamber of environmental horrors," as one author put it. Another book declared that "our beautiful state of Florida is being raped, despoiled and polluted." As asphalt and people replaced wetlands and birds, Florida was mocked as "the New Jersey of the South."

Politically, Florida was still the land of laissez-faire. It was the only state that outlawed county zoning; many property-rights-obsessed Florida politicians considered urban planning a form of communism. Florida still used its water bodies as sewers, and gave away its bay bottoms and lake bottoms to developers. In the back-slapping, steak-and-bourbon culture of Tallahassee, lobbyists for corporate interests usually bought the meals, often wrote the legislation, and occasionally ran the committee meetings.

But as Florida's environment suffered, Florida's environmental movement grew stronger. The Audubon Society's statewide membership increased sevenfold in a decade. Citizens mobilized against the Cross-Florida Barge Canal—and helped shut down the project. A coalition of scientists, activists, and politicians called Conservation 70s began lobbying the Florida legislature—and helped pass forty-one eco-friendly bills in its first year. Activists also stopped plans for an oil refinery and a new Miami Beach–style city along Biscayne Bay, prevented a nuclear plant from dumping hot water into the bay, and led a campaign to create Biscayne National Park to prevent future threats to the bay. When it became clear that the Aerojet Canal was creating a saline superhighway into Everglades National Park, environmentalists filed a lawsuit that forced the Corps to plug it.

There was a new breed of green activist in south Florida, typified by Audubon's abrasive but effective southeastern representative, Joe Browder, whose horn-rimmed glasses, youthful swagger, and ferocious intensity gave him the air of Buddy Holly on a tight deadline. Nature had been the one constant in his childhood; his father's work as a CIA operative and freelance gun-runner kept his family moving around the hemisphere, and Joe sometimes missed entire years of school. So he spent his afternoons studying ant colonies, or reading about coyotes, or watching birds with the slack-jawed fascination that children usually reserve for TV. He earned an ornithology scholarship to Cornell, but dropped out to get married. He eventually landed a job as a TV news reporter in Miami, while volunteering for Audubon on the side. But he had trouble pretending to be objective about the ongoing rape of south Florida, so he gave up journalism for full-time activism, fighting to save Biscayne Bay, spearheading the lawsuit to plug the Aerojet Canal.

Even Browder's friends called him a tiger, a bulldog, a zealot. But he channeled his passion into calculated political strategy; he loved building grassroots support for environmental causes almost as much as he loved wading in swamps. He appealed to hunters and Hispanics, labor unions and Indians; he used his media savvy to stir up publicity that stirred up the public. Browder believed that Americans genuinely cared about the fate of the earth, and that environmentalists could win if they got their message out.

FLORIDA HAD its own version of Rachel Carson to help carry that message, Arthur R. Marshall Jr., another visionary Fish and Wildlife Service marine biologist who seemed eternally disappointed with the human race. He looked like a taller version of coach Vince Lombardi, and he could be just as relentless and inspirational. Marshall became the apostle of the Everglades, preaching the gospel of ecology in his rich baritone, thundering that south

Florida was on the road to hell while pointing out the road to redemption: "It is time—well past time—that we abandoned the centuries-old belief that man's dominion over the earth includes its willful destruction."

Marshall was born in South Carolina, then spent his teenage years in West Palm Beach, fishing, swimming, and canoeing the waters of natural south Florida. He then served as an Army captain under General George Patton, leading 250 men of the 81st Chemical Mortar Battalion onto Omaha Beach on D-Day, fighting in five bloody campaigns, and helping to liberate a Nazi concentration camp. Marshall brought a similar determination to his battles to save his boyhood paradise. "I once offered all I could to America because I wanted it to survive," he wrote to a friend. "I am going to do it again—this time, intellectually instead of militarily."

When Marshall took over Fish and Wildlife's south Florida office in 1955, he began protesting dredge-and-fills and mangrove removals that had always been accepted practice, challenging the region's developers and the politicians who coddled them. But he was usually ignored, and when he was noticed, he was vigorously attacked. After he warned that a bridge to Sanibel Island would wipe out a scallop fishery, local officials tried to get him fired for promoting hysteria. Then they built the bridge—and sure enough, the scallops vanished. Senator Holland also sporadically tried to get him transferred for trying to stifle development, but Marshall's bosses protected him. "Certainly some find my views disputatious; others fear or despise them," Marshall wrote. "These attitudes do not disturb me; in fact they partially fortify me in the sense of the biblical declaration: Woe unto me when all men think well of me."

The criticism may have fortified him, but it also enraged him; Marshall became increasingly distraught about south Florida's war against Mother Nature, and his inability to broker a truce through rational science. He channeled his frustration into messianic advocacy, proselytizing to garden clubs, rotary clubs, hunting clubs, and any other group that would listen to his premonitions of doom. He became the most obsessive defender of the Everglades since Ernest Coe, spreading word that wetlands were wonderlands to be treasured instead of wastelands to be conquered, and that south Florida was careering toward environmental bankruptcy. He bluntly warned that the green revolution was a simple matter of self-preservation, that the Gold Coast would not survive without a healthy Everglades, that Florida's mania for growth was societal suicide. "In Florida, it has always

The Everglades

South Florida's modern march into the Everglades

3

OSCEOLA (c. 1804–38), the heroic
Seminole Indian warrior who was captured
by U.S. soldiers under a white flag of truce,
was painted shortly before his death by
the frontier artist George Catlin.

4

HAMILTON DISSTON (1844–96),
the heir to a Philadelphia saw-making
fortune, launched the first major effort
to drain the Everglades.

5

HENRY MORRISON FLAGLER
(1830–1913), John D. Rockefeller Jr.'s right-
hand man at Standard Oil, enjoyed a second
career as a railroad and resort builder, laying
the foundation for modern south Florida.

6

NAPOLEON BONAPARTE BROWARD
(1857–1910), a former St. Johns River steamboat captain, was the driving force behind
Everglades drainage as Florida's governor,
declaring that "water will run downhill!"

The Dredge "Everglade," Draining the Everglades.

7

In 1906, Governor Napoleon Bonaparte Broward's first dredge, *Everglade,* began digging a canal from Fort Lauderdale to Lake Okeechobee—the first of four diagonal canals through the Atlantic coastal ridge and the River of Grass. Broward was confident that millions of acres of the Everglades would be reclaimed almost instantaneously, but he was wrong.

8

A Seminole Indian poles a dugout canoe through a canal in the Everglades, where land companies hyped waterlogged swampland as an agricultural paradise. "I have bought land by the acre, and I have bought land by the foot; but, by God, I have never before bought land by the gallon," one dismayed buyer said.

9

10

After Lake Okeechobee blasted through its flimsy muck dike during the hurricane of 1928, killing as many as 2,500 south Florida pioneers, authorities worried about disease outbreaks stacked corpses like cordwood and burned them at roadsides in the Belle Glade area. The calamity inspired the construction of the Hoover Dike, which has imprisoned the lake ever since—and cut it off from the Everglades.

SPESSARD HOLLAND (1892–1971), known as Mr. Florida during his three decades as the state's governor and U.S. senator, was responsible for the creation of Everglades National Park as well as the massive flood-control project that seriously degraded the park.

TENTATIVE
REPORT
OF

FLOOD
DAMAGE

FLORIDA
EVERGLADES
DRAINAGE
DISTRICT

1947

11

The "Crying Cow" report chronicled how the floods of 1947 inundated just about all of south Florida, and helped persuade Congress to authorize the flood-control project that tamed the Everglades.

12

13

MARJORY STONEMAN DOUGLAS
(1890–1998), the influential author of *The
Everglades: River of Grass*, became an activist
in her eighties and kept fighting for the
Everglades until her death at age 108.

ARTHUR R. MARSHALL JR.
(1919–85), an impassioned ecologist
who got his start at the U.S. Fish and
Wildlife Service, was the intellectual
godfather of Everglades restoration.

14

In the late 1960s, Dade County officials bragged that their "Everglades jetport" would
attract the next generation of supersonic jets to the Big Cypress Swamp, and that a city
of half a million people would soon sprout around it. They were oblivious to the green
revolution that was brewing around them. Environmental activists persuaded the Nixon
administration to kill the jetport in 1970, even though one runway had already been
built. Hundreds of thousands of acres of Big Cypress have been preserved as a result.

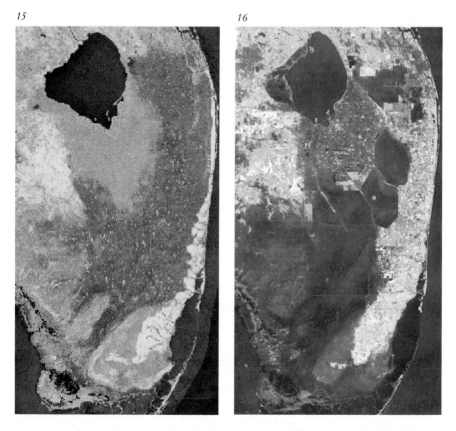

In natural south Florida, pictured in the simulated satellite image to the left, Lake Okeechobee spilled over its lower lip during summer storms, sending a broad sheet of shallow water on a leisurely journey south. That was the Everglades, and it trickled all the way down to Florida Bay and the Gulf of Mexico without interruption. But the satellite image of modern south Florida to the right shows how the region has changed. The Hoover Dike has cut off Lake Okeechobee from the northern Everglades, which has been transformed into the agricultural area below the lake. The eastern Everglades has been overwhelmed by suburban development, and the rest of the Everglades has been fragmented by highways and canals.

Arthur R. Marshall Jr. Loxahatchee National Wildlife Refuge

A red mangrove in Florida Bay

19

Big Cypress National Preserve

20

A lone egret in an Everglades swamp

been said that if we can just get a bigger population, we'll get more business and more dollars and solve all our problems," he said. "That's a bunch
of crap."

Marshall conducted well-regarded studies of south Florida's declining
snook populations, but his real genius was synthesis, connecting the dots of
esoteric studies by other scientists into grand unified theories of the Everglades ecosystem. He considered himself a theoretical ecologist, just as
Albert Einstein was a theoretical physicist, a comparison he did not dis-
courage. "If you don't synthesize knowledge, scientific journals become
spare-parts catalogues for machines that are never built," he explained. "I
am as good a diagnostician of ecosystems as any doctor is of human beings,
and I'm not on any damn ego trip when I say that. Sometimes I wish I
didn't have the knowledge that I do, because I can get pretty damn glum."

Marshall was prone to depression, and he drank too much at night; his
gruesome wartime experiences were always with him. He often assumed
that anyone who disagreed with him was stupid, evil, or out to get him,
especially fellow environmentalists he considered insufficiently hard-nosed.
And he could be exasperatingly judgmental; he once told an interviewer
that an Audubon activist ought to be fired because he wasted gasoline by
driving too fast: "Ignoring the principles of the environment—that's what
he's doing!" Marshall took the abuse of nature intensely personally. Every
new subdivision west of I-95 seemed to jab a dagger into his heart.

But Marshall did not just moan about Florida's ecological collapse. He
did all he could to prevent it—not only by driving slowly and growing his
own organic food, but by diving into environmental controversies throughout the state. He served as an adviser to Conservation 70s; he helped halt
the barge canal and create Biscayne National Park; he later led the Coalition for Water Resources and the Coalition to Repair the Everglades. Douglas called him "the leading man . . . in all of the organizations." When
Florida was about to let U.S. Sugar farm the "Holey Land"—a former bomb-
ing range in the Everglades Agricultural Area that was still relatively pristine—Marshall testified as a private citizen against the giveaway, warning
that the state would need that land someday to help repair the Everglades.
"I am here today," he said, "to represent myself and my two sons."

Marshall often seemed to be shouting into a void. He lambasted south
Florida's car-dependent culture long before it was fashionable, warning that
"the people in Dade County will get to work next week if the Arab chieftains

want them to get to work." But counties kept building more roads that attracted more strip mall sprawl. Marshall also forecast the deterioration of Lake Okeechobee and the collapse of Florida Bay, provoking almost no reaction whatsoever. But Douglas began to champion Marshall as a voice in the wilderness, and her credibility and fame helped make him a voice of authority. "Do not treat Art Marshall lightly. He is your Paul Revere!" Douglas chastised a panel of Florida bureaucrats. "We've come to new times. This is a new voice. You can't afford to be contemptuous!"

Marshall believed there was still time to save the Everglades, but not much time. He thought the ecosystem was rapidly approaching its tipping point—or, as he put it, "a snowballing degeneration of major resources." And he was not the kind of scientist who perennially advised more study. He proposed solutions. Then he demanded action.

In the late 1960s, politicians began to listen. Sometimes, they even acted.

In 1967, two political earthquakes shook up Florida.

One was a U.S. Supreme Court decision ordering the reapportionment of the state legislature, which had been so skewed in favor of small, rural counties that 18 percent of the population could have elected majorities in both houses. With the arrival of one man, one vote, Florida's politics caught up to its demographics, ending the century-long stranglehold of the arch-conservative Pork Chop Gang, essentially expelling the state from the Deep South. Power shifted from north Florida's good ol' boys to south Florida's northeastern and midwestern transplants; the Miami area's delegation expanded from one senator and three representatives to nine and twenty-two. Many of them were young progressives like Bob Graham, a Harvard-educated lawyer who was the son of the Everglades pioneer Cap Graham, but who did not share his father's desire to subdue nature. The legislature instantly became more urban, more reform-minded, and more environmentally conscious; two legislators who had labored in obscurity under Pork Chop rule, Reubin Askew and Lawton Chiles, enacted a bill creating Florida's first real environmental agency, the Air and Water Pollution Control Commission. Democrats like Graham, Askew, and Chiles—all future two-term governors—and some of their new Republican counterparts began pushing to protect beaches, bays, and lakes, questioning whether the god of growth was destroying the resources that made the state so attractive, portraying natural Florida as an asset to be treasured instead of a commodity to be sold.

The other tectonic shift in Florida politics in 1967 was the ascension of Claude Roy Kirk Jr., a little-known insurance salesman who looked like a mob boss, partied like a frat boy, and stunned the state's political establishment by becoming its first Republican governor since Reconstruction. Kirk only received the GOP nomination because no serious candidate wanted it, but he exploited the growing rift within the Democratic Party, winning votes from north Florida conservatives by attacking his opponent as a Miami liberal. Kirk squired a mystery blonde he dubbed "Madame X" to his inaugural ball, and continued his flamboyant antics for the next four years, hiring a private security firm to lead his war on crime, upstaging the black militant H. Rap Brown at a stadium rally, planting Florida's flag on the ocean floor, mobilizing the power of his office to save a child's lemonade stand. "Claudius Maximus" had an insatiable thirst for publicity and a poorly disguised lust for higher office. But he also had a passion for rattling powerful cages; he called himself "a tree-shakin' son of a bitch." And with the help of an energetic young aide, he became Florida's first environmental governor.

The aide was Nathaniel Pryor Reed, a blue-blooded outdoorsman whose family had developed Jupiter Island, America's wealthiest community, the winter retreat for old-money names like Whitney, Harriman, Ford, Duke, Doubleday, and Bush. Growing up in Greenwich, Connecticut, summering in Maine and wintering in Florida, Reed fell in love with all things natural; he collected bugs in his bedroom, memorized the names of birds, trees, and butterflies, and spent so much time chasing snook and redfish that his mother, Permelia, the legendary doyenne of Jupiter Island society, used to say he came out of the womb casting a fly rod. After a stint as an Air Force intelligence officer overseas, Reed returned to Florida to run his family's real estate business. When he saw the way ticky-tacky development was overrunning the region, he began to promote green causes as well. He also agreed to help Kirk's campaign on a lark, never dreaming his candidate might win. But on inauguration day, the unpredictable new governor pulled him aside and pointed out a cramped office: "If you want to change the things you've been hollering about, there's your desk." Reed accepted a salary of $1 a year, and Kirk set him loose.

Reed's angular six-foot-five, 170-pound frame evoked a great blue heron, but his ruddy complexion, unfailing sense of noblesse oblige, and soaring rhetoric began drawing frequent comparisons to John F. Kennedy. Reed launched an all-out battle to save natural Florida, turning his office into a war room reminiscent of his days in military intelligence, with maps dotted by pins representing his environmental allies around the state. Reed regularly took on Florida's most powerful interests—developers, business-men, manufacturers, and their friends in government—but his boss backed him all the way. Kirk didn't care that much about the environment, but he loved antagonizing the powerful and defending the underdog. Back then, the environment was the underdog. Audubon's Joe Browder, Fish and Wildlife's Art Marshall, and all the other pins on Reed's maps knew that they could always take him problems to take to Kirk. And as the governor later put it, "I would take up whatever crusade [Reed] thought we ought to do for the day."

Their first crusade was emblematic of their assault on Florida's Augean stables, an unprecedented effort to clean up the state's sewage. Reed briefed Kirk that Tampa was Florida's only city with decent sewage treatment, and that Gold Coast communities had to sweep their beaches every morning for condoms, tampons, and other "floatables" that washed ashore overnight. "You mean when I take a crap in Palm Beach it goes straight into the ocean?" Kirk gasped. "My God! We've got to do something!" Just like that, Kirk handed the new Pollution Control Commission over to his aristocratic young aide, who began firing off letters ordering thousands of industrial and municipal polluters to treat their waste, and pressuring communities through the media to raise taxes for sewage projects: "Anybody who says you can achieve environmental quality without paying for it is a liar and a fool!" Reed was pilloried as an out-of-touch tree-hugger, but Kirk threatened to strip power from any local official who ignored his edicts. "The chamber of commerce types would raise holy hell," Reed later remembered. "But Claude would say: 'Gentlemen, Reed's right. It's time to clean up this state.'"

Kirk rarely started out on the green side of an issue, but thanks to Reed and the new politics of the environment, he usually ended up there. Dur-ing his campaign, for example, Kirk opposed Biscayne National Park as a crimp on free enterprise. But after the election, Reed arranged for a friend in the Marine Patrol to guard Kirk during a sailing getaway with Madame X; the governor was so enthralled by the bay's beauty, and the officer's pleas to protect it, that he agreed to fight for the park. Similarly, Candidate Kirk supported the Cross-Florida Barge Canal, but Reed persuaded Governor Kirk it was a Democratic boondoggle and an ecological nightmare, and Kirk helped persuade President Nixon to kill it. Reed also urged Kirk to crack down on the sale and destruction of state-owned wetlands. Dredge-and-fill permits plummeted 90 percent during his administration.

Kirk did not get off to a green start in the Everglades, either. During a drought in 1967, Kirk complained that activists who were demanding water releases to Everglades National Park cared more about gators than people. And when ground was broken for the world's largest jetport in the middle of Big Cypress Swamp, Kirk was on hand to hail the project's "new vision" for south Florida. But Kirk was a publicity hound with national ambitions, at a time when the fight for the Everglades was drawing national publicity. The governor had his faults, but foolish consistency wasn't one of them.

"The End of the Park As We Know It"

ONE OF THE ENDURING MYTHS of the Everglades is that the Army Corps was merely following orders from Congress when it presided over the ecosystem's decline. In fact, the law authorizing the C&SF flood control project specifically directed the Army Corps to protect Everglades National Park, and designated "fish and wildlife preservation" as an official project purpose. The Corps just didn't pay attention to that particular purpose. Senator Holland and his colleagues did care more about flood control than the environment, but the Corps barely cared about the environment at all.

By the late 1960s, Corps leaders in Florida had at least learned to pay lip service to the environment, decorating their C&SF reports with egrets and gators. But their overriding goals were still flood control and economic development, and their most valued "customers" were still farmers, builders, and pro-growth politicians. One Corps hydrologist who spent several years working on the C&SF later admitted he hadn't even known of the park's existence: "Nobody ever mentioned it." It was a Corps of Engineers, after all, not a Corps of Ecologists. "You didn't join the Corps because you wanted to save the earth," explained Richard Bonner, the agency's longtime deputy in Jacksonville. "You joined the Corps because you wanted to build something."

The Corps did build four sluice gates along the Tamiami Trail called the S-12s, which were supposed to let water flow from the conservation areas into the park. But the S-12s were slammed shut after their dedication ceremony, and remained shut for several years. It never occurred to the Corps or the flood control district's board—which was dominated by farmers, ranchers, and businessmen—*not* to divert water that people needed to irrigate their tomatoes, bathe their children, or wash their cars. So the park rustled dry. "Instead of a lush wetland wilderness, we were flying over a parched desert," wrote a sportsman who toured the park by air. The slough that had drawn millions of birds and birdwatchers to Anhinga Trail became a silent mud flat, littered with gar carcasses. The alligator, as much a symbol of Florida as the bald eagle was of America, also neared the brink of extinction; the life-sustaining gator holes that served as dry-season oases for fish and wildlife became so scarce that park scientists launched "Operation Survival" to blast artificial holes with dynamite. A boat company purchased full-page ads in magazines urging readers to "Take a Long Last Look at a Famous National Park—Or Wake Up the Army Engineers." The Corps finally agreed to open the S-12s, but only opened one gate a few inches for one week, a response one critic compared to spitting on a forest fire.

The park scientists were desperate, and decided to calculate the minimum amount of water the park needed to survive. They worked night shifts on the Cape Canaveral supercomputer that NASA used during the day to send a man to the moon, and they eventually came up with a figure of just over 100 billion gallons per year. Their bosses then urged Congress to guarantee those deliveries. "If we don't get water," the park's superintendent said, "it will mean the end of the park as we know it." But the Corps and its supporters in Florida insisted that people were more important than the park. If the park needs water so badly, they said, its managers ought to conserve rainfall more efficiently within its borders, by building levees to block the flow to Florida Bay. Even Lamar Johnson, a relatively progressive engineer who worked for the park as a consultant, complained that "the park sits there like a fledgling egret on its nest, mouth open and squawking, waiting to be fed." Johnson called it "inexcusable" that the park would allow even a drop of water to dribble out to sea unmolested, even though the C&SF was flushing billions of gallons to sea out of Lake Okeechobee.

Nathaniel Reed was tired of water managers treating the national park like a foreign usurper. He knew that blocking the overland flow out of the park with coastal levees would turn the brackish bay into a saltwater lagoon, wiping out some of the most biologically productive mangrove swamps on earth. So when a state official, an unreformed Pork Chopper named Randolph Hodges who considered environmentalism a passing fad, slammed shut the floodgates during the 1967 drought, Reed persuaded Governor Kirk to overrule Hodges and fling the gates open. There wasn't enough water in the marsh to make much of a difference either way, but Kirk wanted to send a message that he cared about the Everglades, just as Hodges wanted to send a message that he cared about farmers.

Under pressure from Holland, the Corps devised a typical *essayons* solution to the water crisis: a $70 million expansion of the C&SF project. The goal was to store enough water to meet all the needs of current as well as future urban and agricultural users—and in most years, the minimum needs of the park—through 2000. "In short, we can have our cake and eat it, too," an engineer wrote in an internal memo. But the Corps refused to guarantee that any of the new water would go to the park, raising the specter that farms and cities would take the entire cake.

This set the stage for the first water war over Everglades National Park. Joe Browder, Nathaniel Reed, and several conservation-minded congressmen insisted that any changes to the C&SF project should at least assure the park the water it needed to survive before supplying new water to subsidize new growth, but Hodges, Senator Holland, and other conservative congressmen objected to any congressional water guarantee for the park as a violation of state sovereignty. The stalemate was broken after Reed persuaded Governor Kirk that a guarantee was the only way to stop reactionaries like Hodges from killing the park, and Kirk persuaded Senator Holland that one of his greatest legacies was in danger. Publicly, Holland still claimed to oppose the guarantee, but behind the scenes, he persuaded his colleagues to make a one-time exception for the Everglades, one of his last achievements before retiring in 1970.

As it turned out, the guarantee didn't do much to help Holland's beloved park. The computer models dramatically underestimated its needs, and water managers generally provided the guaranteed deliveries when the park least needed them. The Senate tried to make it clear in its legislative report that Congress did not intend the C&SF project to be a water supply boondoggle for south Florida, but even the guarantee's supporters foresaw that "the pressures for making this water available to people rather than wildlife in case of a drought will be overwhelming." Once a canal was in place, it was hard to persuade an engineer to use it to supply water to wading birds.

Art Marshall, geologist Garald Parker, and ecologist Frank Craighead, three of the most perceptive Everglades scientists, all warned that the park needed much more dramatic changes to the C&SF than a flimsy water guarantee, and Reed and Browder would both come to regret that they did not pay closer attention at the time. "In my opinion, many of the expensive structures will be obsolete, and many of the drainage canals will be refilled by another generation who will be better informed," Craighead advised park leaders. In a letter to Marshall, Parker declared that the Everglades was doomed if men continued to exploit it: "The only 'out' I see, and one that probably will not be politically practical, is to buy out the farmers, close up the big drainage-canal outlets, and let nature take over restoration of this misused land."

Still, the guarantee was more than a Pyrrhic victory, because it established the precedent that the Everglades had rights. Congress had declared that the park was entitled to water, because the nation believed that the park was entitled to survive.

An Airport in the Everglades

DROUGHT AND DRAINAGE were not the only threats to the park's survival. It was also imperiled by development, especially a proposed airport four times the size of Miami International in the Big Cypress Swamp, just six miles upstream of the park.

The Dade County Port Authority bragged that its billion-dollar, thirty-nine-square-mile "jetport of the future" would attract 50 million passengers a year, and that a city of half a million residents would sprout around it. There would be a takeoff every minute, and runways as long as six miles for supersonic jets. There would be a "super-train" linking the terminals to the Gold Coast, probably across the water conservation areas, and the Tamiami Trail would be widened for airport traffic. This "Everglades jetport" would have cut off the flow of fresh water into the northwestern section of the park and the Ten Thousand Islands, but the Port Authority's analysis only mentioned the park once, to note that its existence assured the absence of neighbors who might oppose the project.

Audubon's Joe Browder loved Big Cypress. Its dark and mysterious bogs evoked the coal swamps of the Carboniferous Era, dominated by 500-year-old bald cypress trees with massive trunks flared out like bell bottoms on steroids. It sheltered some of the last Florida panthers and a spectacular collection of orchids. Browder saw his first mud snake there, with a red face that seemed to be smiling at him; he spent hours crouched in cypress ponds with only his head above the waterline, studying the shafts of light streaming through the canopy. Despite Gulf American's hawking of Golden Gate Estates, almost all of Big Cypress was still wet and wild country, and Browder was determined to prevent supersonic jets from drowning out the *kuk-kuk-kuk* of its pileated woodpeckers. This was no place for a new megalopolis.

Some environmentalists figured the jetport was a done deal, and Nathaniel Reed wasn't paying attention to the issue. But Browder frantically mobilized opposition, exhausting his annual phone budget in three months, bringing together interests as diverse as the Miccosukee Tribe, hunting groups, airline unions, and Everglades National Park's most notorious gator poacher. And after Kirk attended the groundbreaking in 1968, Reed finally agreed to fly over the site, where he saw bulldozers gouging a scar the size of Miami in the heart of the Everglades watershed. He immediately realized he had dropped the ball, and called Art Marshall at Fish and Wildlife. "I don't need you next week," Reed said. "I need you now!"

Marshall and Browder quickly assembled a list of 119 questions about fuel spills, air and noise pollution, road and rail access, bird impacts, spin-off development, and other environmental issues regarding the jetport. Dade County officials grudgingly agreed to answer them. On the appointed date, at a Miami restaurant packed with bureaucrats, scientists, and activists, an airport manager read the county's responses.

What will be done to mitigate runway pollution? "The answer to that question is under study."

What will be done to control land use outside Dade County? "The answer to that question is under study."

The charade continued for several more questions, as county officials exchanged smug grins. Reed, unaccustomed to such dismissive treatment, finally jumped to his feet and berated Dade County mayor Chuck Hall for wasting everyone's time. Browder added that these were serious questions, deserving serious answers. Mayor Hall, decked out in a white suit, white tie, and white shoes, shouted that Reed and Browder were "white militants," and insisted the jetport would be built regardless of their petty complaints.

"I guess they want war," Marshall whispered to Reed.

THE JETPORT'S BACKERS thought they had already won. They had built one runway, and another was under construction. They had lined up state permits and federal funding. So they didn't bother to disguise their hostility to the Everglades, dismissing its endangered species as "yellow-bellied sapsuckers" and its defenders as "butterfly-chasers," quipping that "alligators make nice shoes and pocketbooks," describing Big Cypress as "typical south Florida real estate." Florida's transportation secretary said he would miss the gators no more than he missed the dinosaurs. One port official argued that "Hollandizing" the Everglades would be a good thing, since "the Dutch are some of the best-adjusted, most prosperous, happiest people today." Another piously proclaimed that "we will do our best to meet our responsibilities, and the responsibilities of all men, to exercise dominion over the land, sea and air above us." That kind of sentiment had once inspired pioneers to remake the continent, but in the era of *Silent Spring* it just sounded silly. The jetport men were oblivious to the revolution in their midst, and by the time they tried to mount a defense, their pronouncements that "no pollution is anticipated" and "the operations are not expected to create excessive noise" had no credibility.

After the Miami showdown, Reed told Kirk the jetport was another Democratic land swindle, and easily persuaded the governor to withdraw his support. Kirk had even less trouble persuading President Nixon's interior secretary, Walter Hickel, a former Alaska developer and governor who wanted to burnish his green credentials after disparaging trees during his confirmation hearings. Hickel camped out in the Everglades with Kirk on his first official trip, and the two gregarious politicians bonded over booze and canoes, spending several hours imitating alligator mating calls. By the end of the trip, Hickel had agreed to make the Everglades his signature conservation issue.

Nixon's transportation team supported the jetport, but Hickel was not much of a company man; Nixon would later fire him for publicly denouncing the Vietnam War. Now he began agitating for an environmental study of the jetport, and when Democratic senator Henry Jackson of Washington, a potential Nixon challenger, announced a public hearing on the issue, the White House ordered the study to steal Jackson's thunder. Overseen in Washington by hydrologist Luna Leopold, the son of Wilderness Society founder Aldo Leopold, and coordinated in Florida by Art Marshall, the Interior Department's study was released in September 1969. Its first sentence made the department's position abundantly clear: "Development of the proposed jetport and its attendant facilities . . . will inexorably destroy the south Florida ecosystem and thus the Everglades National Park."

While Kirk and Hickel led the inside war against the jetport, Browder led the outside war. He ginned up a flurry of publicity—including features in *Look, Time,* and NBC's *Today* show, as well as an influential *Life* article by the mystery novelist John D. MacDonald, who warned that the jetport would "kill what is left of the Everglades, kill Everglades National Park [and] upset the water tables and the water supply in all south Florida." Browder also served as the coordinator for a new group called the Everglades Coalition, a partnership of twenty-one local and national green groups that opposed the jetport, raising the issue's national profile. And he helped persuade Miccosukee Indian chairman Buffalo Tiger, who initially supported the jetport after his members were promised jobs, that saving the Big Cypress was more important. The 300 or so Miccosukees felt a deep connection to the Everglades; they had split off from the Seminole tribe because they refused to accept money as reparations for past injustices, only land. (The U.S. government had refused to recognize the Miccosukees, but relented after Tiger flew to Havana to meet with Fidel Castro.) The white man's drainage canals had already withered the Everglades, and Tiger could only imagine the impact of an airport in the swamp. "It happens to Indians year after year: progress wasting the hunting grounds," Tiger told the *New York Times.* The Miccosukees still called the Everglades their Mother, and still credited it with saving their forefathers from genocide. "When it's gone, it's gone forever," Tiger said.

Finally, Browder visited the most famous Everglades advocate, the seventy-eight-year-old Marjory Stoneman Douglas, and asked her to issue a ringing denunciation of the jetport. She replied that people wouldn't listen to a ringing denunciation from an old lady; they only listened to organizations. "Well, why don't you start an organization?" Browder replied. So Douglas founded Friends of the Everglades, and began a new career as a tart-tongued and uncompromising activist, the living symbol of her beloved River of Grass. Douglas issued hundreds of ringing denunciations of the jetport, delivering speeches around Florida in floral dresses, dark glasses, and floppy hats that one writer said "made her look like Scarlett O'Hara as played by Igor Stravinsky." She spoke with precise Victorian diction—e-lo-cu-tion, she called it—and she knew how to exploit her moral authority as the grandmother of the Glades. "Nobody can be rude to me, this poor little old woman," she once confided. "I can be rude to them, poor darlings, but they can't be rude to me." Douglas informed her audiences that America had a choice to make: It could have a fancy Big Cypress jetport, or it could have a River of Grass, but it couldn't have both.

Nixon chose the River of Grass. Shortly after declaring in his New Year's Day message that the 1970s would be the decade of the environment, the president scuttled the jetport. In a *Times* article titled "Against All Odds, the Birds Have Won," novelist Philip Wylie celebrated the decision as a signature victory for the earth: "Natural assets and wildlife preserves have been rescued before, but what was new here was the magnitude of the work already done, the money spent, the solid expectations suddenly rejected."

BIG CYPRESS WAS STILL in private hands, still vulnerable to development. But Joe Browder, who was now a Washington lobbyist for Friends of the Earth, drafted a bill for Congress to buy more than half a million acres of the swamp. The newly retired Senator Holland then lobbied his old colleagues to support the bill, one of his last acts before his death in 1971— although he first insisted on the exclusion of the Okaloacoochee Slough, which his friends in the Collier family still hoped to develop someday. The Nixon administration also wanted to save Big Cypress, thanks in large part to Nathaniel Reed, who was now an assistant interior secretary. But the administration wanted to rely on modest land-use restrictions rather than an expensive buyout, which would have left the swamp at the mercy of the same zoning officials who had welcomed Golden Gate Estates.

Once again, politics rescued the swamp. Browder secretly tipped off the White House that Senator Jackson planned to kick off his presidential campaign with a hearing on Big Cypress in Miami. Nixon domestic adviser John Ehrlichman, a former land-use lawyer who had devised the zoning compromise, promptly called Reed to announce a change in strategy. "We're buying the Big Cypress," he said. "We're going to knock Jackson out of the box in Florida!" Hunting, off-roading, and even oil drilling would still be permitted in Big Cypress National Preserve, but development would be banned forever.

Saving Big Cypress made economic and scientific as well as political sense. The jetport could have required a drainage effort as expensive as the C&SF; its high-speed rail link could have cost even more; supersonic jets were not really the wave of the future. And as Ernest Coe had recognized decades earlier, water flows from the Big Cypress were vital to Everglades National Park.

But there was also something spiritual about this rejection of progress and growth. Man had been exercising his power to subdue nature for centuries, but here he had renounced that power. It was an act of mercy, a retreat from the Empire of the Everglades. And for those who demanded more tangible benefits, Wylie suggested that this newly humble approach could save the human race. Who knew which weed or pest or swamp might turn out to be indispensable for man's survival? The Seminoles used various Everglades plants against ringworm, diarrhea, and even impotence; perhaps some wildflower would unlock a cure for cancer someday. "We would probably continue to live and thrive, to the extent we are thriving, if we paved over the Everglades," he concluded. "But the emphasis is on 'probably.'"

Building the Peace

THE DEFEAT OF THE JETPORT eliminated one mortal threat to the park, but the Everglades was still in critical condition, ping-ponging between too-wet and too-dry. South Florida hit rock bottom in 1971, its worst drought on record. Three inches of rain fell in Miami in six months, and government meteorologists used B-57 bombers to try to seed the clouds. Fires in the Everglades spewed black clouds as far east as Miami Beach, and drinking wells turned salty as far west as Miami Springs. Wood storks abandoned their nests, and gators cannibalized their young. "Drought-Ravaged South Florida Faces an Environmental Disaster," blared a front-page *Washington Post* headline. The good news was that south Florida's problems with water and growth were becoming increasingly obvious. And Florida's new governor was a problem-solver.

Running as a sober voice of reason after four years of zaniness, Democrat Reubin Askew of Pensacola easily unseated Governor Kirk in 1970. In many ways, he was Kirk's polar opposite—a conscientious Presbyterian elder who didn't drink, smoke, or gamble, an earnest policy wonk who believed in telling people what they needed to hear instead of what they wanted to hear; his 1984 campaign for the presidency would flame out after he challenged the nuclear freeze movement, the labor movement, and other traditional Democratic constituencies in the Democratic primary. In Florida, his legacy would include a corporate income tax, statewide busing, and sweeping open-government laws. His nickname was "Reubin the Good."

But if Askew lacked Kirk's flair for razzle-dazzle and demagoguery, he was just as committed to the green revolution. He not only kept Nathaniel Reed in the same job, he appointed Art Marshall to the flood control district's board. Askew was no outdoorsman and didn't pretend to be. But he understood that water was south Florida's lifeblood, and that the region—already home to more than two million people, nearly half of them in the original Everglades floodplain—was growing out of control. Askew liked to quote Pogo, the opossum from Walt Kelly's cartoon swamp: We have met the enemy, and he is us.

If Kirk had been governor during the 1971 drought, he probably would have flown one of those cloud-seeding B-57s himself. Askew hosted a conference. But it was no ordinary conference; the governor invited all of south Florida's experts and interest groups to meet in a Miami Beach hotel, ordered them to stay until they all agreed on solutions to the region's problems, and pledged to support whatever they proposed. In his keynote address, Askew warned that unless the conference produced a dramatic shift in south Florida's relationship with nature, the region would become a paradise lost, "the world's first and only desert which gets 60 inches of annual rainfall." It was the first time a Florida governor had publicly questioned the goodness of growth.

Led by John DeGrove, Florida's leading growth management expert, along with Art Marshall, Florida's leading water management critic, the conference's participants produced a slew of blunt recommendations for inaugurating growth management and improving water management. Their fourteen-page report dripped with Marshall's influence, declaring that "there is a water crisis in South Florida today," and that "there is a limit to the number of people which the South Florida basin can support and at the same time maintain a quality environment." Its proposals included strict land-use restrictions for sensitive lands, strict water-quality protections for the Everglades and Lake Okeechobee, and comprehensive state planning. "I'm 81. I won't live to see this through," Douglas snapped during one panel discussion. "But get on with it!"

ASKEW DID. He converted many of the conference's recommendations into a sweeping growth management package, and called a special legislative session in 1972 to push it. He declared that runaway development was destroying Florida's quality of life as well as its environment, overwhelming its roads, schools, hospitals, sewers, trash dumps, and aquifers: "It is not offbeat or alarmist to say that continued failure to control growth and development in this state will lead to economic as well as environmental disaster." State Senator Bob Graham of Miami Lakes and House Speaker Richard Pettigrew of Miami shepherded his package into law, and even though real estate and farm interests weakened some of its reforms, it was still acclaimed as a national model.

The new growth management laws gave the state the authority to regulate large developments, designate sensitive areas where development would be discouraged, and oversee local land and water planning. It realigned the state's flood control districts—called "water management districts" from then on—along watershed lines, required farms, factories, and communities to seek "consumptive use" permits before receiving public water, and set up a legal process for reserving water rights for environmental purposes. It also authorized a $240 million bond issue for acquiring ecologically sensitive private lands, which Florida's voters resoundingly approved.

Skeptics snickered that no laws could stop growth in south Florida, and in some ways they were right. More than 100,000 people kept moving to the region every year; fast-food joints and tract houses kept sprouting in the Everglades floodplain. Developers scrambled to plat zero-lot-line subdivisions and open trailer parks before the new laws could take effect, and quickly devised ways around the laws once they did take effect. Florida's water management districts handed out consumptive use permits to just about anyone who asked, and Florida's water would be reserved for an environmental purpose only once over the next three decades. A follow-up law requiring local growth plans proved similarly porous; the various plans would have generated a statewide population of 100 million, and a cottage industry of lawyers and fixers quickly emerged to help developers secure "variances" to the plans whenever they wanted. Money still talked in south Florida.

But that was the beauty of Askew's $240 million bond issue: It talked back. Developers could maneuver around land-use restrictions and plans, but public ownership locked up land forever. Since statehood, Florida had given away more than half its landmass for a pittance, but since 1972, Florida has led the nation in land acquisition, buying back more than one million acres of environmentally sensitive real estate. Askew's "Lands for You" program inaugurated a bipartisan tradition of land buying which every subsequent Florida governor has continued through programs such as "Save Our Rivers," "Preservation 2000," and "Florida Forever."

The rest of Askew's package had less visible impact. But like the water guarantee for the park, it sent an important symbolic message, even if it was only observed in the breach. Unlimited development was no longer officially condoned, and politicians could no longer pretend that growth was an unqualified good. Growth management has been a state policy since 1972, even though there has been a lot more growth than management. Every governor since Askew has proclaimed his commitment to the cause, even as one commission after another has declared growth out of control. "Managing growth in Florida is like trying to nail Jell-O to the wall," Askew said in a recent interview. "But just think where we'd be if we hadn't done anything."

THE STATE THAT Harriet Beecher Stowe had described as "a prey and a spoil to all comers" was now an environmental model for the nation. One poll found that Florida was the only state where concern about the environment overshadowed the economy, with three-fourths of its residents supporting strict limits on future growth. There was certainly an element of selfishness to this backlash against helter-skelter development, as newcomers tried to slam Florida's door behind them once they had secured their own slice of paradise. But whatever their motives, environmental politics became smart politics in the Sunshine State.

Open season on the Everglades was over. A few extremists still grumbled about yellow-bellied sapsuckers, and one Big Cypress property rights group distributed flyers calling for the assassination of Askew, Browder, and Reed, suggesting that "any citizen killing these people trying to take our land would receive all the blessings [of] God and the Host of Heaven." But from now on, all politicians would at least pretend they cared about saving the Everglades. Once dismissed as a wasteland, "suitable only for the haunt of noxious vermin," the Everglades was now embraced as a national treasure, an International Biosphere Reserve, a United Nations World Heritage Site. It had become the ultimate symbol of "the environment," a wind gauge for man's stormy relationship with nature. "If man cannot live with a living Everglades, he may be incapable of continuing success as a species," Browder wrote. "If man can choose to try and save the Everglades, perhaps he can save himself as well."

Man had already saved Big Cypress and Biscayne Bay from development, amazing substantive achievements. Man had secured a legal promise of water for Everglades National Park and a legal process for managing growth in Florida, important symbolic victories. But the Everglades was not saved yet. The green revolutionaries had successfully treated some potentially fatal symptoms, but their patient was still dying. Now they would have to heal the underlying disease: the C&SF project.

FIFTEEN

Repairing the Everglades

The Everglades is trying to tell us something.
—Florida ecologist Arthur R. Marshall Jr.

Going Down the Drain

HALF THE EVERGLADES was gone, drained for agriculture or paved for development. The rest was an ecological mess—sometimes too wet, sometimes too dry, always obstructed and convoluted by highways, levees, and canals. More than 90 percent of its wading birds and alligators had vanished. Most of its canary-in-the-coal-mine "indicator species" were at risk of extinction, including the Florida panther, Everglade snail kite, Cape Sable seaside sparrow, and American crocodile—the barometers of the ecosystem's uplands, wetter marshes, drier marshes, and Florida Bay, respectively. Melaleuca trees from Australia were invading disturbed wetlands and crowding out all other vegetation; they had no natural enemies in south Florida, and fires only spread their seeds. Similarly invasive Brazilian pepper bushes were so pervasive that they became known as "Florida holly," and were such masters of regeneration that they could only be eradicated from an area by scraping soil down to bedrock and hauling it away.

To Art Marshall, the root cause of all the problems was clear. "The Everglades is not just stressed," he said. "It is distressed—a condition brought about to a major degree by past works of the flood-control project."

It was Marshall who popularized the image of the Everglades ecosystem as a unified organism, connected from head to toe by clean, fresh, slowly flowing water. And it was Marshall who warned the world that the C&SF project was eviscerating that organism. Before drainage, water had spilled down the Chain of Lakes into the meandering Kissimmee River, which had emptied into the liquid heart of Lake Okeechobee, which had overflowed into the sawgrass Everglades, which had dribbled through Everglades National Park to its estuaries. That flowing sheet of shallow water had driven the ecosystem, filtering through wetlands, percolating into aquifers, rising and retreating with the rain. In the Everglades, Marshall observed, the difference between water flowing and water standing was like the difference between being old and being dead.

The C&SF project had sliced up the organism into a disjointed marionette, and the water that was its lifeblood no longer flowed. The Kissimmee was wrestled into a ditch that no longer meandered or flooded its floodplain. Lake Okeechobee was imprisoned by its dike, so it no longer flowed into the upper Glades, except when sugar growers demanded water for irrigation. The upper Glades was becoming a sugarcane monoculture, so it no longer flowed into the central Glades, except when the growers dumped water during storms. The central Glades was divided into five compartments that only flowed into the southern Glades when water managers decided to flood the park; the water in these "conservation areas" barely flowed at all, collecting in stagnant pools where sawgrass decomposed into ooze instead of muck. Inside the park, Shark Slough and Taylor Slough no longer carried much fresh water to the Ten Thousand Islands and Florida Bay, which began mutating from brackish estuaries into saltwater lagoons.

The entire system was broken. Wetlands that had filtered and stored water were gone, or so distressed that they made alluring targets for exotic vegetation. Aquifers were parched. The natural rise and retreat of the River of Grass was overwhelmed by artificial pulses and drawdowns that befuddled wildlife—drowning deer and gator nests, while diluting the fish-saturated wetlands that wood storks had counted on to feed their young.

To Marshall, the plight of the deer, gators, and storks—as well as cash-strapped fishing guides—were all symptoms of a mutilated, mismanaged watershed. The dying park at the bottom of the basin got most of the attention, but Marshall knew its problems began upstream. The head, heart, and body of the ecosystem had been sliced, diced, and bled dry; no wonder its feet had stopped dancing.

Still, Marshall believed the Everglades could be repaired—not *restored,* which would have required moving half a million homes out of its flood-plain, but *repaired.* His first recommendation, the best advice to anyone in a hole, was to stop digging. Even while fires were raging across hundreds of thousands of acres of marshland in 1971, the Corps had dredged more ditches that carried away more water. Marshall knew that the disconnected Everglades could not afford more Aerojet canals. "The Everglades ecosystem as we know it is literally going down the drain," he wrote. "Man has played Russian roulette with the Glades for a very long time. One day soon he may pull the trigger on a loaded chamber."

But Marshall also knew that preventing additional mistakes, while nec-essary, would not be sufficient. It was also time to start fixing past mistakes. And there had been no mistake more egregious than the destruction of the Kissimmee River, at the headwaters of the Everglades ecosystem.

Undoing the Ditch

ONE NINETEENTH-CENTURY VISITOR described the natural Kissimmee as "the most crooked stream in the world," a madcap squiggle fringed by live oaks, cypress domes, and "vast swamps covered with water lilies and beau-tiful flowers." Its basin also included a web of lush marshes known as the Little Everglades, where a *Harper's* writer enjoyed "grasses and vines as graceful as Nature's hand could fling abroad," as well as fire-swept prairies that looked so much like African savannas that one naturalist thought they cried out for antelope. Even after Hamilton Disston sliced off a few of the river's hairpin turns in 1880s, it was still distinguished by "its narrowness, the rampant growth of the water plants along its low banks . . . the variety and quality of its birdlife . . . and above all the appalling, incredible, bewil-dering crookedness of its serpentine body."

The natural Kissimmee basin had attracted 320 species of fish and wildlife—including heavy-billed caracaras that patrolled its prairies; shore-birds that nested on its sandbars; ring-necked ducks, blue-winged teal, and other winter waterfowl that visited its marshes; and one of the world's rich-est bass fisheries. As late as 1958, Fish and Wildlife described the flood-plain as "a fast food factory and nursery ground for sport fishes." But the Corps saw the basin as Disston had seen it, "the source of all the evil," and devised a plan to straighten and confine the 103-mile-long, 6-foot-deep river into a 56-mile-long, 30-foot-deep canal that would never overflow its banks or flood its floodplain.

Environmental agencies failed to predict most of the C&SF's ecological damage, but anyone could see that jamming the Kissimmee into a ditch would kill it. Fish and Wildlife protested that the river "furnishes an unusually valuable and unique bass fishery which will be lost if the present plan for flood control is carried out." And putting ecological concerns aside—as the Corps tended to do—its analysts calculated that the project's cost would outweigh its economic benefits, which should have stopped the channelization in its tracks.

But the canal was Senator Holland's gift to cattlemen in the Kissimmee basin who wanted year-round pastures, and to homeowners around the Chain of Lakes who wanted year-round flood control. So the Corps manipulated its economic analysis—by double-counting and inflating benefits, and using an artificially low interest rate to deflate costs—to claim $1.38 in benefits to landowners for each dollar spent by taxpayers. Now the project was "justified," so Congress could fund it. A battalion of suction dredges and hulking draglines descended on the Kissimmee, bullying it into a ramrod-straight engineering marvel that was christened the C-38 Canal. "I've just returned from the deathbed of an old friend," a *Herald* outdoors writer reported after visiting the project. "Although it has been definitively established that the death will be a boon to something called progress, the sight was a most depressing one."

The Corps spent ten years and $35 million manhandling the Kissimmee into a "wide, broad superhighway," building five dams and moving three million truckloads of dirt. By the time the work was finished in 1971, almost everyone agreed that it never should have been started. Marjory Stoneman Douglas called it the crowning stupidity of the C&SF project; *Time* magazine called it "one of the most disastrous projects ever undertaken by the Army Corps of Engineers." The C-38 did drain runoff from Disney World and the booming Orlando area, and nearly quadrupled the basin's pasture land to 576,000 acres. But as the canal became a national symbol of engineering folly, and unsightly dredge spoil began piling up along the canal's banks, even some cattlemen began to miss the rambunctious river. "The Kissimmee Valley was fantastic country," one wrote. "It is hard for me to understand how man could have the audacity to think he could improve on the Kissimmee River."

The "improvement" created an ecological catastrophe. As Marshall and other critics had predicted, the bass fishery crashed. Waterfowl declined 92 percent, bald eagles 74 percent. The river's sandbars disappeared, along with its shorebirds. Oxbows filled with silt. Common cattle egrets, an invasive species in Florida, replaced snowy egrets, white ibis, and other native wading birds. Dissolved oxygen levels plunged, until gar and bowfin were the only fish surviving in the ditch. The Kissimmee floodplain, previously wet more than half the time, was now dry almost all the time; the basin lost enough wetlands to cover Manhattan twice. "One year they found eight ducks—one, two, three, four, five, six, seven, eight!" recalled Johnny Jones, a hunter and conservationist who was Marshall's close friend and political ally.

To Marshall, the C-38 was not just a murder of a vibrant river, but a disaster for the ecosystem. In the natural Kissimmee basin, water had ambled down the river and dallied in the floodplain; now it whipped down the ditch in violent bursts. Even routine storms now discharged more water into Lake Okeechobee than the 1928 hurricane, forcing water managers to empty the swollen lake down the St. Lucie and Caloosahatchee Canals to prevent a dike collapse, ravaging the balance of fresh and salt water in their rich estuaries. Water managers reversed the natural flood regime to ensure year-round flood control; nature had moved the most water out of the basin in summer and fall, but man removed the most water in the spring.

The C-38 also became a glorified sewer pipe, whisking waste from the upper basin's toilets, citrus groves, and cattle pastures into Lake Okeechobee and south Florida's water supply. (Fishermen joked that when the water got low, they'd ask folks in Orlando to flush twice.) Marshall was concerned that dung from the basin's new cows was washing straight into the ditch, and that the wetlands that used to filter such runoff had disappeared. He warned that without a revolutionary new approach to the C-38, the only question was when—not if—Lake O would become a dead zone.

THIS WAS MARSHALL'S PLAN for the C-38: Get rid of it.

As in: Turn it back into the Kissimmee River. As in: Backfill the canal, blow up its dams, and let the river flow again. Marshall's plan was as simple as it was radical: Get out of nature's way, and let it heal itself. "The river is still there," he wrote in a letter to Douglas. "It's the water that's been taken away. The river is still there."

Johnny Jones decided to convert his friend's vision of a resuscitated river into law. Jones grew up in West Palm Beach—not the part along the beach, but the part in the former Everglades, where he developed a passion for hunting and fishing. He dropped out of school in tenth grade to get married and work as a plumber, but as the marshes where he hunted and fished became condos and turnpike interchanges, Jones developed a new passion for conservation. He was a natural lobbyist, a backslapper with the persistence of a bulldog—he even looked like a bulldog—and he soon gave up his plumbing business to lobby full-time for the Florida Wildlife Federation. The federation was a stodgy hook-and-bullet coalition that had helped limit the size of Everglades National Park, but Jones turned it into a spirited environmental force, forging alliances with Marshall and Marjory Stoneman Douglas. She hated hunting, but she liked Jones, especially after he got the purple gallinule removed from the state's game bird list. She liked people who got things done.

Jones was particularly anxious to get something done for the Kissimmee, his favorite Florida getaway before the engineers turned it into a "dirty ditch." Jones was a plumber, and he knew lousy plumbing when he saw it. But his Kissimmee restoration bill was a legislative long shot. No one had ever tried to reverse a Corps project, and even many conservationists doubted the river could be revived. "It was a sad mistake to tamper with the Kissimmee originally, but now that she is what she is, our view is to make the most of it," *Florida Sportsman* editorialized. "Simply stated, man can easily destroy, but can recreate only with immense difficulty, if at all." Jones had to settle for a compromise bill that failed to define "restoration" and assigned yet another commission to study what to do next. Cattle lobbyists then deleted its funding. But Marshall was unusually cheerful about it. "I believe we are on the road to restore that disheveled river," he wrote Douglas.

Marshall yearned to show the world that man could undo his mistakes, and something about the prospect of undoing such a colossal one seemed to dissolve his pessimism. After years of playing defense, trying to limit man's incursions, he loved the idea of winning back ground for Mother Nature. "We shall see the Kissimmee River flowing sweet and beautifully again," he wrote. "We are on the road, Marjory. Not only for the Kissimmee, but for Lake Okeechobee and restoring the River of Grass."

★   ★   ★

IN MARSHALL'S WAR to save south Florida, the C-38 was Omaha Beach. But it was only one battle in a larger war, one link in the Kissimmee-Okeechobee-Everglades chain. Marshall had a holistic vision for repairing the entire ecosystem. It was known as the Marshall Plan, and in some ways it was as bold as its namesake in postwar Europe.

The Marshall Plan revolved around another simple but radical strategy: re-create the sheet flow down the Peninsula, by reconnecting the ecosystem. It aimed to remove levees, refill canals, revamp water management, and buy sensitive land in order to knit together the River of Grass and mimic the watershed's original hydropatterns. The eighteen-point plan called for water managers to let the Chain of Lakes fluctuate naturally instead of holding them artificially low for flood control, so that they could once again spill into an untrammeled Kissimmee River. It then proposed to restore sheet flow in the Holey Land and adjacent Rotenberger wetlands in the southern end of the agricultural area, then down through the largest water conservation area, then down through Shark Slough and Taylor Slough, all the way to the estuaries. Marshall suggested that someday, as soil subsidence rendered the agricultural area less productive, enough farmland might be repurchased to reconnect Lake Okeechobee to the Everglades.

By keeping more of the Everglades wet for more of the year—not with standing water, but with flowing water—Marshall expected to regenerate muck, recharge aquifers, and attract more forage fishes for wading birds, while reducing the impact of droughts throughout the region. He recognized that parts of the Everglades were too wet at times, but overall, the Everglades was much too dry, because rain no longer lingered on the land in the winter. He thought of summer rains as paychecks for the Everglades; his goal was to keep the water flowing on the land for as long as possible, until the next summer's rains could replenish the account.

To Marshall, every link was vital to the chain, from controlling pollution in Lake Toho at the northern edge of the watershed to filling in some of the Aerojet Canal along Florida Bay 200 miles to the south. But there was no doubt about his favorite link. When Douglas printed up the Marshall Plan as a pamphlet, three of its four pages were devoted to the Kissimmee—"the BIG issue," as Marshall put it. This was man's chance to unshackle Mother Nature from human bondage, to bring a river back from the dead. It had never been done before, so there was no way to be sure whether the Kissimmee could ever flow again—and if it could, whether its fish and birds would ever return. But Marshall had a feeling it could, and they would.

"No one really knew if Einstein was correct in his theory—not even he was absolutely certain of its validity—until that bomb went off at Alamogordo," he wrote. "My bomb is the Kissimmee ditch restored."

"OK. I'll Do It."

"THE GOOD NEWS is on the cover, Governor."

It was the winter of 1981, and Estus Whitfield, an environmental aide to Governor Bob Graham, was giving his boss the *Sports Illustrated* swimsuit issue. The supermodel Christie Brinkley, in a fuchsia bikini, seemed like very good news indeed.

"The bad news," Whitfield added, "is inside."

On page 82, the magazine had used the occasion of Brinkley's photo shoot on Captiva Island to investigate Florida's environment. And it had not found much good news to report. The headline read: "There's Trouble in Paradise." It was superimposed on a jarring portrait of the C-38 Canal— still a dirty ditch, still an ecological disaster. "The sad fact is that Florida is going down the tube," the article declared. "Indeed, in no state is the environment being wrecked faster and on a larger scale."

The article included plenty of sad facts: South Florida's mullet catch had dropped more than 90 percent in five years, C&SF canals were swarming with coliform bacteria, and the Chain of Lakes was collapsing. The basic message was that Florida was growing to death. In the 1970s, its population had kept growing four times the national rate, and ecologically, the growth had been in all the wrong places. While builders continued to drain Everglades wetlands to create low-lying suburban outposts up to twenty miles west of the coastal ridge, cities on the high ground were struggling. *Sports Illustrated* noted that Miami was becoming the drug and murder capital of America—an epidemic that would soon be glamorized in pastels on *Miami Vice*—and that race riots had erupted in 1980. That was also the year of the Mariél boatlift, an influx of 125,000 Cuban refugees—including criminals, mental patients, and other "undesirables" released from Castro's prisons—that had fueled Miami's image as a city out of control.

Halfway through the article, Johnny Jones griped that Florida's governor wouldn't even meet with him anymore, probably because he hoped to run for president and didn't want to look like a liberal. Jones told the magazine that Graham had been one of Florida's best environmental legislators ever, "but as governor he has wandered away from us. . . . As a governor, he ain't got it." As he read those words, Governor Graham's face turned a shade similar to Brinkley's bikini. That was the moment, according to Whitfield, when Florida embarked on its journey toward a restored Everglades.

DANIEL ROBERT GRAHAM grew up in a coral-rock farmhouse in the Everglades, playing with snakes, shooting frogs with a BB gun, fishing in the Miami Canal. What the Mississippi River was to Tom Sawyer, he said, the Everglades was to him. One of his first jobs was as an aide to Democratic Congressman Dante Fascell of Miami, a passionate defender of the Everglades who helped create Biscayne National Park and the Big Cypress National Preserve. But Graham's family made its money from sugar, cattle, and development, the three main threats to the Everglades; his father, Cap Graham, was a proud exploiter of the Everglades who fervently believed in man's dominion over the earth. As a boy growing up in the wetlands of Pennsuco, Bob loaded manure, drove tractors, and raised a prize-winning heifer. As a young man in Miami Lakes, he worked as an attorney for the family business, building houses in the Everglades.

But after Graham followed his father into politics, he made it clear that he did not share his father's views of nature. He was no radical, but he could see that Florida was ravaging its most precious resources, and he was determined to try to swing the pendulum back toward moderation. His environmental impulse sprouted in part from his childhood in the Everglades, but mostly it sprouted from the progressivism of his generation—just as the development impulse of drainage boosters like Napoleon Bonaparte Broward had sprouted from the progressivism of theirs. Graham supported racial equality and public education as well as growth management and environmental protection. It all seemed to go together at the time.

It just didn't seem like a formula for getting elected to statewide office. When Graham entered the crowded field to succeed Governor Askew in 1978, his decision was widely ridiculed as an act of political suicide. "Professional pols viewed Graham's candidacy as a darkly amusing blood sport, like feeding a bunny to a pack of starving dogs," one author wrote. Graham was studious, courteous, diligent, and smart, but with his chipmunk cheeks and high-pitched voice, he did not look or sound like a governor. A political almanac would later describe him as "careful, methodical, thorough, hardworking, reliable"—it might as well have added "zzzzzz." He was a whiz at explaining Medicaid reimbursements, but he wasn't much for chitchat; he came off a bit robotic, and a bit eccentric. Today, he is best known for his obsessive-compulsive habit of scribbling the dullest minutiae of his life—that he drank a chocolate Slim-Fast, that he changed into gray pants—into little notebooks that he color-codes by season. His mania for documentation and his inability to emote can approach self-parody; he once dutifully recorded: "2:39 p.m.—pilot announces hydraulic failure, must make emergency landing."

In any case, no south Floridian had ever been elected governor; when Cap came in third in a primary in 1944, one pundit had predicted it would happen on a cold day in hell. And there wasn't much evidence of a hidden groundswell for a multimillionaire Miami liberal named D. Robert Graham; he was polling at 3 percent when he declared, behind a half dozen better-known politicians. But Graham ran as just plain Bob, a husband and father of four daughters, a likable pro-business, pro-agriculture, pro-environment centrist. He also had a brilliant gimmick, spending 100 full days of the campaign working ordinary jobs with ordinary Floridians. The gimmick really worked because Graham *really worked*—as a cop, a bellboy, a construction worker, a manure shoveler. It was clear that he wasn't just slumming, and voters rewarded his efforts to learn about their lives. It snowed in Tallahassee the morning of his inauguration—the cold day in hell after all.

Environmentalists were ecstatic to have one of their own in power, but Graham initially ignored their issues. At first, he struggled to get anything past the legislature; one newspaper dubbed him Governor Jell-O, and the Senate president declared him the worst governor in the history of the world. Once he found his footing—the turning point was his popular decision to send a murderer to the electric chair—he continued to focus on creating jobs, fighting crime, and reforming schools. President Jimmy Carter, an engineer who loved rivers and hated the Army Corps, had ruined his relationship with Congress by unveiling a "hit list" of nineteen water projects he wanted to kill—and he hadn't even succeeded in killing them. Florida's legislative leaders had declared a moratorium on environmental legislation, and Graham didn't want to repeat Carter's mistake.

But then *Sports Illustrated* hit the stands, with its revelations about

Florida's disappearing mangroves, contaminated drinking water, and overdeveloped beaches—and its criticism of a governor who allegedly "sided with the despoilers." Jones soon received an invitation to meet with Graham, and insisted on a full hour of the governor's time. He spent much of the hour complaining that the C-38 study was still dragging on, and that speculators were starting to buy land in the Kissimmee basin. Jones warned that if Graham didn't move forward soon with dechannelization, it might never happen. The governor listened patiently, scribbling in his notebook.

"Okay," Graham finally said. "I'll do it."

Jones then turned the floor over to Marshall, who gave the governor a twenty-minute version of the Marshall Plan, explaining his vision of a reconfigured south Florida.

"Okay," Graham said. "I'll do it."

As they left, Marshall gasped: *Did you hear what that man said?* Jones explained that politicians say things all the time. But Graham was serious.

THE 1980s were not a great time for environmental protection. Americans had replaced the dour Carter with the sunny Ronald Reagan; it was "Morning in America," an age of conspicuous consumption, and eco-complainers were beginning to sound like scolds from a bygone era. Reagan and his conservative Republican base tended to view environmental regulations as handcuffs on free enterprise and environmentalists as tree-huggers who wanted to tell people what to drive and how to live. Reagan's interior secretary, James Watt, described the environmental movement as a "left-wing cult," and avidly promoted mining, drilling, and logging on public land. The Corps continued to manhandle rivers for phantom barges; for example, its Tennessee-Tombigbee Waterway carried 1.7 million tons of cargo in its first year of operation, a mere 25.3 million tons less than the Corps had predicted. And when Marjory Stoneman Douglas—now in her nineties, and legally blind—called for restrictions on development in the east Everglades at a public hearing, landowners booed and yelled at her to go back to Russia. "I've got all night, and I'm used to the heat," Douglas shot back.

Graham was not as brassy as Douglas, but he did swim against the tide. In 1983, he announced his "Save Our Everglades" program, an effort to turn back the clock so that "the Everglades of 2000 looks and functions more like it did in 1900 than it does today." Graham publicly renounced a century of draining and diking, declaring that people were as dependent on the natural flow of the Everglades as swamp lilies or blue herons: "We face an awesome truth. Our presence here is as tenuous as that of the fragile Everglades." The governor acknowledged that "restoring" the Everglades to its pre-drainage condition would be as impossible as restoring a half-eaten omelet to its egg, but just because it couldn't be perfectly natural didn't mean it couldn't become more natural. "Whatever the price," Graham said, "the price of inaction is higher still."

Most of Save Our Everglades came straight out of the Marshall Plan. It aimed to "re-establish the values of the Kissimmee River," and endorsed a pilot project to restore a few miles of the old riverbed, a first step toward dechannelization. Save Our Everglades also proposed to restore sheet flow through the Holey Land and Rotenberger tracts in the agricultural area. It then called for flow ways under Alligator Alley to reconnect the agricultural area to the conservation areas, a study of similar adjustments to the Tamiami Trail to reconnect the conservation areas to Everglades National Park, and the restoration of flows through Shark Slough and Taylor Slough within the park. Overall, it envisioned more than 500 square miles of land acquisition, an area larger than the city of Los Angeles. Former senator Gaylord Nelson, who had become the chairman of the Wilderness Society, called it "the most comprehensive environmental program in the history of this country."

Save Our Everglades was radical, but not overly controversial; Graham mostly avoided confrontations with powerful interests, focusing on generous buyouts and other win-win solutions that didn't gore anyone's ox. The exception was the proposed restoration of the Kissimmee River, which directly threatened pastures in the upper basin. Florida's cattlemen hired the former leader of the state's C-38 study to fight the plan, and a local group called Residents Organized Against Restoration filmed a video tribute to the C-38, with the Beatles song "Let It Be" playing in the background. Some critics argued that it was lunacy for the government, after spending millions of taxpayer dollars to convert worthless marshes into improved pastures, to spend millions more to convert the pastures back to marshes. Who ever heard of a government project designed to *promote* flooding? "The Everglades is a national park down there by Miami, and we have no quarrel with saving it," one rancher told the *Los Angeles Times*. "But that doesn't mean they ought to come up here and turn everything upside down."

Thanks to Art Marshall, Graham now understood that the health of the park was linked to the health of the Kissimmee, and he believed that restoring the river would be more like turning the ecosystem right side up. In the Broward Era, and his father's era, man's faith in his power and his duty to improve nature had been almost absolute. "Now we're less confident in technology," Graham said. "We're more prepared to rely on what God feels is the appropriate way to relate to these problems."

GRAHAM KNEW FLORIDA could not save the Everglades on its own, and he wasn't willing to leave it in God's hands. But the Reagan administration refused to help him purchase buffer lands; it was trying to ease the federal government out of the land-buying business. And the Corps refused to dismantle its C-38 Canal without an economic justification, insisting it lacked the authority to take on purely environmental projects.

Graham decided he needed to thrust the Everglades back into the national spotlight, so he personally revived the dormant Everglades Coalition, the alliance of local and national green groups that had raised the profile of the Big Cypress jetport. His staff organized the coalition's first annual conference in 1986, focusing on the Kissimmee restoration; two decades later, politicians still appear at the conferences to prove their Everglades bona fides. "The national environmental groups had tied a ribbon around the Everglades and stopped paying attention," recalled Charles Lee, an Audubon activist in Florida for the last three decades. "Graham got it back on the national agenda."

Later that year, Graham was elected to the U.S. Senate, and the Kissimmee remained his top environmental priority. In 1990, he managed to slip language into a public works bill to authorize the Corps to take on purely environmental projects—a little-noticed turning point for an agency that had traditionally taken on environmentally disastrous projects. General Henry Hatch, the commander of the Corps at the time, addressed the Everglades Coalition conference the next January, and stunned the crowd by declaring that he considered his agency's new environmental mission was as important as its economic mission. "The Corps didn't have a lot of credibility on the environment back then," Hatch recalls. "But you could see the era of big dams was over. We had to find something else to do."

Senator Graham made sure the Corps began its new mission on a certain appallingly crooked river in Florida. The price tag for the Kissimmee plan would skyrocket to more than $500 million, about fifteen times the original cost of the C-38. And it would only aim to fill a third of the ditch, in order to maintain flood protection for the homes that had been built in the basin while the Corps dithered. But it was still the most ambitious effort to resuscitate a river ever—Art Marshall's Alamogordo after all.

Getting the Water Right

MARSHALL DIED OF LUNG CANCER in 1985, so he never saw the results of his grand experiment. But by the time of his death, he was no longer so melancholy about the future of south Florida. His larger vision of a free-flowing ecosystem was supplanting the *Waters of Destiny* mentality, and the Marshall Plan was becoming the blueprint for policy in the Everglades. "We are making fair headway," he wrote. "There is in all this, I believe, an opportunity to regain some of that most needed natural resource—HOPE."

The Kissimmee restoration was only the most prominent of Marshall's proposed projects in the works by the time he died. There were also on-going efforts to repair Lake Toho and the Turner River, add 250 square miles to Big Cypress, protect the reefs in a Florida Keys National Marine Sanctuary, and restore Taylor Slough in Everglades National Park.

Congress also approved an ambitious initiative to expand the park and restore Shark Slough, yet another exercise in ecological revisionism inspired by the Marshall Plan. After World War II, Senator Holland had sliced some of the slough out of Ernest Coe's original park boundaries, and the C&SF project had diverted the slough west to protect land in the east Everglades. In 1989, Senator Graham helped pass a bill that put 107,000 acres back into the park and that authorized a project to replenish the slough and shove it back east. The bill also authorized flood control for the low-lying east Everglades neighborhood known as the Eight-and-a-Half-Square-Mile Area, which was directly threatened by the restoration work. Graham hailed the plan as another win-win solution, good for the Everglades and good for people. "We have fashioned balanced bipartisan legislation which will help restore an international treasure," he announced.

THESE EARLY RESTORATION EFFORTS were all designed to fix links in the ecosystem. But some environmentalists began dreaming of a megaproject that would fix the whole chain, converting Art Marshall's vision into a multibillion-dollar government mission. Florida's conservation movement had begun by saving birds from hunters; it had progressed to saving habitats from developers, then trying to restore degraded habitats. Now the Everglades Coalition began strategizing about restoring the entire ecosystem, while improving the human environment as well.

The driving force behind the new vision was Jim Webb, a former Marine from an Arizona ranching family, an erudite man with a passion for big words, big ideas, and big open spaces. Webb was a veteran of western water wars as well as national park wars; he had served in Tucson municipal government and President Carter's interior department. He was also an avid outdoorsman, and after he moved to south Florida in 1986, Joe Browder helped persuade him to start a new career as a Wilderness Society activist. "When I hiked through a swamp or a desert with Jim, I saw more than I would have seen alone," recalled Browder, who befriended Webb while serving with him at Interior. "I wouldn't say that about anyone else."

Webb soon recognized that even though south Florida received five times as much rain as Tucson, its problem was the same: not enough water to go around. Communities, farms, and the Everglades all competed for the same supply, and the competition was intensifying by the day. Florida was now the fourth-most-populous state, and newcomers were pouring into Everglades suburbs; Pembroke Pines grew 83 percent in the 1980s, Davie 130 percent, and Hialeah Gardens 186 percent. The C&SF project, originally designed for 2 million people, now supported 6 million. Giant agribusinesses sucked billions of additional gallons out of Everglades aquifers, paying almost nothing for the privilege. And when water managers had to choose among communities, farms, and the Everglades, the Everglades invariably lost.

Webb also noticed that the C&SF project was continuing to pour billions of gallons of fresh water out to tide, in order to prevent Lake Okeechobee from blasting through its dike again. These releases were not only terribly wasteful, they were terribly destructive to the Caloosahatchee and St. Lucie estuaries. But that got Webb thinking: What if the C&SF project were managed as an environmental project as well as a flood control and water supply project, as its original law had directed? And what if the project could store more fresh water for people as well as nature, instead of blasting so much into delicate estuaries? "Jim knew it was going to be tough to get billions of dollars for sawgrass," recalled George Frampton, his boss at the Wilderness Society. "But if you could help the residential areas with water supply, maybe you're in business." Webb, Browder, and other Everglades Coalition activists translated his vision into a strategic plan titled "Everglades in the 21st Century," and the coalition began pushing for a comprehensive "Restudy" of the C&SF project.

Webb also found an unlikely supporter for his dream of revamping the C&SF: the Army Corps colonel in Florida, Terrence "Rock" Salt. Salt was a disciple of General Hatch, part of a new vanguard of baby-boom engineers who came of age in the Earth Day era. A jowly, bulky soldier with a strong Christian faith and a bureaucratic penchant for "talking through issues," Salt had served in the Pacific northwest, where he fought the ecological wars over salmon and developed a strong preference for ecological peace. Salt wasn't exactly an environmentalist, but he was a realist, and he knew the Corps had to adapt if it hoped to survive. In the nineteenth and twentieth centuries, it had kept busy maiming ecosystems; to stay busy in the twenty-first century, it would have to learn to heal some of the wounds it had inflicted. "Look, we're engineers. We like to turn dirt," Salt later said. "But we started thinking, 'Hey, let's see if we can turn some green dirt.'"

Man had made south Florida safe for gated communities and golf courses. But the consequences of man's manipulation of nature were obvious in any satellite photo. The eastern Everglades and the coastal ridge were now a gray block of development. The upper Glades was a pink block of farmland. The central Glades was crisscrossed by levees and highways. The tree islands that once all pointed in the same direction as the flow of the River of Grass now pointed every which way—where they were visible at all.

Now, man was thinking about the replumbing of the Everglades, to try to make the ecosystem look more like it did before he started messing with it. He was finally making an effort to Get the Water Right in the Everglades—not by squirting it wherever it served his economic interests, but by letting it flow as it naturally flowed, in the right amounts to the right places at the right times.

There was just one problem.

The water was dirty.

SIXTEEN

Something in the Water

*Tragically, the ecological integrity and ultimately the survival
of the Park and the Refuge are today threatened by the inflow
of nutrient-polluted water.*

— U.S. v. South Florida Water Management District

The Other Evil Empire

ONE DAY IN THE EARLY 1980S, Nathaniel Reed took an airboat ride through the sawgrass of Water Conservation Area 3 with Walt Dineen, the South Florida Water Management District's top scientist. Dineen cut the engine just south of the Everglades Agricultural Area, in front of a dense thicket of head-high cattails. Then he began to cry.

Cattails are attractive plants, but they didn't belong in the Everglades. "What's the problem?" asked Reed, who was serving on the district's board.

"I don't know," Dineen sobbed. "We better find out fast."

When he was Governor Kirk's pollution czar, Reed had written Florida's water quality standards, outlawing any discharges that changed the flora or fauna of a receiving body. Now he could see that something in the water of the Everglades was changing the River of Grass into a sea of cattails, crowding out the native sawgrass, unhinging the native food web, making the marsh smell like rotten eggs. Unless that something was removed, restoring more natural water flows and quantities would only accelerate the poisoning of the Everglades. "Where do you think it's coming from?" Reed asked.

Dineen pointed north, toward the immense green carpet of sugarcane dividing Lake Okeechobee from the Everglades.

SAVE OUR EVERGLADES was mostly pain-free environmental restoration, except for the taxpayers who footed the bill. But if sugarcane runoff was polluting the Everglades, that would be a harder fix. No one had ever gored the sugar industry's ox.

In the 1980s, Big Sugar joined Big Tobacco and Big Oil among the ranks of reviled American industries—a symbol of corporate welfare, backroom politics, and greed, attacked in countless editorials about "sugar daddies" and "sweet deals." The stereotype was crude, but it wasn't really unfair. Thanks to lavish campaign donations and A-list lobbyists, Big Sugar was one of the most powerful industries in Washington, and was rivaled only by real estate as the most powerful industry in Tallahassee, even though it accounted for less than one-tenth of the state's agricultural output. Its profits were a direct result of its outsized clout—especially in the Everglades, where the industry owed its existence to government support.

Big Sugar received no direct subsidies, as its army of spokesmen constantly pointed out, but it depended on federal import quotas, tariffs, and price supports that cost American consumers as much as $2 billion a year. Florida's growers also relied on a federal program to import their labor pool of 10,000 impoverished West Indian cane-cutters; the industry was notorious for mistreating them, withholding their wages, and deporting any who dared complain. The growers also reaped the benefits of the C&SF project, which irrigated their fields in the dry season and drained their fields in the rainy season. They received more than half the project's water releases, while paying less than one percent of the district's taxes. Meanwhile, state and federal research scientists all helped the industry conserve soil, eradicate pests, and breed more profitable cane at taxpayers' expense.

The recipients of all this largesse were not exactly the small family farmers that Willie Nelson and Bob Dylan were singing about in their Farm Aid concerts, and their farms were a far cry from the ten-acre tracts originally planned for the Everglades. The agribusinesses U.S. Sugar and Flo-Sun controlled more than half the Everglades Agricultural Area's 450,000 acres of cane fields, and raked in more than $100 million a year from Uncle Sam. Their executives—especially the Fanjul brothers, the jet-setting Cuban exiles who ran Flo-Sun—became black-hat emblems of the so-called Decade of Greed. "The closest most of them got to the actual crop were the cubes they dropped in their coffee at the Bankers' Club," the novelist Carl Hiaasen wrote in *Strip Tease,* a thinly veiled satire of the Fanjuls and the politicians who did their bidding. "The scions of sugar growers wouldn't be caught dead in a broiling cane field."

Critics also lambasted "sugar barons" for rotting America's teeth, expanding America's waistlines, and even fueling America's drug habits. (The argument was that U.S. protection of domestic sugar induced foreign farmers to grow coca and marijuana instead.) The grandiose Palm Beach mansions of the Fanjuls—and their posh resort in the Dominican Republic, Casa de Campo, where Michael Jackson married Lisa Marie Presley—were contrasted with the squalor of sugar towns such as Belle Glade, which was so racked by poverty and AIDS that foreign-service trainees were sent there to prepare for the Third World. Sugar growers joked that their industry had become the second Evil Empire, just behind the Soviet Union, and that their neighbors were shocked they didn't have horns and tails.

It was in this context that environmentalists began to accuse the sugar industry of damaging the Everglades. Marjory Stoneman Douglas wrote Governor Graham to express her "violent conviction" that the entire Everglades Agricultural Area should be returned to nature in order to save the River of Grass. Sugar barons became the scapegoats of choice in green-group fund-raising appeals, caricatured as scheming villains who were killing the Everglades. "Those of you who may not know the power of this industry, consider that Florida sugar interests alone contribute more money to political candidates than corporate giant General Motors," one Audubon Society solicitation said. Environmentalists tended to exaggerate the sins of Big Sugar, but the industry really did contribute more than General Motors. And it really was damaging the Everglades.

The Everglades Agricultural Area, after all, was in the middle of the Everglades. Lake Okeechobee had once spilled into the sawgrass, but now the cane fields blocked its path like a giant clot, choking off the original sheet flow. Drainage of the EAA also lowered the region's water table and depleted its soils, while Big Sugar's water demands exacerbated droughts and threw off the ecosystem's natural rhythms. Growers wanted their land dry when Mother Nature wanted it wet, and vice versa.

Big Sugar's impact on water quality was less obvious than its impact on water quantities and flows. But scientists like Art Marshall recognized that one source of Lake Okeechobee's smothering algal blooms, along with Kissimmee valley cattle pastures that poured manure down the C-38 Canal, were cane fields that back-pumped irrigation water into the lake—as Douglas put it, "along with all the pesticides, fertilizers, dead cats and old boots that the water had absorbed." Johnny Jones and his Florida Wildlife Federation filed a lawsuit, and in 1979, the growers were ordered to stop back-pumping their runoff north into the lake. Instead, they started pumping it south into the water conservation areas. Into the Everglades.

MALCOLM "BUBBA" WADE, a stocky U.S. Sugar executive with a handlebar mustache that makes him look like a politician in a Thomas Nast cartoon, readily acknowledges that before the 1980s, sugar growers paid almost no attention to the environment, because they had no economic incentive to do so. U.S. Sugar was partly owned by the Charles Stewart Mott Foundation, a charity that focuses on environmental issues, but in 1981, the company's official history of its first fifty years did not even mention the environment. After all, government agencies had created the Everglades Agricultural Area, and had promoted its conversion to sugar fields to reduce American reliance on Cuba. The government provided the industry's irrigation water, flood control, and research scientists, and never bothered to regulate its runoff. Big Sugar was under no pressure whatsoever to rein in its pollution.

Of course, Big Sugar's political contributions had a lot to do with its lenient treatment; the water management district's sugar-friendly board had been blocking pollution studies for years. Nathaniel Reed finally got the studies funded after his eye-opening visit to the cattails, and the science soon demonstrated that phosphorus from the sugar fields was polluting the Everglades. But the board still refused to do anything about it. In any case, phosphorus wasn't a toxic chemical; it was a natural nutrient, so vital to plant development that it was found in most commercial fertilizers. And it wasn't filthy sludge; even the dirtiest sugar runoff looked relatively clear, with only about 200 parts per billion of phosphorus, less than a thimble in an Olympic-sized pool. Sugar growers still argued that phosphorus was good for the environment, because it made things grow. At public events, they sometimes drank their runoff to prove it was harmless.

But that runoff wasn't harmless to the Everglades, because the things that extra phosphorus made grow generally didn't belong in the marsh. The Everglades was "phosphorus-limited," with flora and fauna peculiarly adapted to a nutrient-starved environment, and ill-suited to compete when even minute amounts of phosphorus became available. And those thimbles added up; the agricultural area pumped 100 tons of phosphorus a year into the Loxahatchee refuge, fertilizing the march of the cattails.

The scientist who best documented this was Ronald Jones, a nerdy young Florida International University microbiologist who was a devout adherent of an Amish-style sect called Apostolic Christianity, and believed God had sent him to Florida to save the Everglades. (He also believed God had chosen his wife; he asked a church elder to arrange their marriage before they went on a date.) To Jones, the lessons of microbiology were the lessons of God: Everything visible depends on the invisible. In the Everglades, he believed the explosion of cattails was foreshadowed by subtler shifts in bacteria and other microscopic organisms hidden in periphyton mats, the microbial mush at the bottom of the Everglades food chain. It looked like pond scum, but to Jones it was the real canary in the coal mine, and the path to salvation for the Everglades.

Even before he finished his phosphorus studies, Jones declared the maximum concentration acceptable in the marsh: a mere 10 parts per billion (ppb), the equivalent of a penny to a millionaire. He infuriated sugar growers and their political supporters with his cocky pronouncements, but he proved that when infinitesimal amounts of phosphorus were added to the Everglades, soils became saturated, periphyton mats disintegrated, spinach-like algae proliferated, and dissolved oxygen was sucked out of the marsh. The diversity of invertebrates and tiny fish plummeted, and sawgrass grew abnormally tall. Eventually, the sawgrass marshes were displaced by cattails so thick that fish and birds couldn't swim or land in them, much less feed in them. It all began with the unseen microbes.

The science also proved that the "nutrient front" of phosphorus-saturated soils directly followed the path of sugar runoff from the Everglades Agricultural Area. That front was advancing like the Blob, spreading about five new acres of cattails a day. Soon, one-fifth of the Loxahatchee refuge was infested with cattails, while Jones documented microbial changes as far south as Everglades National Park. And once phosphorus became entrenched in the soil, it was almost impossible to get out. "Cattails were the grave markers on the Everglades, and the periphyton changes were the warnings that death was on the way," Jones says. "But nobody wanted to fight Big Sugar—except Dexter."

Dexter Lehtinen was the new U.S. attorney in Miami, a forty-two-year-old Vietnam veteran with the ferocity of a bull shark. Lehtinen was not the kind of prosecutor who shied away from fights. He went looking for them.

No Guts, No Glory

LEHTINEN WAS ANOTHER CHILD of the swamp, a hypersmart and hyper-intense Eagle Scout who had grown up near Homestead, hunting, fishing, and camping in Everglades marshes and hammocks. Buffalo Tiger, the Miccosukee Indian leader, was one of his neighbors, and used to store milk in his family's icebox when Dexter was a boy. From his father, a carpenter who had inspected Liberty and Victory ships during World War II, Lehtinen inherited a lifelong belief in hard-and-fast rules. "Sometimes, Dad would reject a ship because the welding wasn't right, and his bosses would say: 'Come on! We need that ship. There's a war on,'" Lehtinen says. "But Dad wouldn't budge. His job was to say if the welding was right. It was black and white. There were rules, and you followed them."

Lehtinen served as an Army paratrooper in his own generation's war, leading a platoon of Rangers in Vietnam. He lost a chunk of his face to shrapnel during the Laos invasion of 1971, then spent the next eighteen months in a hospital bed while doctors reconstructed his jaw with bone from his hip. He lost his sight in one eye, and the left side of his face is still sunken and inert. But he remained as combative as ever, erupting in fury when he heard the future senator John Kerry's congressional testimony accusing his fellow veterans of war crimes. More than three decades later, Lehtinen would spend his own money to produce his own ads attacking Kerry's campaign for president.

Lehtinen's war wounds healed slowly, but they never dampened his ambition. He earned two master's degrees from Columbia, finished first in his class at Stanford Law School, then worked as a federal prosecutor in Miami. In 1980, he was elected to the Florida legislature as a law-and-order Democrat, although he became a Republican after marrying one of his GOP colleagues, Ileana Ros, another firebrand who would become the first Cuban-American congresswoman. Lehtinen looked like a cannonball, sounded like a machine gun, and made waves like a tropical hurricane, crusading against bail bondsmen and lenient judges with a righteous anger rarely seen in go-along, get-along Tallahassee.

Lehtinen was also enraged by the pollution of the Everglades, and he decided to write a law to stop it. But after researching the issue, he decided that Florida already had strict water quality laws. They just weren't being enforced. When he asked state officials why at committee hearings, they claimed that enforcement was too expensive, too labor-intensive . . . it was always too something. "I never understood that," Lehtinen says. "I thought the whole point of the rule of law was that you had rules, and people followed them. When rules are ignored, you get a My Lai massacre."

In June 1988, President Reagan named Lehtinen the top federal prosecutor in south Florida. He informed his staff that its new motto would be "No Guts, No Glory," and posted a quotation that summarized his philosophy on his office wall: "On the plains of hesitation lie the blackened bones of countless millions who at the dawn of victory lay down to rest." He carried around a plastic AK-47 to emphasize his commitment to the drug war, and personally took over the prosecution of Panamanian dictator Manuel Noriega. And on his first day of work, he told a deputy he intended to do something about the Everglades. He began meeting secretly with Everglades hunters, Everglades environmentalists, and especially the superintendent of Everglades National Park. "I had always said the laws ought to be enforced," Lehtinen says. "Well, now I was the law. It was up to me."

Lehtinen knew his new bosses in the Reagan administration would never approve legal action against Florida and Governor Bob Martinez, a Republican whose commerce secretary was Vice President George H. W. Bush's son Jeb. He figured there was even less appetite for a war against the sugar barons; Flo-Sun president José "Pepe" Fanjul was a top donor to Vice President Bush's presidential campaign. So he waited until a few weeks before election day, when Bush was pledging to become "the environmental president," and was running ads attacking Massachusetts governor Michael Dukakis over the filth in Boston Harbor. Lehtinen then filed a lawsuit in Miami without informing his superiors in the Justice Department, an unprecedented October Surprise.

Lehtinen's complaint accused Florida of failing to enforce its own water quality laws by allowing Big Sugar to pollute the Everglades, and breaking contractual promises to help protect Everglades National Park and the Lox-ahatchee refuge; it did not cite the Clean Water Act, which would have required formal approval from Washington. The political fallout was swift and intense. Governor Martinez, who had just recommended Lehtinen for his new job, petitioned the Justice Department to overrule him and drop the case, arguing that litigation would hamstring Florida's efforts to repair the Everglades. Big Sugar also lobbied to get the lawsuit withdrawn, and Lehtinen fired. Even Army Corps officials urged Justice to drop the suit, saying it would damage their agency's close relationship with the water management district.

Lehtinen was summoned to Washington, where his bosses savaged him as a rogue prosecutor and a disloyal Republican. But Lehtinen had forced them out on a limb, and sawing it off would have been a political fiasco. Lehtinen was never formally nominated for his job, and he was hounded by internal Justice Department investigations for the next four years, but he was allowed to pursue his phosphorus case.

To LEHTINEN, *U.S. v. South Florida Water Management District* was a clear-cut case about dirty water. Two valuable federal properties were being damaged, because Florida was shirking its responsibility to prevent the fouling of the Everglades. With noxious cattails rampaging across the marsh, and with Ron Jones serving as his expert witness, Lehtinen thought Case #88-1886 was a slam dunk.

But Florida officials were furious about being dragged into court like criminals, and they had no intention of admitting guilt. They did not see how the U.S. government could blame Florida for the decline of the Everglades, when the U.S. Army had reconfigured the ecosystem and created the agricultural area in the first place. So the state hired a New York law firm with a reputation for scorched-earth litigation—Skadden, Arps, Slate, Meagher & Flom—and began to contest every allegation and stonewall every document request. In response, the feds tried to out-Skadden Skadden, demanding fifteen years' worth of memos and calendars, deposing scores of state employees. The result was a nasty legal quagmire funded by the public, with dozens of attorneys bickering over half a million pages of documents, and quibbling over issues as mundane as where to hold depositions. By 1991, Skadden had billed Florida taxpayers $5.5 million, including $35 an hour for its copy boys, and one $17 charge for a courier to pick up the morning paper. "The litigation thus far has failed dismally to stem the decline of the Everglades," *Legal Times* reported. "Instead, it has descended into a petty slugfest between two powerful, stubborn armies of lawyers." Even Reed—who had clamored for a cleanup of the Everglades after his visit to the cattails, and had denounced his own board as "a hand-maiden to agriculture"—called the lawsuit "an absolute disaster."

"Millions of dollars have been spent over issues that could have been set-tled by men and women of good intentions," he said.

This was the heart of the matter: Could well-intentioned public servants be trusted to fix the Everglades? Well-intentioned public servants such as Reed and Senator Graham believed they could. In fact, they believed that cooperation and consensus was the only path to solving the ecosystem's problems, since politicians and sugar growers would have to implement the solutions. There was no way a judge could wave a wand and clean up the Everglades; ultimately, there had to be a deal. And the Martinez adminis-tration was already floating vague cleanup plans to try to get rid of the law-suit, proposing some phosphorus-reduction requirements for sugar farmers, and the conversion of some sugar fields into artificial phosphorus-filtering marshes. Reed and the Audubon Society's Charles Lee—a lobbyist whose penchant for compromise earned him the nickname "Let's Make a Deal" Lee among fellow environmentalists—endorsed one plan that would have lim-ited the sugar industry's liability to $40 million.

But Lehtinen did not believe in consensus. He believed in the rule of law, and in Ronald Reagan's arms-control dictum: Trust, but verify. He was glad to hear well-intentioned public servants say they wanted to clean up the Everglades, but well-intentioned public servants had let it get dirty in the first place. Lehtinen would accept a deal only on his terms, with strict man-dates and deadlines, backed by a court order. And he couldn't help but notice that in court, Florida's well-intentioned public servants still refused to admit there was anything to clean up. Even the litigious carpet-bombers at Skadden had publicly advised the state to pursue a settlement, but after Big Sugar protested, the district's board—led by one of south Florida's best-connected lawyers for agricultural interests—had rejected their advice.

Lehtinen's case was about dirty water, but his subtext was dirty politics. Big Sugar had a stranglehold on the state, and he wanted to break it. He often quoted Mark Twain: "Water flows uphill—towards money." Florida was never going to get serious about protecting the Everglades as long as

Big Sugar was calling the shots, because even the best-intentioned Florida politicians went wobbly with sugar. When the industry's subsidies came under fire in 1990, Senator Graham—the son of a sugar grower—argued that they were necessary to finance the cleanup of the Everglades. "Because the health of the Everglades is inextricably linked to the sugar industry's economic stability," he said, "Florida sugar cane fields are an integral component of the Everglades ecosystem." He might as well have called the *Exxon Valdez* an integral component of the Prince William Sound ecosystem.

Big Sugar fought off congressional attacks on its subsidies, but the growers recognized Lehtinen's challenge as an equally grave threat to their bottom line. As the lawsuit dragged on, they launched a public relations blitz, warning that 38,000 jobs were being jeopardized to save birds, grass, and microbial slop—even though Flo-Sun, tired of fighting lawsuits over its labor abuses, was preparing to replace all its cane-cutters with mechanical harvesters, and U.S. Sugar would soon follow suit. Growers complained that runoff from cattle ranches and vegetable farms contained more phosphorus than theirs, which was true, and that Evian water would fail the Ron Jones phosphorus standard of 10 parts per billion, which was also true. They trumpeted the findings of their own scientists, who concluded that the Everglades could tolerate phosphorus levels as high as 50 ppb, which was not true at all. They warned that strict phosphorus enforcement would mean the end of agriculture in the Everglades, shattering the dreams of Hamilton Disston, Napoleon Broward, and Spessard Holland. "I don't believe it," Lehtinen told the *Herald*. "But if those are the terms, I say it's an easy choice. Save the Glades and let agriculture move to Wisconsin."

ON MAY 21, 1991, a new lawyer appeared in a Miami courtroom to represent the state of Florida, a lanky sixty-one-year-old with a folksy drawl. "If it please the court, for the record, I am Lawton Chiles," he said. "I am the governor of Florida."

Chiles had a lot in common with Senator Holland, his boyhood idol and political mentor. Like Holland, he was an avid turkey hunter from a piney-woods central Florida town. He was also a consensus-builder who believed that reasonable people could always find common ground, and a wily politician who positioned himself as a friend of nature as well as a friend of agriculture and development; he had succeeded Holland in 1970 after walking the entire length of the state during his campaign. But "Walkin' Lawton" never got used to the partisan nastiness that had fractured Washington since Holland's departure, and quit the Senate in 1988 after a bout with depression. With the help of Prozac, Chiles returned to politics to unseat Governor Martinez in 1990. He hired Carol Browner—a thirty-five-year-old Miami environmentalist who was serving as an aide to one of America's greenest politicians, Senator Al Gore of Tennessee—to be his top environmental official, and promised the Everglades Coalition that he would end the war over the swamp. He also fired the take-no-prisoners litigators of Skadden, Arps, which is how he ended up in court that May morning.

The governor's closing remarks in that Miami courtroom have become the stuff of legend in Florida, but his initial arguments were straight out of the Skadden playbook. He assured Judge William Hoeveler that he had come to make peace, but he wanted peace on the state's terms. He urged the judge to suspend the lawsuit for a year, so that Florida's scientists could start cleaning up the Everglades instead of wasting time in depositions. Chiles claimed he had already "demonstrated beyond any shadow of a doubt" that his administration could be trusted to do the job: "Obviously, our task is far from over, but I am absolutely committed to finishing the task of saving the Everglades."

Finishing the task? Chiles had barely started. He had floated a cleanup plan with 17,000 acres of filter marshes, but that was even less than the last Martinez plan. And it was only a plan. Lehtinen rose to deliver his rebuttal, brandishing a glass of water he had scooped out of an Everglades canal. This case, he said, was not about good faith, or trust, or even plans. It was about the water in his glass: "We sued over that water because it is *dirty water.* We will be only satisfied when all of the plans have an effect on the water that's in that glass." Lehtinen welcomed the governor's promises, but Martinez had made similar promises. If the state's lawyers were so eager to clean up the water, he asked, "why won't they stand up at this podium and say that this water is dirty?"

Judge Hoeveler had clearly decided to deny the stay, but he asked whether anyone had anything to add. Like a quarterback calling an audible, Chiles suddenly abandoned his strategy. "I am ready to stipulate today that the water is dirty!" the governor blurted. "I am here and I brought my sword. I want to find out who I can give that sword to!"

Jaws dropped throughout the courtroom, especially around the defense table. "What I am asking is to let us use our troops to clean up the battlefield now, to make this water clean," Chiles continued. "We want to surrender!"

Two months later, Lehtinen and Chiles announced a settlement, establishing strict phosphorus limits and mandating the largest nutrient removal project in history. Florida would have to build 35,000 acres of filtration marshes in the Everglades Agricultural Area by 1997, and probably a lot more by 2002. By 1997, runoff from the EAA would have to meet an interim standard of 50 parts per billion, and by 2002, it would have to meet a final limit that no longer altered the Everglades—presumably around 10 ppb. The deal did not specify who would pay for the cleanup, or what the precise numerical limit would be. But it set up a science program to determine that limit, and Judge Hoeveler approved the settlement in a consent decree, retaining his right to intervene in case of state backsliding. It seemed like the end of the water quality wars, or at least a ceasefire. "We are no longer going to spend millions of dollars litigating this," Chiles said.

BUT THE LITIGATION wasn't over yet. Big Sugar hadn't surrendered its sword. And its lawyers began hacking away at the settlement like canecutters at harvest time.

The agreement would have required the growers to reduce their own phosphorus releases at least 25 percent. But they filed at least thirty legal challenges in state and federal courts, alleging violations of everything from the National Environmental Policy Act to Florida's open-meeting laws. They claimed the closed-door settlement had short-circuited the political process, which would have required environmental reviews, public participation, and industry input for such a massive initiative. They argued that Florida politicians should decide their fate, not lawyers, scientists, and bureaucrats.

Of course, they got along well with Florida politicians. George Wedgeworth, the son of Everglades pioneers and the founder of an Everglades sugar cooperative, never forgot the advice Senator Holland once gave him: There are 200 million Americans who care about the Everglades, and only 100 sugar growers, so solve your problems in Florida. The growers were skeptical of the filter marshes, which had never been tried on this scale, and they were afraid the settlement would allow government officials to keep seizing their land until the filter marshes produced results; in their subsidized industry, every acre was a virtual license to print money. The growers were also dubious that they would be able to slash their own phosphorus outputs significantly with on-farm management reforms. Wade says that when the industry's friends at the state agricultural institute suggested that a 25 percent reduction was a reasonable target, "we were ready to string them up."

Big Sugar had a lot to lose. Economists estimated its annual profits at $238 per acre, and the industry's own scientists predicted the state would need 100,000 acres for filter marshes. The lawsuit was also emboldening environmentalists, who stepped up their attacks on federal sugar subsidies, and called for a buyout of the entire agricultural area. "The sugar barons aren't going to let it go easily or without a fight," the Audubon Society told donors. *"It will surely be the biggest fight in the history of saving the Everglades."* During a public debate against a frail but feisty Marjory Stoneman Douglas, still crusading against sugar at 102, one executive held a paper bull's-eye over his heart to illustrate his industry's predicament. "The environmentalists wanted a virgin, pristine, no-people Everglades," Wedgeworth recalled. "That didn't leave a lot of room for us." One executive warned an Audubon leader that the assaults on sugar subsidies reminded him of an old organ grinder's saying: You can play with the chain, but don't fuck with the monkey. "You guys are fucking with our monkey," the executive said.

So the farmers took the Skadden approach to litigation, demanding thousands of documents, deposing hundreds of witnesses, grilling one district manager for two weeks, waiting for the government to tire of the expensive legal battle. By 1993, the cleanup plans were still tied up in court. Cattails were spreading, and the water of the Everglades was as dirty as ever. Florida was spending millions of dollars to defend a deal it had once spent millions of dollars to avoid. So far, the settlement hadn't settled anything.

"Bottom line: Gridlock reigns," reported the magazine *Florida Trend*.

"The Clinton Administration Delivers!"

THE DEATH OF THE EVERGLADES ECOSYSTEM, after years of rumors, was less of an exaggeration than ever. The cattails, as Jones said, were just the markers on the grave.

The most pathetic symbol of decline was the vanishing Florida panther, the official state mammal, the star of Florida's best-selling license plates. The last thirty or so panthers were so inbred that males were being born without testicles, and wildlife officials were importing Texas cougars to try to diversify their gene pool, raising questions about whether the next generation of panthers would really be panthers at all. The leading cause of panther mortality was car crashes, but the underlying problem was habitat loss: The cats needed room to roam, as much as 200 square miles for an adult male, and that room was being ripped up for treacherous four-lane highways, limestone mines that produced the raw materials for those highways, and master-planned communities with ironic names such as Wildcat Run and The Habitat.

The southeast coast was continuing its westward expansion into Everglades suburbs such as Weston, which jutted into the edges of Shark Slough about twenty miles west of Fort Lauderdale, attracting 50,000 suburbanites in less than a decade. And now the long-dormant southwest coast was sprawling into the panther country east of Naples and Fort Myers. Regulators at the Army Corps, the unlikely guardians of the Clean Water Act, approved more than 99 percent of all applications to develop in Florida wetlands. Biologists at Fish and Wildlife, the overseers of the Endangered Species Act, tried to warn that a number of projects would jeopardize the panther's survival—including several subdivisions, a massive rock mine, and Florida Gulf Coast University—but they were always overruled by their bosses. "There was no place left for the cats to go," said Andy Eller, a Fish and Wildlife biologist who was fired after he tried to object to several gated communities in panther habitat. "It was just a development free-for-all."

The entire ecosystem was stressed. Lake Okeechobee's sandy bottom turned to mud. Floods and fires erased nearly half the tree islands in the Everglades, and health officials had to post mercury warnings for fish caught in the Everglades. There were now 1.4 million acres of invasive melaleuca, Brazilian peppers, and Australian pines across southern Florida, along with nonnative animals like Nile monitor lizards, Burmese pythons, and so-called evil weevils that devoured rare bromeliads. Along the Gulf Coast, red tides the size of small states were killing dolphins and manatees, which were also under siege from the region's fast-growing flotilla of speedboats. The St. Lucie estuary was buried in a black gunk that scientists called "flocculent ooze."

The most telegenic disaster was the collapse of Florida Bay, the magical estuary at the bottom of the ecosystem, the favorite fishing hole of novelist Carl Hiaasen, Red Sox legend Ted Williams, and the first President Bush. Vast swaths of its sea grasses, sponges, and mangroves died, and its gin-clear waters turned a slimy pea green. Its pink shrimp, spiny lobster, and stone crab catches crashed. The lush coral reefs at its edges, the only living reefs in the continental United States, decayed into gray moonscapes, stripped of their colorful vegetation and glittering fish. A seventh-generation Floridian named George Barley—a hard-nosed Orlando developer who had a fishing getaway in the Keys, and became an indefatigable Everglades Coalition activist after his beloved bay began to deteriorate—called it "an environmental collapse unprecedented in Florida history." Ron Jones compared the bay to a baby strangled in its crib. "Mother Nature is having a nightmare," lamented *Outside* magazine, "and the nightmare is the Everglades."

STILL, WHEN THE COALITION met in Tallahassee in February 1993, many Everglades activists were euphoric. That's because Bill Clinton had moved into the White House. The Man from Hope was the first Democratic president in twelve years, and the first baby boomer president ever. He was a pro-business centrist who had ignored the environment as governor of Arkansas, but he had spoken eloquently during his campaign about the links between a healthy environment and a healthy economy. And his vice president was his fellow boomer Al Gore, the darling of the environmental movement, a politician so green that President Bush had labeled him "Ozone Man." In his best-seller *Earth in the Balance,* an ecological manifesto inspired by *Silent Spring,* Gore had not only predicted the demise of the internal combustion engine, he had argued that sugar price supports accelerated the destruction of the Everglades. Now Gore was the administration's point man on the environment, and two top administration officials were south Florida environmentalists. Carol Browner had left the Chiles administration to run the Environmental Protection Agency. And Janet Reno—a true swamp rat who had grown up in the palmetto scrub west of Miami, where her pet dog was killed by a rattlesnake—was now the attorney general; her sister, Maggy Hurchalla, was one of the most respected activists in the Everglades Coalition. Nathaniel Reed, now sixty years old, told reporters he had never felt so upbeat about the River of Grass. "Christmas has come early for protectors of the ailing Everglades. At least it seemed that way at the annual Everglades Coalition conference," one observer wrote in the *St. Petersburg Times*. "Long accustomed to dealing with hostile politicians and no-can-do bureaucrats, Glades lovers walked around wearing surprised grins."

The star of the conference was the new secretary of the interior, former Arizona governor Bruce Babbitt, the former head of the League of Conservation Voters. He came to Tallahassee on his first official trip, and immediately decided to make Everglades restoration one of his top priorities. Babbitt sensed an opportunity to make history after Jim Webb of the Wilderness Society, a friend of his from Democratic circles in Arizona, introduced him to Colonel Salt of the Corps. Together, the environmentalist and the engineer lobbied Babbitt to push the Restudy of the C&SF project, describing it as a unique chance to revive a dying ecosystem and replenish dwindling water supplies all at once. Babbitt then delivered an inspired speech about ecosystem-wide restoration, declaring the Everglades "the ultimate test case," a referendum on man's ability to live lightly on the land. He was interrupted three times by standing ovations.

Babbitt had grown up in a prominent ranching family in Arizona's wide-open canyon country, spending his boyhood fishing, rafting, hiking and collecting rocks; he was deeply influenced by the writings of conservationists such as Aldo Leopold. He later marched for civil rights in Selma, studied geophysics as a Marshall Scholar, and earned a law degree at Harvard. He was a vision guy, and often talked about thinking "on a landscape scale." But for all his environmental passion and intellectual firepower, Babbitt was at heart another consensus politician. As governor, his favorite part of the job had been bringing together divergent interests and hashing out complex deals. In his new job, he was looking forward to sitting down with loggers and conservationists in the Pacific Northwest to find common ground over spotted owls and old-growth forests. Babbitt sensed that in the Everglades, divergent interests were already starting to come together and were thinking on a landscape scale. They just needed leadership to guide them to consensus. "That's where I'm at my best," he later explained. "I could see how the human complexity of the Everglades reflected the complexity of the ecosystem itself. And I could see the pathway."

The pathway, Babbitt realized, would have to go through the Corps. He didn't trust the dredge-and-dike Corps to restore a piece of furniture, much less an ecosystem. But the C&SF was a Corps project, under the Corps budget, and he couldn't think of a way to revamp it without the Corps in charge. So he decided to try to co-opt the Corps. The interior secretary is not usually a link in the Pentagon chain of command, but Babbitt summoned the agency's leaders to his office. Babbitt informed them that the administration was deeply committed to the C&SF Restudy—which at the time was true only of its interior secretary—and gave them a pep talk about how restoring ecosystems could be the future of the Corps. "You're going to be heroes!" Babbitt told them. They agreed to start the Restudy immediately, with an accelerated timetable. Babbitt also created a task force exclusively devoted to Everglades restoration, a deft bureaucratic move to get the Corps and other federal agencies singing from the same sheet of music—with Interior as the conductor. He assigned Assistant Interior Secretary George Frampton, the former head of the Wilderness Society, to chair the task force. He later hired Colonel Salt as its executive director, his symbolic way of announcing a new direction for the Corps, a new kind of water project for America.

Even though Interior's responsibilities in lower Florida were strictly environmental—four national parks, sixteen wildlife refuges, sixty-eight endangered species—Secretary Babbitt did not want a strictly environmental plan. He had mediated enough western water disputes to know that a strictly environmental water plan would be politically untenable. He wanted a balanced plan, a win-win for people and the Everglades, produced by partnerships among federal, state, and tribal officials as well as south Florida's interest groups. Governor Chiles had the same goal, and set up the Governor's Commission for a Sustainable South Florida to seek consensus solutions for the region, bringing together representatives of development, agriculture, water utilities, Indian tribes, environmental groups, and a slew of government agencies. Senator Holland had insisted on unity when he pushed the original C&SF, and Babbitt and Chiles believed unity would be necessary again to reconfigure the C&SF.

The most obvious obstacle to that unity was the phosphorus battle. A long-term strategy was unfolding in Babbitt's mind, and it didn't include years of litigation over a few parts per billion. He didn't see how there could be progress toward consensus while the players were fighting in court. During the coalition's conference, stenographers had been on hand to transcribe speeches for the legal record, even during cocktail hour. That didn't feel like a formula for success. "I knew we could fight sugar until the end of time, and everyone would feel righteous, but nothing would get done," he said later. "There had to be a solution, and sugar had to be a part of it."

A week after the conference, Babbitt met secretly for ninety minutes with Flo-Sun CEO Alfonso "Alfie" Fanjul Jr., the most powerful sugar baron. Babbitt then assigned Frampton to negotiate a cleanup deal with the industry, using a Flo-Sun proposal as a starting point. Babbitt and Frampton had environmental pedigrees, but when it came to phosphorus, they parted ways with the environmental community. They were highly intelligent and highly confident, and they were convinced that environmental protection worked best when no one's ox was gored. "The enviros were obsessed with phosphorus, because they were obsessed with punishing sugar," said Frampton, a former Supreme Court clerk and Watergate prosecutor with degrees from Yale, Harvard Law, and the London School of Economics. "We saw the pollution lawsuits as a diversion, a distraction from the larger restoration of the Everglades. We wanted to move on."

IN JULY, AN EBULLIENT Secretary Babbitt announced that he was ready to move on. At a news conference in the Interior Department auditorium, he unveiled a framework for a deal to settle all of the phosphorus lawsuits. Florida would build at least 40,000 acres of filter marshes, almost enough to cover Washington, D.C., and growers would contribute $233 million to $322 million of the cost over twenty years, depending on how much they reduced their own phosphorus outputs. Water quality was only the first step in Babbitt's plans to rehabilitate the Everglades, but he hailed the plan as "the largest, most ambitious ecosystem restoration ever undertaken in this country."

"The River of Grass has a new lease on life!" he declared.

In a rare public appearance, Alfie Fanjul was just as exuberant. He hailed the framework as "the end of gridlock," proof that what's good for the environment could be good for business as well. "In November, America cast a vote for change," Fanjul said. "Today, the Clinton administration delivers!"

Does it ever, Joe Browder thought. In the quarter century since he had fought the Big Cypress jetport, Browder had served at Interior under President Carter, worked with Native Americans in the Southwest, then returned to Washington as a consultant. But he still loved the Everglades—he had helped his friend Jim Webb write the coalition's twenty-first-century vision statement—and he sensed it being sold out before his eyes.

Alfie Fanjul was the Democrat in a family that covered its political bases. While his brother Pepe had been the vice chairman of President Bush's finance committee, Alfie had cochaired Clinton's campaign in Florida, and had raised $100,000 for the candidate at a dinner in Miami. Now Alfie and his U.S. Sugar counterpart were beaming on an Interior Department stage, boasting about progress and ecological sensitivity. The only environmental representative on stage was Webb, who was an old friend of Babbitt's and a former employee of Frampton's, but who looked like he had just eaten some bad fish.

As Browder listened to the details of the "Babbitt Agreement," he could see why the sugar executives looked so pleased. They were about to escape a federal court order, extend their cleanup deadlines, and foist most of the bill for their pollution onto taxpayers. During the Q&A for reporters, Browder slipped into a seat a few rows from the stage, and assumed the role of the skunk at Babbitt's garden party. "It's an absolute betrayal, Bruce, and it won't stand," Browder said. Babbitt's aides hastily shut down the news conference, but Browder cornered him before he left. "This will hurt the Everglades," Browder said. "How could you agree to it, Bruce?"

"Well, that's my job, Joe," Babbitt replied. "To find compromise."

The Everglades Coalition attacked the agreement as a terrible compromise, a fuzzy retreat from the firm mandates of Lehtinen's original settlement, a "sweetheart deal" that would allow Big Sugar to use the Everglades as a sewer for the foreseeable future. The developer-turned-activist George Barley, outraged that sugar would escape the full cost of cleaning up its mess, launched a campaign for a penny-a-pound tax on sugar sales. Even Webb soon turned against the deal. "Somebody listened to sugar," he said.

The Miccosukee Indians—who still lived in a slice of Everglades National Park below the Tamiami Trail and retained hunting rights in Conservation Area 3—also blasted the compromise. The Miccosukees did not trust the government that nearly wiped out their ancestors, and they hired Dexter Lehtinen, who had resigned from federal service, to defend the Everglades against his old employer. Lehtinen noted that the Babbitt Agreement had plenty of specifics about money and marshes, but no guarantees that the water of the Everglades would ever be clean. "It has the potential to be the Munich of the Everglades, buying 'peace in our time' with Big Sugar," he said.

Babbitt was surprised and annoyed by the criticism from environmentalists, who had hailed him as a hero just a few months earlier, and had even urged President Clinton not to appoint him to a Supreme Court vacancy because he was supposedly indispensable at Interior. Now he was suddenly a sellout for jump-starting the largest cleanup in history? Some environmentalists had been willing to accept $40 million from the sugar industry a few years back; shouldn't they be ecstatic about a deal that was six to eight times better? Senator Graham and Governor Chiles supported the deal; were they sellouts, too? Babbitt claimed he hadn't even realized that Fanjul was involved in politics before their meeting. "I just thought he was a big grower," Babbitt recalled. "He didn't tell me he was the Evil Fanjul, cradling Florida's destiny in his hands."

In January 1994, Babbitt returned to the Everglades Coalition in Miami, and again called the Everglades "the single most important test case of whether we can restore an ecosystem." This time, there were no standing ovations, and Babbitt barely pretended he was among friends. His aides had promised he would announce a major initiative, but he didn't. President Clinton had just awarded Marjory Stoneman Douglas the Presidential Medal of Freedom, but Babbitt refused to meet with her group, Friends of the Everglades. He did make a point of chatting with migrant workers at a Farm Bureau rally protesting the conference, and promised to try to protect their jobs. After six months at Interior, Babbitt was already sick of activists who fixated on ideological purity, who made the perfect the enemy of the good, who saw concessions as proof of conspiracy.

But operatives like Browder and Barley didn't think they were making the perfect the enemy of the good. They didn't see anything good about the Babbitt Agreement. Big Sugar was illegally polluting the Everglades; why cut a deal with a scofflaw? Interior's primary mission was protecting national parks and refuges; how could an interior secretary compromise on their biological integrity? The federal consent decree had ensured clean water by 2002; the Babbitt Agreement didn't seem to ensure it at all. "We are dealing with people who have shown their ill intentions," a Clean Water Action activist wrote in a memo to Browder. "They have consistently demonstrated that their intentions are about seeking compromise, AT ANY COST."

* * *

Everglades Forever

GOVERNOR CHILES AND Secretary Babbitt had both staked their reputations on a compromise, Chiles with his courtroom surrender, Babbitt with his tentative agreement. But the sugar growers walked out of the final settlement talks. The parties could only agree to seek a deal based on the Babbitt Agreement in the Florida legislature, Big Sugar's home court, a venue where lobbyists joked that cane juice flowed out of the drinking fountains.

In February, Chiles introduced a cleanup bill called the Marjory Stoneman Douglas Act, and the sugar industry began spreading even more campaign cash than usual around Tallahassee. It hired three dozen lobbyists to work the bill, including the governor's former chief of staff and two former House speakers. Determined to beat sugar at its own game, Barley approached Florida's best-connected law firms about representing the Everglades Coalition, but most of them had sugar clients, and when he found one that didn't, a grower snagged the firm a few days later. The sugar industry also lined up support from agriculture groups, business interests, and labor unions, which warned legislators that strict phosphorus standards would endanger machinist jobs in sugar mills. Barley was an in-your-face multimillionaire with sharp elbows and powerful friends; he wasn't used to losing. But he had never faced a juggernaut like Big Sugar.

It was clear that the legislature would not pass anything without sugar's approval, and that the Chiles and Clinton administrations were determined to pass something. Despite the vitriolic anti-sugar views of its namesake, much of the Marjory Stoneman Douglas Act was written by sugar's hired guns. And even though the state's failure to enforce its own water quality laws had inspired the original lawsuit, the Clinton administration let the state drive the final settlement. "National Democratic Party operatives . . . tell the White House there is no real controversy, just another case of the environmentalists demanding 100% and getting mad when only 98% is delivered," Browder warned the coalition.

In fact, they were getting less than 98 percent. The Douglas Act delayed the final phosphorus standards until 2006, and included loopholes that left it unclear whether there would even be final standards. The bill included $700 million—less than one-third of it from sugar—for the 40,000 acres of marshes that were expected to cut phosphorus levels to 50 parts per billion, but it had no money and no plan to get to 10 ppb. Lehtinen warned that the bill would gut the water quality agreement he had negotiated in court, relying again on the state's good faith and promises. The scientist Ron Jones, now a consultant to the Miccosukees, believed it would allow the state to declare victory once it hit 50 ppb—which would destroy the Everglades a bit more slowly, but just as surely. "Six bullets kills you, but three bullets also kills you," Lehtinen said.

The environmentalists did have one secret weapon. A month after she was honored at the White House, and President Clinton compared her to Mother Nature, the 103-year-old Douglas wrote to Chiles to request that her name be removed from the cleanup bill: "I disapprove of it wholeheartedly." It was a public relations bombshell, the Everglades equivalent of George Washington demanding that his name be taken off the Washington Monument. But the Clinton administration still endorsed the bill, which was renamed the Everglades Forever Act, and the Florida legislature overwhelmingly passed it.

The Everglades Coalition unanimously denounced the bill, declaring that "the Clinton administration has joined with the Florida Legislature in surrendering to the big sugar corporations that pollute the Everglades." Ron Jones called it "a total and complete disaster." Joe Browder, a lifelong Democrat, told a top White House aide that he had dealt with every presidential administration since Lyndon Johnson's and had never seen one so quick to accept damage to the Everglades. George Barley, a staunch Republican, warned that friends of sugar and enemies of the Everglades would soon pay a political price. "I'm determined to make this not only a national issue but an international one, and I have the money, resources, connections and determination to do it," he wrote to Browder. "I do not want to reveal all of our strategies in this campaign, but there are unpleasant political surprises in store for those on the wrong side of this issue."

Everglades Forever was so unpopular that Chiles's signing ceremony was closed to the public, even though it was held in Everglades National Park. Protesters from Friends of the Everglades were threatened with arrest unless they moved to a nearby parking lot, where they held up a "Killing the Everglades Forever Act" banner, and denounced the bill as "a death sentence for the Everglades." Secretary Babbitt approached the protesters afterward and tried to explain that Florida was launching a cleanup for the ages. But they kept interrupting him, calling the bill a sham. He finally walked away, rolling his eyes in disgust.

EVERGLADES FOREVER DID NOT turn out to be a death sentence for the Everglades, as many environmentalists had predicted. It made the Everglades much cleaner. The sugar industry's gloomy predictions about the inevitable failure of filter marshes were also wrong. So far, the marshes have kept more than 2,000 tons of phosphorus out of the Everglades.

The vigilance of the Miccosukees has been responsible for some of this success; the Everglades saved their ancestors, and they have helped to return the favor. Dexter Lehtinen persuaded Judge Hoeveler to keep his consent decree in place, and the tribe has replaced the U.S. government as the defender of Everglades water quality, litigating to force Florida to live up to its obligations. As a sovereign nation, the Miccosukees also set a powerful precedent by enacting a phosphorus limit of 10 parts per billion for their own slice of the Everglades. Florida eventually followed suit for the rest of the Everglades, although there is still controversy over how the limit will be enforced.

But the Indians were not the only ones doing their part for clean water. The sugar growers, required to cut their phosphorus releases by 25 percent, slashed them by more than 50 percent by using less fertilizer, releasing less water, and cleaning their ditches more often. And the water management district built the first round of filter marshes on time and on budget, winning national engineering awards. "I could never understand why environmentalists weren't cheering," recalled Sam Poole, the district's director under Chiles. "We were cleaning up the Everglades. Why were they making my life so miserable?"

One reason was that it remained unclear how the district intended to achieve 10 ppb by 2006. But the Clinton and Chiles administrations were not thinking that far ahead. They drew two lessons from Everglades Forever: Cooperation was the path to restoration. And the Chicken Littles of the Everglades Coalition could not be trusted. "We didn't take them too seriously after that," Frampton recalled.

But state and federal officials remained committed to the larger goal of restoring the hydrology of the Everglades, while assuring a stable water supply for south Florida's population. With the phosphorus battle settled, they were more eager than ever to devise a consensus plan to Get the Water Right—for people, farms, and the Everglades.

The Everglades Forever cleanup and the Kissimmee River dechanneliza-tion were already two of the largest environmental repair jobs in history, with the Shark Slough and Taylor Slough restoration projects not far behind. But replumbing the C&SF would be a different order of magni-tude. It would be the ultimate test of humanity's ability to make peace with nature, the ecological equivalent of the moon mission. "Everglades Forever was a defining moment," Babbitt said. "But we had our eyes on the bigger prize."

SEVENTEEN

Something for Everyone

The model is consensus rather than confrontation.
—Florida governor Jeb Bush

"Time Is of the Essence"

SOUTH FLORIDA IN THE MID-1990s was everything its founders had imagined and more—a winter playground, a retirement home, a sugar bowl, a melting pot. It was the home of the citrus industry and the cruise industry, Little Havana and Little Haiti, the Professional Golf Association and *The Golden Girls*. Sawgrass Mills Mall, at the edge of the Everglades in Sunrise, was the state's number-two tourist attraction after Disney. Fort Lauderdale was the "Yachting Capital of the World," Islamorada the "Sport-fishing Capital of the World." South Florida was still a magnet for celebrities like Gianni Versace, Rush Limbaugh, and Madonna, but its lure transcended the rich and famous. The Naples area, for example, not only had America's highest concentration of millionaires, including Steve Case and Larry Bird, it had the second-highest growth rate, behind Las Vegas. More than 300 newcomers moved to south Florida every day.

But the region was also on the verge of loving itself to death. The Florida Panthers ice hockey arena was built so close to the Everglades levee that an errant slapshot could almost land in the swamp, neatly illustrating the wall-to-wall sprawl that was wiping out the actual Florida panther. The rapid declines of Lake Okeechobee, Florida Bay, and the coral reefs were pummeling bait shops, dive shops, and motel owners. Aquifers were overtapped, floodplains were overbuilt, and schools were so overcrowded that students in Broward County were lining up for lunch at 9:50 A.M. And Hurricane Andrew's rampage through the Homestead area drove more than 250,000 people out of their homes, a painful reminder of the perils of living in harm's way. As bulldozers plowed deeper into the Everglades, downtowns were dying and suburbanites were idling in traffic; Miami was America's poorest city, and the average south Florida commute doubled in a decade. The entire Miami–Fort Lauderdale–West Palm Beach metropolitan area was becoming an indistinguishable glob of gated communities, Jiffy Lubes, strip malls, Comfort Inns, RV parks, Taco Bells, and clover-leaf interchanges. The regional economy was a kind of ecological Ponzi scheme, dominated by low-wage tourism and construction jobs that relied on the constant pursuit of more people and more development that put more stress on nature. There was agriculture, too, but most of the region's farms were basically prebuilt real estate, subdivisions awaiting their zoning variances. One newspaper published a cautionary vision of south Florida in 2015, featuring dead lawns, $50-a-pound shrimp, government restrictions on washing machine use, and a dead Everglades; six million eco-tourists would no longer visit the region's parks and refuges every year, and water bills would soar. "Even if you could not care less about wood storks, warblers or wood rats, the environmental damage will touch you," the article said.

In the 1960s, Art Marshall had been a lonely voice bellowing that south Florida was going to hell. But in October 1995, all forty-two members of the Governor's Commission for a Sustainable South Florida—including developers, farmers, and bankers, as well as environmentalists—agreed that the region was just about there. "It is easy to see that our present course in South Florida is not sustainable," they warned in a unanimous report.

The destruction of south Florida's ecosystem, the Governor's Commission concluded, was destroying south Florida's quality of life, creating lower water tables, higher flood risks, gridlock, "mind-numbing homogeneity and a distinct lack of place." South Florida was already living well beyond its means, and every new resident was demanding an additional 65,000 gallons of water per year. The region's highway mileage had quintupled in two decades, and could now circle the earth twice; one study concluded that merely maintaining current congestion levels would cost $26 billion over two decades. Hurricane researchers calculated that a repeat of the 1926 storm would cause $80 billion in damage, and warned that evacuation

BY GENE THORP—CARTOGRAPHIC CONCEPTS INC.

Today, half the Everglades is gone, drained for agriculture or paved for development. South Florida has been transformed from a watery wasteland into a fast-growing megalopolis of 7 million residents, 40 million annual tourists, and one-fifth of America's sugar production. Disney World, the Sawgrass Mills Mall, Florida International University, and Burger King corporate headquarters were all built in the natural Everglades ecosystem. But millions of acres have been preserved in dozens of parks and refuges.

routes were dangerously clogged. "Time is of the essence," the commission warned. "If we are to curtail the deterioration and evade further catastrophe, urgent strategic action is needed."

The action the Governor's Commission had in mind was the Restudy, the effort to revamp the C&SF project to make the region ecologically and economically sustainable. The Army Corps was already working on it, but Colonel Terry Rice, a Ph.D. hydrologist who had succeeded Colonel Salt in Jacksonville, asked the commission to design its own conceptual plan first. Rice promised that if the plan made sense, the Corps would adopt it. The Corps was notorious for devising projects behind closed doors, then foisting them upon the public as done deals, but Rice and Stuart Appelbaum, the Corps planner in charge of the Restudy, were convinced business as usual wouldn't fly. "This couldn't be your father's Corps of Engineers," said Appelbaum, who had overseen the agency's first environmental study on the Kissimmee River. "The Corps couldn't tell Florida what to do. The political consensus had to come first."

IT DIDN'T SEEM LIKE a very good time for political consensus.

In Washington, partisanship had become so venomous that the federal government shut down for a week over a budget dispute. Republicans had seized control of Congress in 1994, and House Speaker Newt Gingrich was pursuing his "Contract with America" as a new conservative mandate, while Senate Majority Leader Bob Dole was preparing to run against President Clinton. The environment emerged as one of the nastiest battlegrounds, as the new GOP majority began crusading to roll back environmental regulations—House Majority Whip Tom DeLay compared the EPA to the Gestapo—and Clinton began ripping Republican leaders as anti-environmental zealots. The Republicans showed little inclination to compromise, which was fine with Clinton, who planned to portray them as paradise-paving extremists in his reelection campaign. The one belief the parties seemed to share was that, as Clinton put it, "the era of big government is over," which did not bode well for a multibillion-dollar restoration effort.

The atmosphere in Florida was even more toxic, as the environmental community and the sugar industry went to war over the proposed penny-a-pound tax. Passions were further inflamed after the anti-sugar crusader George Barley was killed in a charter jet crash; his widow, Mary Barley, was so suspicious of Big Sugar that she hired private investigators to probe his death. At Barley's graveside, his fishing buddy and Keys neighbor, the billionaire commodities trader Paul Tudor Jones II, asked the mourners: "Who will pick up the flag?" Mary took over the penny-a-pound fight, and Jones decided to finance the battle to save the Everglades. "Rest Easy," Jones wrote in Barley's funeral book. "I Got Everything Under Control."

Like Barley, Jones was a high-energy, ultracompetitive alpha male. He had been a welterweight boxing champion at the University of Virginia, and E. F. Hutton's youngest vice president at twenty-five. He then launched one of Wall Street's most successful hedge funds, becoming a legend in financial circles when he made $100 million betting against the market before the 1987 crash. He married an Australian model and bought estates in Connecticut, Colorado, Maryland, the Keys, and the Bahamas; he also bought the sneakers Bruce Willis had worn in *Die Hard,* and wore them to make big trades. Now Jones intended to bring his killer instinct to Everglades activism. He thought most environmentalists were nice people with sensitive souls, but useless in a back-alley knife fight. So he shelled out $10 million for a bare-knuckle ad campaign against Big Sugar and any politician who opposed penny-a-pound, including Dole, Graham, and his Republican counterpart from Florida, Connie Mack, a former Fort Myers banker who was the grandson of the Hall of Fame baseball manager. The most brutal ad suggested that Congressman Mark Foley, a freshman Republican whose district included the sugar fields, was drowning deer in the Everglades. It was a cheap shot, but Jones believed the only way to beat a self-interested force like Big Sugar was to play by its rules, which meant no rules.

Growers fought back with their own no-holds-barred campaign, spending at least $25 million attacking the proposed tax and the "environmental extremists" who backed it. They went after Jones as a hypocrite who had paid a record $2 million fine for destroying wetlands on his Maryland duck-hunting reserve. They took out full-page ads in Spanish-language papers comparing Nathaniel Reed to Fidel Castro, and encouraged black pastors in sugar towns such as Belle Glade and South Bay to spread the word that environmentalists were racists. "What's next: Special taxes on golf courses?" one industry ad asked. "Special taxes on toilets that use water?"

Penny-a-pound was the ugliest Everglades battle since the Big Cypress jetport. Senator Graham had always been sympathetic to sugar, but the Everglades Coalition's leaders were so angry when he opposed penny-a-pound that they disinvited him from their annual conference, even though he had personally revived the coalition a decade earlier. Al Gore had lambasted Big Sugar in his book, but Alfonso Fanjul was so angry when the vice president endorsed penny-a-pound that he called the White House an hour later to complain. At the time, President Clinton was in the Oval Office telling an intern named Monica Lewinsky that he no longer felt right about their sexual relationship, but he interrupted the breakup to speak to Fanjul for twenty-two minutes.

"I think it's fair to say that tensions were high," Graham recalled.

MEANWHILE, A NEW ECO-WAR was erupting over a new plan for an Everglades airport, this one at Homestead Air Force Base, nestled between Biscayne and Everglades National Parks. For Joe Browder, history was repeating itself—as tragedy, not farce. It was depressing to fight another airport, but Browder believed the Clinton administration was selling out the parks.

During the 1992 campaign, Clinton had visited the ruins of the base after it was leveled by Hurricane Andrew's 175-mile-an-hour gusts, and had pledged to support its redevelopment. A group of Cuban-American developers led by Carlos Herrera, president of south Florida's powerful Latin Builders Association, then devised a plan to convert the base into a commercial airport. And in January 1996, Dade County awarded the group a no-bid lease, even though it refused to release its financial statements; in classic Miami fashion, the vote to approve the deal was held at 5:40 A.M. Herrera, a Republican who had become a generous donor to Clinton and the Democratic Party, received audiences with the president and key administration officials, and his plan began sailing through the federal bureaucracy. The environmental review approving the airport was so slipshod that its maps did not even identify Biscayne Bay, much less the national park established to protect Biscayne Bay.

The airport seemed like a done deal. Senators Graham and Mack were so eager to fast-track it that they squelched a congressional investigation of its potential impact on Everglades restoration, even though it called for more than 600 flights a day over two imperiled parks. Alan Farago, a Sierra Club activist who was leading the fight to stop it, had to wait four hours outside a $1,500-a-plate Clinton fund-raiser to hand a letter to White House chief of staff Thomas "Mack" McLarty. When he was finally allowed in, he saw the airport's backers scattered all over the hotel ballroom; McLarty assured Farago that he knew all about the airport, and the administration was working on it. "No, I'm on the *other* side," Farago explained.

The airport battle further strained relations between the Everglades Coalition and the Clinton administration, as groups such as the Sierra Club, Friends of the Everglades, and the Natural Resources Defense Council threatened to sue. But the controversy also exposed fault lines within the coalition, with advocates of moderation—especially the leaders of the Audubon Society, Florida's oldest and largest green group—reluctant to waste energy and political capital fighting the administration and Florida's politicians over a lost cause. They did not want to alienate the leaders they hoped to work with to restore the Everglades, or fuel stereotypes of conservationists as litigious people-haters who reflexively opposed economic growth. The Homestead area had struggled even before Hurricane Andrew, and some Audubon leaders believed anti-airport dogmatists were making the coalition look like "eco-terrorists."

But the hard-core airport opponents saw little difference between moderation and capitulation on such a clear-cut, yes-no issue. They saw the airport as a corrupt deal for a handful of well-connected land speculators, the Big Cypress jetport all over again. Browder, who had worked for Audubon during the jetport fight, sent a memo to his old ally Nathaniel Reed, predicting that "when this chapter of Everglades history is written, Audubon will be shamed." Browder blamed the influence of Paul Tudor Jones, who was a family friend of Gore's from Tennessee, and was giving Audubon nearly $1 million a year. In fact, Jones helped fund some anti-airport advocacy as well, but it was true he wasn't looking for a knife fight with the Clinton administration. When it came to politicians and the environment, Jones didn't think anyone was better than Al Gore.

But grassroots activists like Farago and Browder believed it was time to start slashing. The administration was about to green-light an airport—and the inevitable pollution, sprawl, and noise that would follow it—at the edge of two national parks, after a laughable environmental review. Enviros weren't supposed to seek reasonable compromises on disasters like that. They weren't supposed to worry about relationships with friends in high places, or console themselves with the knowledge that other politicians might be worse. They were supposed to raise hell. After listening to Audubon's ever-pragmatic Charles Lee argue for the middle ground at a coalition meeting, a young Natural Resources Defense Council intern whispered: "I thought these were environmentalists!"

Ultimately, the coalition agreed to push for a more thorough environmental review of the airport. And even though Audubon's leaders refused to sign letters threatening lawsuits, others continued to send them.

The Everglades Consensus

DESPITE ALL THE CONFLICT swirling around the Everglades, it was still a powerful symbol, and politicians still wanted to be seen as its defender. The election year of 1996 inspired a particularly healthy competition to save the Everglades, with both parties jumping back on the bandwagon.

The lovefest began when lobbyists for Paul Tudor Jones persuaded Senator Richard Lugar of Indiana, a Republican presidential candidate, to float a 2-cents-a-pound sugar tax; Jones rewarded him with campaign donations and TV ads. Senator Dole opposed the tax, but as the Florida primary approached, he came under heavy pressure to show he cared about the Everglades as well, so he proposed $200 million for land acquisition.

Vice President Gore did not intend to let Republicans claim the Everglades high ground; his staff was already pulling together the administration's efforts into a marquee White House plan. Gore started discussing the Everglades with Clinton every week at their lunch meetings, and they agreed that a big push for Everglades restoration would be good politics as well as good policy. In February, Gore announced that the C&SF Restudy would be a top administration priority—which was a bit irritating to Secretary Babbitt, who thought it already was—and vowed to double government spending on the Everglades. He also pledged to purchase at least 100,000 acres of sugar fields for restoration, a key demand of the Everglades Coalition.

Speaker Gingrich soon embraced the Everglades as well. It was becoming clear that the GOP assault on environmental regulations was alienating moderate voters, and Gingrich decided that the Everglades could help his party look green. The speaker did not usually participate in floor debates, but he made a rare speech to support Dole's $200 million plan. "Newt told me: 'This is great politics!'" says Congressman Mark Foley, who had begged

Gingrich and Dole for help after the drowning-deer ads made him look like a Bambi-killer. "Republicans could still be against overregulation of business, but we could be for the Everglades."

THE COMBATANTS ON the Governor's Commission also carved out common ground, setting aside their differences to build consensus for a south Florida megaproject. They recognized that they would never agree on zero-sum issues like penny-a-pound and the Homestead airport, but the C&SF Restudy offered potential benefits for all of them.

The main problem with the flood control project was obvious: 1.7 billion gallons of fresh water were bleeding out to sea every day. The solution seemed obvious as well: storage. If more fresh water could be captured and stored in the wet season, it could be redistributed in the dry season to farms and cities as well as the Everglades. That didn't have to be divisive; sugar growers, home builders, water utilities, and environmentalists all had an interest in preventing shortages. They were all battling over a limited "water pie," but commission chairman Richard Pettigrew, the former Florida House speaker, figured that if the Restudy could "expand the pie," the water wars could end. Pettigrew was a consensus-building virtuoso, and in August 1996, his commission approved a conceptual plan to revamp the C&SF— again by a unanimous vote. The primary thrust of the plan was maximizing storage in deep wells and huge reservoirs, including at least 50,000 acres of sugar fields that already had a willing seller, but not necessarily the 100,000 acres proposed by Gore. At the same time, the commission emphasized that "flood protection and water supply for all users are critical components of sustainability." It was a win-win compromise, a perfectly balanced plan; no one would get hurt, and everyone would work together. "We had to come up with something for everyone," Pettigrew said.

Three weeks before election day, President Clinton signed a bill directing the Corps to convert the Governor's Commission's something-for-everyone vision into a restoration blueprint. "When we . . . preserve places like the Everglades, we are standing up for our values and our future," he said. Clinton didn't know much about hydrology, but he knew a lot about politics. "This is a great issue!" he enthused at the bill-signing ceremony. "You know what? We're going to win Florida!" And he was right.

Florida's voters narrowly rejected penny-a-pound, and Congress rejected sugar taxes as well. But Congress did approve Dole's $200 million gift to the Everglades, which financed the purchase of those 50,000 acres of sugar fields. And after the election, Vice President Gore intervened to hold up the Homestead airport deal, ordering the more stringent environmental review that the Everglades Coalition had demanded. "The Everglades is one of the greatest environmental treasures on the planet, and it's in the custody of the United States of America," Gore recalled. "We took that seriously."

It's Not Brain Surgery

NOW THE EVERGLADES was in the custody of the Army Corps of Engineers, the agency that had helped put the ecosystem on life support in the first place.

The Corps was a behemoth, with more employees than the Departments of Energy, Labor, and Education combined; it was overseeing 12,000 miles of navigable waterways, operating 2,500 recreation sites, producing one-fourth of the nation's hydropower and managing enough land to cover the entire states of Vermont and New Hampshire. But General Joe Ballard, the agency's new commander, was concerned about its future. America wasn't damming rivers anymore, and the Corps budget was barely keeping pace with inflation. Ballard decided he needed to "grow the Corps," approving a secret "Program Growth Initiative" designed to extract more money from Congress, and a dot-com-style business plan that designated "Seek Growth Opportunities" as one of the taxpayer-funded agency's three core principles. He encouraged local commanders to pester congressmen for new work building schools and sewage plants, replenishing beaches, and cleaning up hazardous and radioactive waste. And while Ballard was no environmentalist, he saw that ecosystem restoration could be a huge growth opportunity for the Corps—not only in America, but in ninety countries where it had a presence. "The Corps has nothing going on as big and complex as Everglades restoration," he exhorted his underlings. "If we do it well, our success will open the door wide for similar work around the world. We have no choice! We must put our best foot forward. The future of the Corps depends on it!"

But many critics doubted the Corps could fix man's mistakes in the Everglades. It was a multipurpose agency, not an environmental agency, and it had always put people ahead of nature. Its culture still revolved around ecologically destructive navigation and flood-control projects; in fact, subsequent investigations would reveal that the Corps was manipulating economic analyses in order to justify a billion-dollar lock project on the Mississippi River, deepen entrances to the ports of Philadelphia and Baltimore, and prevent the demolition of the agency's salmon-killing dams on the Snake River. Ballard and his aides seemed to view restoration as a new way of moving dirt and keeping busy, not a new way of thinking. The agency had some biologists now, but it was still a Corps of Engineers. And Congress still steered it toward ecologically destructive boondoggles. For example, the Corps spent $2 billion taming the rambling Red River into slackwater pools for barge traffic—four of the new dams were named for members of Congress, and the new waterway was named for Senator J. Bennett Johnston—but the barges never came.

Exhibit A for the skeptics was the much smaller and simpler Corps effort to restore Shark Slough, which was stalled after a decade. Two hulking floodgates the Corps had built to let water flow into Everglades National Park had never been opened; they loomed above the Tamiami Trail, concrete monuments to bureaucratic paralysis. Above the gates, water stacked up on the Miccosukee Indian hunting grounds, drowning deer and washing away tree islands; below the gates, the park continued to die of thirst. "Glades Plan Turning into River of Morass," the *Herald* reported.

The stalemate centered on the Corps flood control plan for the Eight-and-a-Half-Square-Mile Area, the rural Cuban-American community at the edge of the slough. The park's ecology-oriented scientists concluded that the Corps plan would dry out 30,000 acres of marshes, so they pushed for a buyout of the area's 350 homes instead. They argued that the objective of the Shark Slough project was to restore Shark Slough; it seemed crazy to sacrifice so much of the slough to protect a development—the only development—on the wrong side of the protective levee. But development-oriented Corps officials stuck by their plan for levees and pumps, pointing out that Congress had specifically directed them to protect the Eight-and-a-Half, accusing the park of launching a "jihad" for environmental purity. "They think they are fighting a holy war against the infidels," one Corps hydrologist wrote in an internal e-mail. "It's going to take strong leadership and possibly a chopped-off hand or firing squad to get out of this."

The bureaucratic sniping soon devolved into a litigious quagmire. Environmentalists sued the Corps to protect the Cape Sable seaside sparrow, an endangered bird found only in the park, known as "the Goldilocks bird" because it could not tolerate extreme floods or extreme droughts. The Eight-and-a-Half residents sued to protect their homes. The Miccosukees sued to protect their land. Conflict resolution experts were hired just to ease the tensions among government agencies, but the stalemate continued. One congressman complained that "we will all be pushing up daisies" before Shark Slough was restored.

The parties traded accusations of stupidity, insensitivity, extremism, and racism, but there was one point on which they all agreed: This was a bad sign. If an $80 million Everglades restoration project that affected 350 families over eight and a half square miles was hopelessly bogged down, what would happen to an $8 billion Everglades restoration project that affected millions of people over 18,000 square miles?

BUT IF GENERAL BALLARD saw Everglades restoration as a necessary evil—"We have no choice!"—some Corps leaders saw it as an opportunity for redemption, a chance to make amends for the agency's environmental mistakes in the Everglades while continuing to promote economic development in south Florida. The Pentagon official overseeing the Restudy, Michael Davis, was a former Corps biologist and the son of a Corps engineer; he had always dreamed of turning the agency's technical expertise into a force for environmental good. Colonel Rice was so environmentally conscious that the Everglades Coalition gave him its annual public service award; he later married the president of Friends of the Everglades. Stu Appelbaum, the amiable planner in charge of the Restudy, was determined to lead the most inclusive study in Corps history, and to avoid the bitterness of the Shark Slough fiasco. He welcomed the water management district as an equal partner, and invited thirty other agencies to help develop a consensus plan. He put all the Restudy's biologists, hydrologists, and engineers on one team, and urged them to leave their agency hats at the door so often that they printed up "Agency" hats they could actually leave at the door. Over Labor Day weekend in 1998, as the team raced to complete the plan in record time, every cubicle in the Restudy office was occupied.

In October, the team submitted its $7.8 billion draft plan to transform the C&SF, the largest and most expensive environmental initiative in history. The 4,000-page plan aimed to resuscitate Lake Okeechobee and the estuaries as well as the Everglades, while bringing back panthers, crocodiles, and wading birds. At the same time, it pledged not to reduce anyone's flood control, while expanding south Florida's water supply to serve an astounding 12 to 15 million people. With its new partners in the water management district, the Corps essentially adopted south Florida's regional water supply plan as a national goal, which helped ensure the support of water utilities, home builders, and agricultural interests. The Corps explained that providing new water for people would reduce their reliance on the Everglades, promoting sustainable growth for the region.

Vice President Gore unveiled the plan at a ceremony in West Palm Beach, describing it as a tribute to Marjory Stoneman Douglas, who had died that year at 108. The vice president boasted that it would not only "restore the precious Everglades," it would "protect and preserve south Florida's water supply for farmers, families, and future generations." In south Florida, he said, the environment *was* the economy. And if Douglas had announced the "Eleventh Hour" for the Everglades in 1947, it had to be getting close to midnight. "This is an ambitious and aggressive plan," Gore said. "But this much we know: The cost of inaction cannot be afforded."

RESTORING THE EVERGLADES, Appelbaum liked to say, was not brain surgery; it was much more complicated. Richard Ring, the superintendent of Everglades National Park, likened the C&SF overhaul to converting a twin-engine Cessna into a 747 in midflight.

But for all the complexity of the Comprehensive Everglades Restoration Plan—known as CERP—it was mostly an effort to expand the water pie. It proposed eighteen aboveground reservoirs covering 180,000 acres, an area nearly the size of New York City, including 60,000 acres in the Everglades Agricultural Area. It called for two belowground reservoirs that would be retrofitted from limestone quarries in the eastern Everglades, as well as 330 "aquifer storage and recovery wells" that would inject water a quarter mile into the earth to be withdrawn in droughts. There would also be "seepage management" along the perimeter levee to stop water from escaping the Everglades underground.

By capturing new water and reducing seepage losses, CERP would add nearly a trillion gallons a year to the water pie, creating benefits throughout the ecosystem. Water managers would no longer have to rely so heavily on Lake Okeechobee and the water conservation areas as reservoirs; they could maintain lower and healthier lake levels, and let the conservation areas function as part of the Everglades again. They could also stop blasting coffee-colored lake water into the Caloosahatchee and St. Lucie estuaries, where stressed-out snook and pompano were developing lesions so wide that their entrails were dragging behind them. And south Floridians would have a stable source of water to meet the demands of future growth, so that they wouldn't have to suck water out of Everglades aquifers when they washed their SUVs and irrigated their crops.

CERP also proposed 35,000 additional acres of filter marshes to improve water quality, almost as much as the state's Everglades Forever Act. It called for operational changes to help mimic natural water patterns and eliminate extreme water fluctuations; after decades of manipulating canal levels to protect tomato fields and subdivisions, water managers would have to consider wood storks and oyster beds as well. The plan included one component designed to restore the section of Big Cypress where the Rosen brothers had dug their canals for Golden Gate Estates, and another designed to restore some of the tidal flows cut off by Henry Flagler's railroad down the Keys. Of course, CERP did not exactly aim to "restore" the Everglades; that would have required the relocation of several million people west of I-95. But the plan did aim to improve 2.4 million acres of wetlands—not in the old sense of improving them for human use, but improving their ecological health. The goal was to Get the Water Right—the quantity, quality, timing, and distribution—across the south Florida landscape.

The reengineered Everglades would not be a natural Everglades; it would still be intensely managed and tightly controlled. Corps officials described it as "a Disney Everglades," and warned that it would have to remain on life support. But at least it would be alive, with more water available for farms, people, and the Everglades during droughts, and less water dumped on estuaries and the Everglades during storms. And once the hydrology came back, the biology would follow: "The entire south Florida ecosystem, including the Everglades, will become healthy. . . . The numbers of animals—crayfish, minnows, sunfish, frogs, alligators, herons, ibis and otters—at virtually all levels in the aquatic food chain will markedly increase."

AT LEAST THE CORPS HOPED SO.

The entire Restudy depended on intricate models of the natural Everglades that took a week to run through a computer. But only a few engineers understood how they worked and what assumptions drove them,

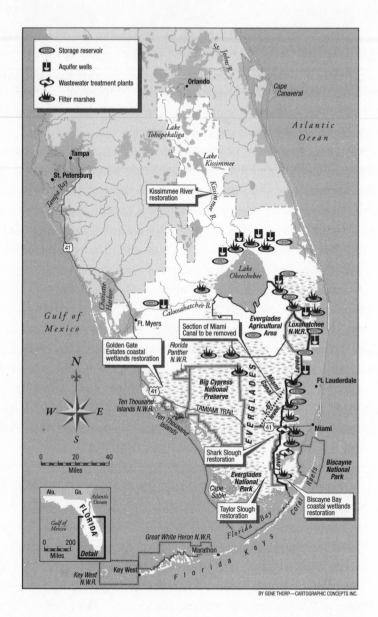

BY GENE THORP—CARTOGRAPHIC CONCEPTS INC.

The Comprehensive Everglades Restoration Plan was extremely complex, but it was essentially a storage plan. The goal was to store more than 1 trillion extra gallons of water for cities, farms, and the Everglades in massive reservoirs and high-tech wells. The plan also included seepage controls to keep water from escaping the Everglades, as well as filter marshes to help ensure that the water would be clean.

and no one was sure they accurately simulated the original hydrology and topography of the Everglades, where a few inches of water depth or ground elevation could transform the landscape. Even the inventor of the models thought they were given far too much weight in the Restudy. "There are unique and significant uncertainties that remain with these models," the plan acknowledged. And even if the project did replicate some semblance of the natural Everglades, no one could be sure the fish and birds would return: "The ways in which this ecosystem will respond . . . almost certainly will include some surprises." The cochair of the science team for CERP thought a better acronym would be SWAG—Scientific Wild Ass Guess.

While some uncertainty was inevitable, the restoration plan was highly dependent on four technological gambles that accounted for nearly $4 billion of its cost. The plan's aquifer wells were expected to hold twenty times more water than the world's largest storage site of its kind; it was unclear how much of the water would be recoverable, and whether overloading the aquifers would fracture them. CERP also entered unknown territory with its plan to convert mined-out limestone pits into eighty-foot-deep reservoirs; many geologists doubted the quarries would hold water, and some feared they could contaminate Miami's drinking water with deadly bacteria. The plan's "seepage management" strategy included a subterranean barrier to stop groundwater from flowing out of the eastern Everglades, another unprecedented structural interference with Mother Nature. And when modeling revealed that the plan was cutting off flows to Biscayne Bay, the team hastily added two wastewater treatment plants designed to scour urban runoff and divert it to the bay. That had never been tried, either, and even the project's leaders considered the idea impractical.

John Ogden, Appelbaum's counterpart at the water management district, joked that CERP was exactly like the Apollo mission—except no one was sure where the moon was, or how to find it, or whether it was made of cheese. But the plan at least recognized these uncertainties, and included $100 million for pilot projects that would test the four speculative technologies before they were deployed. And if one of them didn't work, the Corps intended to adjust the plan. CERP called for "adaptive management," a scientific way of saying the plan would be flexible.

Previous engineers in the Everglades had assumed they knew all the answers, and the result had been a century of unintended consequences.

The Restudy's leaders tried to approach the Everglades with humility. They knew their plan was imperfect and would have to change over time. "Maybe this plan is premature, but I don't want to do a post-mortem on the Everglades," Appelbaum said. He recognized there would be bumps along the road to restoration. But he was confident he was on the right road.

"These Are Deep, Systematic Problems"

THE FIRST BUMP was a doozy. On December 31, 1998, Everglades National Park's scientists submitted their comments on CERP. These were the critics who had scuttled the Corps plan for Shark Slough, and they were even less impressed now. The restoration plan, they wrote, "does not represent a restoration scenario for the southern, central and northern Everglades." The park scientists used the plan's own projections to show that it offered swift, sure, and lucrative benefits to south Florida's homeowners, developers, and agribusinesses, while its benefits for the Everglades were riddled with uncertainties and delayed for decades. Perhaps the plan offered something for everyone else, but it didn't offer much for nature. "There is insufficient evidence to substantiate claims that [the plan] will result in the recovery of a healthy, sustainable ecosystem," their forty-four-page comments concluded. "Rather, we find substantial, credible and compelling evidence to the contrary."

The Restudy team was shell-shocked. How could it sell a $7.8 billion Everglades restoration plan that was being trashed by Everglades National Park? Park scientists had taken potshots at the Restudy for months, and had complained that their concerns were being ignored. But no one had expected this public New Year's Eve stink bomb. Park officials were already considered poor team players, and their harsh rhetoric only cemented feelings among rival agencies that the park only cared about the park.

The park's scientists did care about the park—that was their job—and the plan, they pointed out, almost completely stiffed the park. For $7.8 billion, CERP would barely increase flows to the southern Everglades—from about 60 percent of predrainage levels to 70 percent, and not until 2036, and then only if the influential rock-mining industry was finished excavating 20,000 acres of the Everglades, and even then only if the risky effort to convert its limestone pits into reservoirs panned out. For the first decade of the plan, there would be no new water for the park in dry years, with only a modest increase in wet years, which would be achieved by reducing vital flows to Biscayne National Park. The park's staff agreed with the Restudy team that CERP should aim to restore the entire ecosystem, not just Everglades National Park. But they had expected at least some help for the park. It was the 1.5-million-acre public face of the Everglades, and its decline had inspired the demands for restoration. Protecting a national park was not supposed to depend on broad consensus; it was the law. It seemed presumptuous to ask Congress to spend so much money on an Everglades plan that did so little for the federal government's main property in the area.

And the park's environmental concerns were not limited to the park. Restoration was originally supposed to expand the spatial extent of the Everglades, but CERP actually called for the sacrifice of more than 30,000 acres of Everglades wetlands outside the park for reservoirs. The plan did next to nothing to address invasive species, or the runaway development that was whittling away the Everglades every day. And the plan's few promising ecological components were back-loaded. The park scientists concluded that "it is difficult to identify any significant environmental benefits" from CERP's first decade of projects—not just for the park, but for the entire ecosystem.

By contrast, the plan was expected to meet all its urban water supply targets for 2050 in that first decade, storing enough water to subsidize at least six million more south Florida residents. That sounded less like a restoration project than a federally subsidized water supply project for the unborn. "The Corps gave the cities and the ag guys all the water they needed up front," recalled hydrologist Robert Johnson, the director of the park's science staff. "Then they said: Okay, if there's anything left, we'll try to get it to the Everglades someday, as long as nobody gets flooded. How is that an environmental plan?" It was nice that everyone finally recognized that south Florida's economy was linked to its environment, but they were not the same thing, and park scientists believed that environment-is-the-economy rhetoric was being used to dress up an economic boondoggle in environmental clothing. Johnson was afraid that after paying for the front-loaded water supply components and observing little ecological progress, Congress would conclude that Everglades restoration was a Florida scam, and would stop funding it before the greener components could even begin.

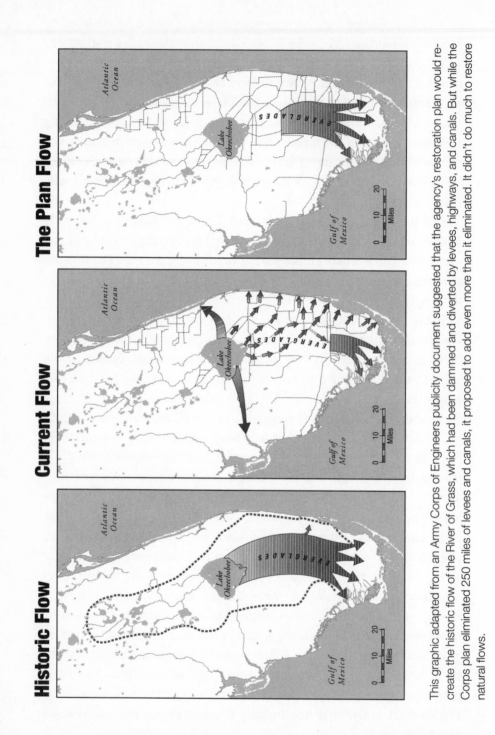

Historic Flow

Current Flow

The Plan Flow

This graphic adapted from an Army Corps of Engineers publicity document suggested that the agency's restoration plan would re-create the historic flow of the River of Grass, which had been dammed and diverted by levees, highways, and canals. But while the Corps plan eliminated 250 miles of levees and canals, it proposed to add even more than it eliminated. It didn't do much to restore natural flows.

The park also had a philosophical problem with CERP: It wouldn't achieve Art Marshall's dream of a reconnected, free-flowing ecosystem. The Corps boasted that it would remove up to 240 miles of levees and canals in an effort to "reestablish the natural sheet flow through the Everglades." But the scientists noted that the plan would add more levees and canals than it would remove, as well as numerous pumps and control structures, and that it "largely retains the fragmented management and compartmentalization characterizing today's Everglades." In reality, the plan aimed to mimic the historic water depths and durations of the Everglades—at least the ones predicted by the computer models—but not the historic flow. The park's scientists thought a $7.8 billion operation should do more than keep the Everglades on life support. They wanted to remove as many man-made barriers as possible, and let the waters of the Everglades flow as naturally as possible.

The Restudy's leaders found this critique particularly annoying. They insisted they had no love for levees. They had hoped to devise a more natural plan with more connectivity and flow. But millions of people now lived in the path of the original Everglades. Their models suggested that removing more barriers and restoring more north-to-south flow—the Corps called this strategy "let it rip"—would increase flooding in Weston and degrade tree islands in the conservation areas. Weston jutted into the Everglades like a pie wedge; it probably never should have been developed in the first place. But it had been developed, and it was hard to imagine a successful restoration that required the evacuation of Miami Dolphins quarterback Dan Marino and 50,000 of his neighbors. A reconnected Everglades felt right, but the Restudy's leaders saw no evidence that getting the water to flow again was as important as getting it to the right places at the right times. Appelbaum often asked: Do you want it natural, or do you want it like the Everglades?

The park scientists understood that the Everglades could never again be purely natural, and that water managers would have to retain some control over the ecosystem. But sheet flow and connectivity were two of the central features of the historic Everglades. Water depths and durations were a matter of guesswork, but it was a matter of fact that the original Everglades was unfettered. CERP seemed to reflect an engineer's bias for Rube Goldberg–style structural engineering, relying on uncertain technologies and human manipulation instead of simply getting out of Mother Nature's way and letting her reclaim her territory. A $7.8 billion Everglades plan that virtually ignored sheet flow seemed as bizarre as a $7.8 billion Everglades plan that virtually ignored Everglades National Park.

THE PARK'S COMPLAINTS were soon seconded in a letter signed by six of the nation's best-known ecologists, including Harvard's E. O. Wilson, Stanford's Paul Ehrlich, and the University of Tennessee's Stuart Pimm, the group's brash ringleader and the author of an ecological manifesto called *The World According to Pimm.* "There are serious failings in the plans being considered," the scientists wrote. "These are deep, systemic problems." Pimm, an endangered-species expert who was studying the Cape Sable sparrow, scoffed that only a structure-loving, control-minded Army Corps engineer could coin a phrase like "let it rip" to describe the almost glacial flow of an unobstructed River of Grass. Pimm believed that CERP violated the basic principles of ecology: Connected is better than fragmented, and natural is better than managed. "It's not that there are gaping holes in this plan," Pimm wrote in an e-mail to conservationists. "It's that we scientists are having trouble finding even a thread of restoration upon it."

The Everglades Coalition was again divided about how to respond, with Audubon's leaders and their allies typically reluctant to criticize the plan or the administration. They considered CERP a once-in-a-lifetime opportunity, and feared that any hint of negativity from the environmental community could doom the plan in Congress. "I urge everyone to be very careful with how you choose your words to describe the plan's shortcomings," Audubon lobbyist Tom Adams e-mailed coalition members. When Pimm and his fellow ecologists called for an independent science review of the Restudy's environmental performance, Audubon's Charles Lee publicly dismissed their criticism as "cries from the fringe."

Thanks to Paul Tudor Jones, Audubon had more scientists and staff working on the Everglades than the rest of the coalition combined, but its leaders focused on promoting the plan, not critiquing it. Adams was the only full-time CERP lobbyist, and his marching orders were simple: Pass the Plan. Another Audubon activist went to work for the Corps as the plan's chief spokeswoman. Audubon's Florida leader, Stuart Strahl, served on the Governor's Commission, and he believed the coalition would only marginalize itself by attacking a consensus plan. If Everglades activists couldn't support an effort to pump $7.8 billion—four years' worth of spending on all national parks—into the south Florida ecosystem, what could they support? Strahl believed that constant negativity was a guaranteed path to irrelevance.

Jones stayed behind the scenes, but he was now the major force behind Everglades advocacy, funding the National Resources Defense Council, the World Wildlife Fund, and his own Everglades Trust and Everglades Foundation as well as Audubon. He also set up an alliance known as the "Barley Group" that became the locus of power within the coalition, setting policy at weekly conference calls. And while Jones could see that the restoration plan wasn't perfect—he was paying the salaries of quite a few scientists who confirmed that—he believed it was the only hope of averting a total ecosystem collapse. "Regardless of my concerns and those of experts about the details of parts of CERP, I have learned that a bird in hand is worth a thousand in the bush," Jones later explained. "If CERP failed, it could not be recreated. No CERP, no Everglades restoration. Frankly, it was as simple as that."

The rest of the coalition wanted an Everglades restoration project, too, but some groups argued that it ought to restore the Everglades, instead of just subsidizing more of the growth and development that was killing the Everglades. A consensus plan that left out Everglades National Park didn't sound like much of a consensus plan. Joe Browder fired around e-mails complaining that Jones and Audubon were shilling for Clinton and Gore, and that the administration would have no incentive to improve its plan if it could count on blind loyalty from environmentalists: "We are going to lose what really counts about the Everglades unless someone gets beyond cheerleading for the Administration and mounts a genuine campaign to force the federal government to meet its responsibilities to the National Parks." Browder's criticism of fellow environmentalists made him a bit of a pariah within the coalition, but grassroots activists in Friends of the Everglades and the Sierra Club agreed that the restoration plan was unacceptable.

So did two of the nation's smartest environmental litigators—Tim Searchinger of Environmental Defense, who had represented conservationists in the water quality lawsuit, and Brad Sewell of the National Resources Defense Council, who was handling the legal battles over the Cape Sable sparrow and the Homestead airport. Channeling the criticisms of their allies inside the park, Searchinger and Sewell led a push to improve the plan, arguing that congressmen from outside Florida would never approve it without clear benefits for the Everglades in the first decade and $4 billion. They toned down their rhetoric to appease Audubon and its followers, removing overt threats to oppose CERP from most of their letters. Still, they made it clear to the Clinton administration—and to Audubon—that they would create a ruckus unless the plan included front-loaded environmental progress. "We didn't think it was so radical to expect an $8 billion Everglades project to, you know, help the Everglades," Searchinger recalled. "The good news was, you could produce much better results much faster for much less money."

The critics envisioned a more natural, less structural CERP that would provide more water to the park and faster environmental benefits to the entire ecosystem. Instead of trying to store water at the side of the Everglades and have water managers squirt it wherever and whenever they thought it was needed, the critics wanted to store more water at the top of the Everglades and let it flow south in an uninterrupted sheet. They envisioned a plan based on three strategies: Build even larger reservoirs in the Everglades Agricultural Area, closer to the 100,000 acres that Gore had originally promised. Reduce seepage out of the eastern Everglades. And reconnect the central and southern Everglades by removing a long diagonal levee and elevating the Tamiami Trail. That would provide restoration in real time, while saving taxpayers billions of dollars on unproven technologies. Searchinger's mantra was: Better, Faster, Cheaper. Pimm said: Let It Flow. "It matters not at all who gets credit for this, nor how we get there," Pimm wrote. "I'm not out to embarrass anyone, let alone this administration."

THE ADMINISTRATION WAS EMBARRASSED. It was also distracted by the Lewinsky scandal. But top officials recognized that their seven years of work on Everglades restoration would be wasted unless they could ensure at least tepid support from the Everglades Coalition and Everglades National Park. George Frampton, who was now Clinton's top environmental aide, still didn't trust Everglades activists, but the park's scathing critique, while irritating, persuaded him that the multipurpose plan had drifted too far from its original purpose. In March 1999, Michael Davis, the Clinton appointee overseeing the Corps, promised environmentalists that the administration's final version of CERP would put the Everglades first.

But time was running out: The Corps had to finish its technical plan by April, so that the administration could deliver it to Congress by its July deadline. The Restudy team was sent back to work, and soon discovered an extra 79 billion gallons for Everglades and Biscayne National Parks. In June, the team ran a new modeling scenario with more top-to-bottom flow, which produced "a series of improvements" to Lake Okeechobee and the estuaries within the first decade, and "vast improvements" to the southern Everglades. The new modeling did predict modest reductions in water supply benefits, and some undesirable pooling in one corner of the conservation areas. In any case, the Corps said it was too late to change the ten-volume technical plan. But the new modeling suggested that the plan could be adjusted later to provide up-front environmental benefits, which helped persuade many environmentalists to support it.

To nail down their support, Davis ordered the Corps to insert twenty-seven pages of new commitments into the "Chief's Report" that accompanied the technical plan, including a promise that the extra 79 billion gallons would be delivered to the national parks. Davis also added a pledge that restoration would be the project's "primary and overarching purpose," while flood protection and water supply would only be considered "to the extent practicable." Many Corps officials resented the last-minute political interference, grumbling that the administration was letting environmentalists end-run a consensus process. "There is one more change to the Michael Davis . . . OOPS, SORRY . . . Chief's report that has to be made," one colonel wrote in an e-mail. "Even though I understand that there was a significant need to get these groups onboard with total support, I am uneasy about changing what was in the report," a project manager wrote.

But the administration changed the report anyway. And on July 1, Vice President Gore personally delivered the plan to Congress, with an explanatory pamphlet titled *Rescuing an Endangered Ecosystem: The Plan to Restore America's Everglades*. Its cover was decorated with a panther, a heron, and a smirking alligator, and it promised that 80 percent of the project's water would go to the environment. "The ecological and cultural significance of the Everglades is equal to the Grand Canyon, the Rocky Mountains or the Mississippi River," it said. Senators Graham and Mack stood by Gore's side, and vowed to lead a bipartisan push to save "America's Everglades." Senator John Chafee of Rhode Island, the seventy-seven-year-old chairman of the Environment and Public Works Committee, vowed to pass the plan before his retirement at the end of 2000. "Let's get it done!" Gore shouted.

Ready for Action

THE POLITICAL CLIMATE seemed relatively favorable for the restoration plan. The economy was flush, and the federal budget was in surplus for the first time in decades. Senator Chafee, a genial moderate, was well liked on both sides of the aisle, while Graham and Mack were two of the most respected senators in their caucuses. Graham had once called Mack an "ideological wacko," but now they were close friends, and they marched in lockstep on CERP despite their partisan differences. Graham focused more on details, while Mack handled much of the politics; he was the GOP conference chairman and a confidant of Senate Majority Leader Trent Lott of Mississippi. Mack was retiring in 2000 as well, and most of his fellow Republicans were willing to do anything for him.

Of course, 2000 was also an election year, and election years were generally good for the Everglades. Florida was emerging as a key swing state, so neither party wanted to get on the wrong side of a $7.8 billion Florida issue. And the Everglades plan would be attached to the bill authorizing the biennial preelection smorgasbord of Corps water projects, which meant that scores of congressmen with pork in the bill would have an incentive to vote for the Everglades. House Transportation and Infrastructure Committee Chairman Bud Shuster, the legendary pork dispenser from south-central Pennsylvania, sent word that the Senate could have Everglades restoration, as long as the House could have its usual feast of ports, dikes, reservoirs, and sewage plants. Shuster's aides began calling the bill the Altoonaglades.

But it was unrealistic to expect Congress to approve a project for Florida without consensus support in Florida, and that consensus began to unravel, as the Chief's Report commitments that ensured the support of the Everglades Coalition and Everglades National Park were opposed by just about every other interest group. The Governor's Commission had called for balance; the Chief's Report put nature ahead of people. The commission had also pushed for a guarantee that everyone would retain their existing levels of water supply and flood protection—not just "to the extent practicable," but no matter what. And the Corps had promised that any changes to its technical plan would be submitted for public comment; the Clinton administration had tacked on the 79 billion extra gallons for the park after meeting privately with environmentalists.

In September, Dexter Lehtinen and the Miccosukee Indians filed the first CERP lawsuit, accusing the administration of cutting "back-room, closed-door, secret deals with a few special interests," quoting scenes from George Orwell's *Animal Farm* where the pigs "meet in private and afterwards communicate their decisions to the others." The sugar industry, which had battled the tribe over water quality, soon joined its suit over the Chief's Report, and water utilities, home builders, and state officials attacked the report as well. Sugar growers secretly financed an anti-CERP campaign by a conservative think tank called Citizens for a Sound Economy, which began blasting the Restudy as "A Case of Bad Government."

On October 24, the Everglades plan's prospects nose-dived again when Senator Chafee died suddenly of heart failure. The front-runner to succeed him was James Inhofe of Oklahoma, who was one of the most conservative Republicans in the Senate, and was openly hostile to the Everglades plan. Inhofe saw CERP as a classic boondoggle, an invitation to pour tax dollars into a swamp until the end of time. Inhofe's only challenger was Senator Robert Smith of New Hampshire, who was even more conservative than Inhofe; he was running for president as a militant defender of traditional values, and in July, he had quit the Republican Party to protest its tilt toward moderation. But four days after Chafee's death, Smith ended his presidential campaign and came crawling back to the GOP in hopes of winning the chairmanship. And in November, after a stormy meeting within the Republican caucus, Smith eked out a secret-ballot win over Inhofe by one vote. At the time, it hardly seemed to matter. Smith's last rating from the League of Conservation Voters had been zero on a scale of 100; Everglades activists figured they were in trouble either way.

The new chairman of the subcommittee overseeing the Corps looked like trouble, too. Republican senator George Voinovich of Ohio was a deficit hawk, and he didn't like the idea of a new multibillion-dollar commitment at a time when the Corps had a $30 billion backlog of uncompleted water projects—including flood control projects in Columbus and Cincinnati.

Voinovich was especially unhappy about subsidizing growth in Florida, which was already depleting Ohio's tax base by attracting so many Rust Belt retirees. He saw CERP as pure politics, and he had no interest in contributing to Gore's presidential campaign. "Nobody was looking out for the taxpayers," he says. "I decided I had to bird-dog this thing."

THE OTHER WILD CARD was the new Republican governor of Florida, Jeb Bush, the brother of Gore's likely opponent, Texas governor George W. Bush. Florida would have to pay half of CERP, and it was hard to imagine Jeb Bush—a former Miami developer with a deep belief in small government—pushing a big-government restoration plan associated with his brother's Democratic opponent and his own Democratic predecessor.

Jeb Bush was a conservative policy wonk, much more intellectual than his older brother; he was once considered much more likely to follow their father to the White House. But in 1994, while George rode a Republican wave to the governorship of Texas, Jeb lost a close election to Lawton Chiles. Even in a landslide GOP year, Jeb had tilted too far right, calling himself a "head-banging conservative," ignoring the environment. "If he had just mumbled a few platitudes, he would've won," said Allison DeFoor II, a Florida Republican leader and Audubon board member who had founded the Theodore Roosevelt Society to promote a greener GOP. Jeb learned his lesson, and served on an Audubon committee while laying the groundwork for another run in 1998. This time, he unveiled not only platitudes but actual environmental plans, and cruised to a comfortable victory. He named DeFoor his "Everglades czar," and assigned him to build the broadest possible consensus for a restoration plan. Bush told the Everglades Coalition that if Nixon could go to China, a Miami developer could help save the swamp.

But Bush soon confirmed some of the coalition's fears. He disbanded the balanced Governor's Commission for a Sustainable South Florida, and replaced it with a new commission stacked with builders, farmers, and businessmen. He also infuriated environmentalists by intervening to protect the Cuban exiles of the Eight-and-a-Half-Square-Mile Area. Under Chiles, the water management district had decided to buy out the community in order to help rehydrate Everglades National Park, but Bush's appointees swiftly overturned the decision, sending the Shark Slough restoration back into limbo. The Bush administration also rejected the Chief's Report, especially its elevation of the federal park over Florida's people, farms, and businesses.

In his first year as governor, Bush also presided over a near-disaster for the Everglades, an episode straight out of the environmental community's darkest conspiracy theories. On September 21, 1999, he spent an hour with several representatives from Azurix Corp., an aggressive new player in the $400 billion global water market. Azurix was a subsidiary of Enron Corp., which at the time was still known as one of the world's most innovative firms. And it wanted to control the Everglades.

Accompanied by one of Florida's top Republican fixers, along with another lobbyist who had written a paper on water privatization while working for Bush's private think tank, Azurix's CEO made an audacious proposal: The firm would help pay Florida's $3.9 billion share of the Everglades restoration project, and would build some of its wells and reservoirs, in exchange for the right to sell water captured by the project. Water had always been a public resource in Florida, but Bush was a big believer in private enterprise, and some of his aides were clearly interested in the proposal. "It shows some outside-the-box thinking on a thorny issue," one of Bush's top policy advisers scribbled in a note to DeFoor.

DeFoor was as intrigued by privatization as the next Republican; he knew that as long as water was virtually free, users would have little incentive to conserve it. But he was appalled by the Azurix team, which wanted an immediate no-bid deal, and seemed to think it could skip the courtship and head straight to consummation. DeFoor also saw that Enron's close ties to the GOP and the Bush family could create a political nightmare for the governor. "I want to be perfectly clear on this: No one likes out-of-the-box thinking more than I," replied DeFoor, a former sheriff who was also a divinity student, a Florida historian, and an attorney known for practicing law in shorts and Hawaiian shirts. "However . . . we are going to get our ass handed to us on this."

Azurix continued to meet with high-level state officials, and helped arrange a water privatization conference on Marco Island. But the company soon imploded, Act One of the spectacular Enron collapse. So Everglades restoration was never entrusted to a conglomerate whose name became synonymous with dubious finances, and the people of Florida got to keep their most precious resource. But the Azurix flirtation did not inspire much confidence in the future of Everglades restoration, any more than the new power of chairmen Smith and Voinovich, or the continuing stalemate over

Shark Slough, or the park's scathing comments about the Restudy, or the Chief's Report's unraveling of the fragile Florida consensus.

DeFoor preferred to focus on the bright side: Disaster had been averted. Despite its high-level access, Enron's water grab had failed. And if the Everglades marriage between the state and the feds had the feel of a shotgun wedding, Governor Bush and the Clinton administration still wanted to make it work.

EIGHTEEN

Endgame

We view this as the most important year in our history.
— Everglades Coalition, January 2000 agenda

A Time to Act

THE SLOGAN FOR THE JANUARY 2000 Everglades Coalition conference in Naples was "A Time to Act." The political climate may not
have been ideal, but momentum had been building for eight years, and the
coalition's leaders were convinced that 2000 would be their best
chance—perhaps their last chance—to pass a restoration project. It was
also the decision year for the Homestead airport, the most prominent
threat to the ecosystem in a generation. "Action taken to restore the Everglades in the next year will set the course for the next several decades," the
agenda said.

Over the course of the twentieth century, Florida conservationists had
helped stop plume hunts, preserve millions of acres of wetlands, mandate
minimum flows to Everglades National Park, and secure the largest nutrient
cleanup in history. But the Everglades was still dying. The ecosystem's natural balance was so out of whack that efforts to save the Cape Sable sparrow threatened the survival of the Everglade snail kite. Cattails were still
spreading, tree islands were vanishing, muck soils were shrinking, estuaries
were collapsing, and development was blocking the recharge of the region's
groundwater. The greatest enemy of the Everglades, the coalition's leaders
declared, was further delay.

* * *

SENATOR CHAFEE HAD PROMISED to hold a field hearing on the Comprehensive Everglades Restoration Plan at the Naples conference, and Senator Smith agreed to respect his late predecessor's wishes. Everglades activists were not expecting much from the John Birch Society's top-rated senator; after he took over the committee, the Sierra Club attacked him as "a fox in charge of the henhouse," and one journalist wrote that "many environmental groups are predicting an apocalypse of sorts." They never dreamed he would be one of the Everglades plan's most aggressive champions.

"John Chafee was strongly committed to seeing this restoration effort go forward," Smith said in his opening statement. "I totally agree. You will find no daylight between Senator Chafee's position and my own." The crowd gasped, and then cheered. Smith was a devout Roman Catholic, and he believed in the sanctity of life—not only for unborn children, but for egrets and otters, too. His six-year-old son had seen his first alligator on a vacation in the Everglades, and Smith now saw the swamp as a test for mankind: "When our distant descendants move into the Fourth Millennium, I hope it will be remembered that this generation, at the beginning of the Third Millennium, put aside partisanship, narrow self-interest and short-term thinking by saving the Everglades." Smith was as conservative as it got in American politics, but he figured that part of conservatism meant conserving things.

Senator Smith's witnesses were divided over the details of CERP, especially the Chief's Report's elevation of nature over people. But every key witness supported the Restudy. U.S. Sugar's Bubba Wade distanced the sugar industry from Citizens for a Sound Economy, and said growers now welcomed the restoration plan. Nathaniel Reed, the Everglades Coalition's elder statesman, answered the question of whether the plan would work with "an unequivocal yes!" Even Dexter Lehtinen, who devoted most of his testimony to Miccosukee grievances against the Interior Department, praised the Army Corps technical plan. Governor Bush's environmental secretary, David Struhs, quoted Senator Holland's remarks after the passage of the original C&SF Project: "The whole Florida delegation has stuck together in this matter and will, I am sure, continue to do so. The Florida citizens, industries and public units have also cooperated to the fullest degree, as has the Republican delegation. I want you to remember that this is not a partisan project, and should continue to merit the united efforts of all our people."

"That quote is as applicable in 2000 as it was in 1948," Struhs said.

The Everglades Coalition couldn't have scripted a much better start for its push for action. The restoration plan suddenly seemed sacrosanct; Senator Inhofe was the only politician who publicly opposed it, and he had no power over it. Even Senator Voinovich declared that he supported it despite his concerns about its cost and uncertainty; to be safe, Smith decided to yank the plan out of Voinovich's subcommittee and oversee it himself. "Both parties are sticking to the we-love-the-Everglades script," the *Palm Beach Post* said. Clinton administration officials met with Bush's aides in Naples, and were pleasantly surprised to hear that the governor felt as strongly as they did about swift action. Allison DeFoor, Bush's Everglades czar, called 2000 a "do-or-die year," and vowed that Florida would fund its share of CERP by the end of the spring.

DeFoor sensed that south Florida's interest groups were like drunks at the end of a bar fight. Their arms felt heavy, and they wanted an excuse to stop slugging. DeFoor set up a meeting between Audubon activists and sugar growers at Paul Tudor Jones's estate on the Keys, and both sides agreed over stone crabs to support the governor's funding bill. But the good feelings went only so far; a U.S. Sugar executive could not resist stealing one of Jones's prize orchids before he left.

BEFORE THE NAPLES CONFERENCE ENDED, Secretary Babbitt—back in good graces with his old antagonists in the Everglades Coalition—provided a final jolt of welcome news, announcing his personal opposition to the Homestead airport. A recent draft of the Clinton administration's revised study had suggested that the airport was back on track, but EPA Administrator Browner now came out against it as well. The administration was clearly divided, which meant the decision would be made in the White House.

Alan Farago, the Sierra Club activist leading the airport opposition, always figured the fight would come down to raw politics. Dade County's backroom deals reminded him of the corruption in his hometown of Providence, with Cubans instead of Italians calling the shots. But Farago believed the influence-peddlers could be defeated—not by playing kissy-face with decision-makers, but by building so much public revulsion to the airport that decision-makers would be afraid to approve it. He had quit an Audubon board out of disgust with the group's insider compromises, and he wanted to show that principled grass-roots activism could produce results.

Farago faced an uphill battle. Dade County Mayor Alex Penelas, the most prominent Cuban-American Democrat, was the airport's leading supporter. And Jorge Mas Santos, the leader of the Cuban American National Foundation—the anti-Castro group that dominated Miami exile politics—was one of the airport's key investors. President Clinton had won Florida in 1996 by making new inroads among Cuban voters and donors, and Vice President Gore's advisers feared that alienating Penelas, Mas, and the Latin Builders Association—not to mention Senator Graham—would doom his chances in Florida in 2000. Gore was already scrambling to distance himself from the Clinton administration's handling of Elián González, the five-year-old shipwreck survivor who had become a figure of religious devotion in Little Havana. Some Cuban-American leaders felt just as strongly about the Homestead issue; in fact, rumors were flying that they had offered the boy to the administration in exchange for a guarantee of the airport.

The airport's opponents also faced a serious cash disadvantage. The developers were paying more than $1 million to one of Washington's top lobbying firms, Verner, Liipfert, whose partners included former Senate Majority Leaders Bob Dole and George Mitchell. They also bankrolled an "Equal Justice Coalition," which spread the word that airport opponents were racists who wanted to keep minorities in poverty. The Sierra Club could barely afford buttons and T-shirts. But Farago noticed that Dade County's flight plans for the new airport passed directly over the Ocean Reef Club, a north Key Largo enclave of two thousand of America's wealthiest snowbirds. On his first visit, he met an elderly investor named Lloyd Schumaker, who wrote him a $100,000 check before he could even finish explaining why he was there. When Farago explained that the donation would not be tax-deductible, the crotchety Schumaker said he didn't care; he had already made $30 million that year.

Ocean Reef's residents ultimately decided to tax themselves to provide Farago with a $2 million war chest. That was enough to launch a sophisticated campaign, with pollsters, lobbyists, economic consultants, a Cuban-American community organizer, and slick ads depicting a flock of jets flying over Biscayne Bay, under the caption: Somehow, It's Not Quite the Same. The basic message was that it made no sense for the federal government to green-light a major airport at the edge of the Everglades at the same time it wanted taxpayers to spend $8 billion to restore the Everglades.

The campaign soon converted Senator Voinovich to its cause, partly because Ocean Reef was home to a number of well-connected Ohio Republicans, partly because the senator wanted to prove he cared about the Everglades despite his skepticism about the restoration plan. The usually mild-mannered Senator Mack once yelled at him to mind his own business, but Voinovich believed that if the Everglades was really "America's Everglades," as the Florida senators kept calling it, then a threat to the Everglades was America's business.

The main target of the campaign was Al Gore, who had the power to kill or approve the airport. But the vice president refused to take a stand—even after Babbitt and Browner sided with environmentalists, even after former senator Bill Bradley, his challenger for the Democratic presidential nomination, came out against the airport as well. Gore would only pledge to seek "a balanced solution" that would help the economy without harming the environment. As a public servant, Gore was often far ahead of his colleagues on issues like nuclear proliferation, environmental protection, and the "information superhighway," but as a politician, he had a tendency to straddle.

Gore's aides assumed that Florida activists would forgive him for taking a pass; after all, he had demanded the additional study that held up the airport in 1997, and had spearheaded the plan to restore the Everglades. But the airport's opponents kept up the pressure. In February, they threatened to protest an "Environmental Voters for Gore" rally in Broward County, scheduled to feature Browner with actors Leonardo DiCaprio and Ted Danson. The Gore campaign was afraid of man-bites-dog articles about conservationists attacking the Ozone Man, so the rally was cancelled. "Al Gore spilled blood for these people for eight years, and they were going to protest?" recalled Mitchell Berger, a Fort Lauderdale attorney and Democratic fund-raiser who was Gore's closest confidant in Florida. "Talk about the death of common sense."

Politically, Gore was walking a fine line between Democratic-leaning environmentalists and Republican-leaning Cuban-Americans. But the airport's opponents assumed he would return to his green roots after the predawn raid of April 22, when armed federal agents seized Elián from his Miami relatives so that his father could take him back to Cuba. There was no way Gore could distance himself from the administration now; Nathaniel Reed told the vice president's aides he wouldn't win the Cuban vote if he promised to land the 82nd Airborne in Havana. And it was hard to imagine that Gore still cared about Mayor Penelas, who had made national headlines by declaring that the administration would be responsible if Miami rioted over Elián.

Yet Gore remained on the fence. He wasn't convinced that the airport was central to the plumbing problems that were destroying the Everglades. Neither one of his most trusted Everglades advisers, Berger and Paul Tudor Jones, had raised alarms about the airport; in fact, Berger did legal work for the Mas family, and had once told environmentalists that he could engineer a buyout of the Eight-and-a-Half Square Mile Area if they would back off the airport. Berger helped persuade Gore that the airport opposition had more to do with not-in-my-backyard complaints about noise over Ocean Reef—a vacation getaway for prominent Republicans such as Senate Appropriations Chairman Ted Stevens of Alaska—than ecological concern for the River of Grass. "I didn't think the airport threatened the survival of the Everglades," Gore later recalled. In any case, Gore's advisers figured Everglades activists would back him regardless of Homestead. Shouldn't an $8 billion restoration plan count for something?

"Consensus Was the Only Way to Do This"

BUT MANY EVERGLADES ACTIVISTS remained skeptical of the restoration plan. They had grudgingly agreed to support it after the Chief's Report provided the additional commitments that it would actually restore the Everglades. But it soon became clear that the additional commitments in the Chief's Report were dead on arrival on Capitol Hill. Senators Mack and Graham—as well as Vice President Gore and Governor Bush—believed that passing CERP depended on maintaining a consensus among Florida's interest groups, and there was a consensus among every group except environmentalists that the Clinton administration had unfairly elevated nature over people. "They wrote us a letter," an aide to Mack assured Dexter Lehtinen. "We'll write them back a law."

The nonenvironmental interests all argued that the Chief's Report— especially its guarantee of 79 billion extra gallons for the park—had violated the consensus process that produced the original technical plan.

Even Audubon's Tom Adams, the most active Everglades Coalition lobby-ist, was sympathetic to the accusations of an end-run. Senators Mack and Graham wrote a letter protesting the guarantee, and the Clinton adminis-tration quickly backed off, saying the Corps was only committed to studying whether to provide the extra water. In the spring, Senate staffers agreed that their bill would ignore the Chief's Report, authorizing only the original Army Corps technical plan—the same technical plan that had been lambasted by the scientists at Everglades National Park. After months of cheerleading for CERP, the Everglades Coalition once again had to decide what to do about an Everglades restoration plan with questionable benefits for the Everglades.

The activists who had persuaded the Democratic administration to add environmental commitments to the Chief's Report hoped they could now persuade the Republican-controlled Senate to add environmental assurances to the actual bill, especially legal requirements that would reserve water for the Everglades and ensure ecological progress within a decade. They also wanted to maximize the power of the Department of the Interior, which tended to side with the environment, and minimize the power of the gov-ernor of Florida, who tended to side with his constituents. Unfortunately for the environmentalists, every other key stakeholder wanted the opposite. Sugar growers, home builders, water utilities, and Florida's other economic interests were all determined to make sure CERP did not favor nature over people—by eliminating or weakening environmental assurances, minimiz-ing the power of Interior, and maximizing the power of the state. They had such a common vision for CERP that they shared the same Washington lobbyist, Robert Dawson, a courtly Alabama native who had overseen the Corps during the Reagan administration. Dawson did not mind if CERP was marketed as a pure Everglades restoration plan, but he warned that it would never get out of the bog without solid guarantees for water supply and flood control.

The Seminole and Miccosukee tribes agreed with Dawson's clients that CERP should not favor nature over Floridians. They may have considered the Everglades their mother, but they were Floridians, too, with their own economic interests; the Seminoles ran a $500-million-a-year gaming busi-ness as well as cattle and citrus operations, and the Miccosukees had just opened their own casino overlooking the Everglades. The Miccosukees were especially determined to limit the role of Interior, an institution they despised. They still considered Everglades National Park their rightful homeland, and tensions had flared again recently when park leaders tried to stop them from building homes along the Tamiami Trail. Dexter Lehtinen warned that if CERP gave Interior any power over water management in Florida, "we will put a knife in the heart of this bill."

Governor Bush also sided with the economic interests. Florida's legislature had agreed to pay half of CERP's cost without a single dissenting vote, and Bush was determined to make sure equal money meant equal power. That meant an equal balance between the Everglades and his constituents, and an equal partnership between the Corps and the state; anything less, he told Congress, would be a "master-servant relationship." Senators Graham and Mack, who were in charge of refining CERP to ensure a consensus among Florida's interest groups, tended to agree. With green groups on one side and just about everyone else on the other side, it would be easier to forge consensus by pressuring the green groups to make concessions than by pressuring everyone else.

Even within the Clinton administration, there was only limited support for trying to strengthen the bill's environmental assurances. Vice President Gore had no desire to dive into details; Mitchell Berger had told him the environmental critics were extremists who would only be satisfied if the city of Weston was reflooded. Army Corps officials generally sided with Bush and the Florida interests; they didn't want to share power with Interior, and they didn't want their hands tied by restoration requirements. And George Frampton, the former Interior official who now coordinated policy at the Clinton White House, was tired of the Everglades Coalition's whining. He just wanted to pass a bill. The only administration official willing to fight was Secretary Babbitt, who didn't care too much about the details of the plumbing, but did care about Interior's role. In March, when Frampton was about to agree to strip Interior's power over CERP, Babbitt faxed a heated letter to his former aide threatening to oppose the administration's pet project if the Corps and the state retained full control: "Otherwise we allow a future that repeats past mistakes, with grievous consequences for our children and grandchildren." Frampton backed down, and the internal split never became public.

It was the threat of a public showdown that gave Babbitt his leverage within the administration; Vice President Gore did not want to be blamed for delaying the revival of the Everglades before the election. But the

Everglades Coalition did not have much leverage in the Senate to demand benefits for the environment. The coalition had secured the commitments in the Chief's Report by threatening to oppose CERP, but those threats were a lot less credible now that it had declared 2000 "A Time to Act," and national conservation organizations were clamoring for Congress to pass the bill. Environmentalists who still hoped to improve the bill could see that their chances were shrinking by the day, as Audubon and other groups began jockeying to portray themselves to funders as the saviors of the Everglades. There was intense pressure to stay "on message," to stop quibbling over details, to avoid discrediting CERP. Audubon issued one statement declaring that "we will continue to seek improvements in the bill to increase restoration benefits—as long as they do not endanger its enactment."

"Our feeling was: This isn't perfect, but it's more good than bad," said Audubon president John Flicker. Even CERP's water-supply components would reduce pressure on the Everglades, and several uncontroversial restoration components would benefit Big Cypress, the St. Lucie estuary, and Biscayne Bay's coastal wetlands. The Wall Street wizard Paul Tudor Jones told Audubon leaders that he had spent $5 million on the Everglades; he would consider $8 billion an excellent return. Larry Kast, a brash young water resources lobbyist who joined the Audubon team during CERP, advised environmentalists to stop trying to sweeten the deal. "I was focused like a laser beam on getting this passed, and the key was unity in Florida," Kast recalled. "We had to stop arguing over every frigging detail and every frigging drop of water. We had to get our shit together, or we were going to lose $8 billion."

More skeptical activists such as Environmental Defense's Tim Searchinger and NRDC's Brad Sewell knew that some of their colleagues believed they were threatening a fragile consensus, turning up their noses at an $8 billion restoration plan because it wasn't perfect. But this was the same plan the park's scientists had said "does not represent a restoration scenario for the southern, central and northern Everglades." The latest version of CERP did not even guarantee that the project would do no harm to the Everglades—only that no one's level of water supply or flood control would be reduced. That seemed a lot worse than imperfect.

ON MAY 11, Chairman Smith and Senator Max Baucus of Montana, the ranking Democrat on Smith's committee, held the first hearing on the Everglades bill. It was supposed to be a typical congressional Kabuki show, an opportunity for flowery speeches about the majesty of the River of Grass, with Smith demonstrating the Republican commitment to restoration and Baucus carrying water for the Clinton plan. But as Baucus listened to testimony about the plan's "tremendous amount of flexibility"—and watched witnesses duck questions about its ecological uncertainties—he did something exceedingly rare in Washington. He ditched his script and spoke his mind:

> I'm a little uneasy and I'll tell you why. I worry about seeing the evening news a year or two or three from now, "The Fleecing of America," "It's Your Money," something like that. . . . I have a funny feeling that I might be buying something that sounds good, but down the road, it's going to leave my successors a huge, huge program. And the problem is, we've spent all this money on the Everglades and my gosh, it's not working like it was supposed to work. Oh, we've gone this far, gee, it's like the Vietnam War in a sense, we've got to keep pouring more money into it because it's gone this far. What's our exit strategy?

An aide to the senator kept passing him notes and kicking his chair, but Baucus kept rambling. "Nobody has provided a compelling case that this is going to work," he blurted. "So far, it doesn't totally pass the smell test, if you want the honest truth."

This unplanned outburst of candor offered unexpected ammunition to Searchinger, Sewell, and other environmental critics of the restoration bill. They contended that without strict legal assurances for the natural system, Florida officials would keep giving away water needed for the Everglades to cities and farms, and CERP would never pass the smell test needed to secure national support. Governor Bush's lobbyists argued that assurances were unnecessary, because Florida already had the power to reserve water for the environment. But the state had only used that power once in twenty-eight years, for a marsh in the St. Johns basin. Senate staffers were wary of fixes that would antagonize every lobby except the enviros, but most of them—especially Senator Smith's aides—eventually realized the critics had a point. CERP had to change the status quo that had destroyed the Everglades.

The problem was finding the right language that could nail down the support of queasy environmentalists and avoid "Fleecing of America" exposés without losing the support of the other interest groups. After months of roller-coaster negotiations, Florida's economic interests withdrew their support for the bill in early August, then changed their minds after extracting a few key concessions. In early September, Senators Graham and Mack orchestrated a settlement of every key group, only to see Governor Bush's aides pull out of the agreement. This time, Senator Mack called Bush and explained that the state of Florida was holding up an excellent compromise. The governor, who was campaigning for his brother out west, called his underlings and told them to back down. George W. Bush didn't want to be blamed for scuttling Everglades restoration, either.

Anyway, Jeb Bush had gotten most of what he wanted in the bill. The state would be an equal partner with the Corps, which was already sympathetic to Florida's economic interests. Interior would only have a veto over the rules governing the project, not the sixty-eight project components. Those rules would not define the natural system's water needs, as environmentalists had hoped; they would only set up "a process" to define those needs. (Bush's aides had tried to dilute the rules even further, proposing that they set up "a process to provide procedural guidance" to define those needs.) Senator Voinovich secured a resolution declaring that reuse of the Homestead air base should be compatible with Everglades restoration, but Graham and Mack fought off substantive measures that could have blocked the proposed airport. Nobody's ox would be gored by CERP.

In general, the assurances did not assure much for the Everglades, although they did impose a few restraints on state water managers. The bill stated that "the overarching purpose" of the bill was restoration, but its substantive provisions included much stronger protections for flood control and water supply. And while Senator Smith's aides did jam some extra assurances language into their committee report, highlighting the Army Corps pledge that 80 percent of the water captured by CERP would go to the environment, the report did not have the force of law.

A few environmental groups denounced the consensus legislation, most notably Friends of the Everglades, the grass-roots organization founded by Marjory Stoneman Douglas. Robert Johnson, the head of Everglades National Park's science staff, told the *Washington Post* that the legislation would do almost nothing for the environment: "This is just a situation where the emperor has no clothes." When Audubon, the National Parks Conservation Association, the National Wildlife Federation, and Defenders of Wildlife passed around a draft letter describing CERP as a "must-pass" bill, the ecologist Stuart Pimm wrote a blistering critique:

> Of course, we should all live long and healthy lives; we will need to do so if we are to see this plan's benefits. . . . I can see why the sugar growers like this plan. This is a plan for ecological inaction and that is exactly why I find fault with it. I believe that consensus is fine. I applaud your efforts to work out compromise. But at some level this must fail: just because the policymakers all agree that the sun rises in the west doesn't make it so.

But most green groups went along with the deal—some with trepidation, some with enthusiasm. "This is an historic agreement for the future of America's Everglades," rejoiced Audubon's Stuart Strahl. Johnson's bosses at Interior also endorsed the bill, along with the rest of the Clinton administration. Secretary Babbitt would have preferred solid guarantees for the natural system, but he figured all the hype over "America's Everglades" would at least create expectations of restoration in the future. Perhaps the sugar industry would agree to sacrifice more land for restoration after it exhausted its soils, or after it lost its federal price protections, or after Castro died. Perhaps prolonged water shortages—and the rate hikes that could accompany them—would persuade Floridians to start conserving their most precious resource. Or maybe desalinization or some other new technology would solve south Florida's water problems. CERP would just be a start.

The Everglades Is Coming

THE PLAN WAS NOW IN PLACE, but Congress still had to approve it before adjourning for the election. Army Corps bills tend to pass at the last minute without debate, because Congress prefers to keep its pork platters off C-SPAN. But this one still had to make it through the Senate and House. "The single greatest threat to restoration of America's Everglades is the lack of time left in the congressional session," said Audubon's Strahl.

Behind the scenes, Florida's state officials, economic interests, and tribes had all fought to reduce CERP's emphasis on nature, but they now came together to promote it as a restoration plan for America's Everglades. Audubon lobbyist Tom Adams walked the halls of Congress arm-in-arm with sugar lobbyist Bob Dawson. "If we can agree to support the Everglades," they told members, "then you should, too." Senators Smith and Mack rallied support among Republicans, while Senator Graham and the Clinton administration lined up Democrats. It wasn't too hard. When Senator Inhofe tried to persuade colleagues that CERP was an astronomically expensive, scandalously uncertain exercise in government bloat, they often replied: But it's the Everglades! Senate Majority Leader Trent Lott had no great interest in the Everglades—he joked that he was pretty sure it wasn't in Mississippi—but he fast-tracked the Corps bill as a favor to Mack.

One potential sticking point was a raging debate over "Corps reform." After a Corps economist blew the whistle on the agency's frantic efforts to justify a billion-dollar lock project on the Mississippi River, Corps follies became front-page fodder, and Corps critics called for independent reviews of major projects, setting the stage for an ugly floor fight. But the environmental establishment never pressed too hard for the reforms, because it did not want to endanger Everglades restoration. So the Corps bill went to the Senate floor without them.

Instead, the Senate debate over the bill was dominated by florid tributes to the Everglades, and to the bipartisan consensus that had brought together Florida's Hatfields and McCoys. Senator Smith read a list of endorsers ranging from the Florida Fertilizer and Agrichemical Association to the National Parks Conservation Association. Senator Graham marveled at how much had changed since he launched his Save Our Everglades program to turn back time in south Florida. "In 1983, restoring the natural health and function of this precious system seemed to be a distant dream," he said. "After seventeen years of bipartisan progress, we now stand on the brink of this dream becoming a reality." But Senators Inhofe and Voinovich were not the only voices of caution. Senator John Warner, a Republican from Virginia, complained that the Everglades would dwarf all other water projects, including the restoration of Chesapeake Bay. "All of a sudden, we come along with the romance of the Everglades," Warner said. "Paul Revere called out: The British are coming. I call out: Folks, this is coming. You better go back home and talk to your constituents and say this one is going to be in competition with what I had planned for our state." Senator Baucus tried to defend the bill, but he again betrayed his doubts, acknowledging that part of him agreed with the critics. "This arrangement may not be perfect," Baucus said. "But we are dealing with an extraordinary, special situation, and that is the Everglades. . . . There is a slight tilt in favor of the State of Florida, but the Everglades is really special. It is a national treasure."

The Senate passed the bill by an 85 to 1 margin, with Inhofe the only dissenter. "If you have any doubts about every single 'i' being dotted and every 't' being crossed, take the risk. You'll be glad you did," Smith said. "When the historians look back, they are going to say when it came time to stand up for the Everglades, we did."

Now the House of Representatives controlled the fate of the Everglades. Momentous issues were at stake—the most ambitious ecosystem restoration in history, a new model for dealing with water conflicts, a new direction for the Corps, a chance to prove that man could repair his relationship with Mother Nature. But in the House, only one issue mattered: Clay Shaw was in a tight race. The workmanlike ten-term congressman from Fort Lauderdale was one of the most vulnerable Republicans, and with control of the House hanging on a few contested races, Speaker Dennis Hastert of Illinois was willing to do anything necessary to help the chairman of the Florida delegation. "We knew this could come down to two seats, and if that meant we had to spend $8 billion for Mr. Shaw, that's what we were going to do," one Hastert aide recalled.

In September, Shaw introduced the Senate's Everglades deal in the House, and Chairman Shuster attached it to an Army Corps bill that was so crammed with local water projects it took up forty-five pages of the *Congressional Record*. The bill had been held up all summer in a partisan dispute over prevailing-wage laws, but Republicans now agreed to drop their objections to get the Altoonaglades passed. On October 19, Shaw presided over the debate from the speaker's chair, watched a series of Republicans give him credit for saving the Everglades, and made the final speech before the House approved the bill by a 394 to 14 margin. "We are seeing a rare moment in the closing days of this Congress: both great political parties coming together and doing the right thing," Shaw crowed.

THE CONGRESSIONAL DEBATE over the Everglades was dominated by high-minded rhetoric about the River of Grass being above partisan politics. But it was still election season, and Florida was shaping up as the key battleground between Vice President Gore and George W. Bush. The day before the House voted on CERP, Gore's campaign aides huddled with Everglades activists in Miami, pleading with them to rally their troops behind the vice president. Kathleen McGinty, Gore's top environmental adviser, began the meeting by pointing out that Gore had led the fight to restore the Everglades, taken on the sugar industry over penny-a-pound, and fought for the environment all his life. But all the activists wanted to talk about was his waffling on the Homestead airport.

The Gore campaign had never imagined that they would have to beg Florida environmentalists for support three weeks before Election Day. George W. Bush was the dream candidate of drilling, mining, and logging interests; Gore was their nightmare. When a Democratic operative had tried to warn Gore campaign manager Donna Brazile that Everglades activists were irate about Homestead, the message had come back: "Tell them to go fuck themselves." Where else could environmentalists turn?

The answer, for some of them, had been Ralph Nader, the consumer crusader who was running on the Green Party ticket, attacking Bush and Gore as twin peas in a corporate pod. Joe Browder, a lifelong Democrat, had begun feeding information on Homestead to the Nader campaign and a group called Environmentalists Against Gore, and Nader had started making speeches denouncing the airport plan and accusing Gore of selling out the Everglades. Alan Farago had refused to take Nader's calls, but he knew the airport was costing Gore votes. In September, he had commissioned a Democratic pollster to conduct a survey of Florida voters, which suggested that Gore would gain four points if he came out against the airport. The Sierra Club had given the results to Gore's campaign, but the vice president refused to switch his position.

Even Nathaniel Reed, the ultimate inside player, had grown exasperated after months of behind-the-scenes lobbying against the airport. The vice president's aides had promised Reed that he would make an anti-airport speech, and Reed tended to err on the side of trust, especially with eco-friendly politicians like the Ozone Man. But he eventually realized that Gore had no intention of getting off the fence, and he fired off an e-mail throughout the environmental community, warning that Gore was contemplating the destruction of two beloved parks. "Until the Administration and in particular the Vice President is confronted with opposition, the Administration will continue to ignore the issue," Reed wrote. "From crisis comes opportunity! Force the crisis!"

Now it was three weeks before the election, and the Gore campaign realized it had a Nader crisis in south Florida. McGinty, Gore confidant Mitchell Berger, and former water management district director Sam Poole were dispatched to try to persuade the enviros that Homestead was a crazy litmus test. McGinty argued that Gore had gone to bat for the environment for his entire career; it was time for environmentalists to go to bat for Gore. But the activists just wanted to know why he had stayed in the on-deck circle on the airport. They said his silence was driving their members to Nader.

Berger couldn't believe he was having this conversation. George W. Bush hadn't taken a position on the airport. Neither had Jeb Bush. And Gore at least had an excuse for staying mum; the administration's study was still under way, and taking sides could be construed as interference. Anyway, Gore had intervened to block the initial pro-airport study; didn't that suggest his true feelings? Babbitt and Browner publicly opposed the airport; didn't that suggest where the decision was headed? If Gore made a statement now, it would just look like pandering. "Isn't there any trust in this room?" Berger asked.

There wasn't much. One airport opponent challenged Berger on his work for Jorge Mas; Berger insisted it didn't matter. McGinty said Gore wanted to hold an environmental rally in south Florida to highlight his defense of the Everglades, and asked whether there would be protesters. Absolutely, she was told. "Tell him that only a true friend will tell you what you don't want to hear," one activist said. "And what you don't want to hear is that you are going to lose this election because of Homestead."

MEANWHILE, the Everglades plan was in danger yet again. The House and Senate had to reconcile their Army Corps bills, and Senator Smith objected to a half-billion dollars of "environmental infrastructure" in the House bill. He knew "environmental infrastructure" was a euphemism for water and sewer plants, which were supposed to be local responsibilities. The Corps was already under fire for General Ballard's "Program Growth Initiative," and Smith didn't want to encourage more mission creep. He told Chairman Shuster he would block the bill unless the extra pork was removed.

Shuster was flabbergasted. A committee chairman objecting to the presence of pork in a Corps bill was like a Burger King fry cook objecting to the presence of beef in a Whopper. Environmental infrastructure was especially dear to Shuster's heart; he had invented the concept in a 1992 bill, diverting the first projects to his own district, then authorizing billions of dollars' worth of additional projects for other members. He was appalled by Smith's selfishness, and House Speaker Hastert was even angrier. Several vulnerable Republicans were counting on environmental infrastructure projects to build support at home before Election Day, and Shaw was counting on CERP, but Smith didn't seem to care who controlled Congress.

Only in Washington could an effort to save taxpayer dollars be considered selfish, but Smith's sudden stand on principle did seem odd. He had agreed to a bill with 138 water projects worth $7 billion, not including the Everglades behemoth, which as far as Shuster concerned was just another huge water project. Smith hadn't objected to flood protection for East Saint Louis or the renourishment of Rehoboth Beach or a comprehensive study of the Merrimack River basin in his home state of New Hampshire. Why was he drawing a line in the sand over sewer projects that actually helped people? Smith was hauled into a meeting with Speaker Hastert, and the avuncular former wrestling coach got as livid as his aides had ever seen him, throwing his pen in Smith's direction. "This is bullshit!" Hastert screamed. But Smith refused to budge. He found it hard to believe that a few sewage plants were going to determine the outcome of the election. "Control of the House is in Bob Smith's hands!" one of his aides wrote in a sarcastic e-mail. "Give me a break."

Congress was running out of time, so Senator Mack went to see House Appropriations Chairman C. W. "Bill" Young of Florida, who agreed to tack CERP onto an agriculture spending bill if the larger Corps bill was scuttled. Shuster realized his entire bill was in danger of stalling without its Everglades engine, so he relented and agreed to pass it without environmental infrastructure. Speaker Hastert then forced Young to tack the infrastructure projects onto a health spending bill. Nobody's ox was going to be gored on Capitol Hill.

It wasn't pretty, but four days before the election, Congress finally passed the Altoonaglades, prompting another round of speeches depicting Clay Shaw as the second coming of Marjory Stoneman Douglas. "Governor Broward, for whom my home county is named, ran on the platform that he was going to drain that swamp, the Everglades," Shaw said. "November 3 is the day we took the first step in really restoring this national treasure."

*   *   *

CONGRESSMAN SHAW RACED BACK to south Florida to campaign, and the Everglades headlines helped him edge his Democratic opponent by six hundred votes. Vice President Gore was not so lucky.

Ralph Nader visited Miami for a get-out-the-vote rally on November 5, and used Joe Browder's talking points to blast Gore for "waffling as usual" on the Homestead airport. "Congress and the state of Florida are poised to spend $8 billion to rehabilitate the Everglades," he said. "Why won't the Vice President take a stand against undermining these efforts?" Nader also sent letters to Florida environmentalists, bashing Gore for buckling to real estate interests: "There are no airports situated on the border of national parks in America; the Everglades is the last place to consider changing that."

Gore campaigned in Florida, too, but he never did hold that Everglades rally, and south Florida's environmentalists never did generate much enthusiasm for him. It was frustrating, but Gore always knew that for some Ivory Soap environmentalists, as he put it, "Ninety-nine and forty-four-hundredths percent pure was never good enough." He was more irritated at Mayor Penelas, who was reelected in September, then took off for a vacation in Spain, contributing nothing to Gore's campaign or his fight for a recount. After the votes were counted on Election Day, Gore trailed Bush by 537 votes in Florida. Nader received more than 96,000 votes, and some operatives attributed 10,000 of them to the airport issue. That was more than enough to elect a president who would support oil exploration in the Arctic National Wildlife Refuge, reverse his campaign promise to regulate carbon emissions, and enrage environmentalists like no president since Ronald Reagan. "Oh, I don't think the airport was a major factor in the outcome," Gore said in a recent interview.

Then he paused. "Well, maybe it was."

ON DECEMBER 11, 2000, the Gore campaign's last day in court, Senator Graham woke up in Miami Lakes at 6:05 A.M. According to his characteristically meticulous notebook, he weighed in at 187 pounds, ate some fiber cereal with raisins, and listened to six voice-mail messages. At 9:17

A.M., he flew to Washington on American Airlines flight 1394; he sat in seat 3A, and updated his notebooks for ten minutes at 10:30 A.M. After arriving in Washington, he purchased 11.833 gallons of gas at $1.599 per gallon at a Pennsylvania Avenue Amoco. Then he headed to the White House to celebrate the crowning achievement of his thirty-four-year political career. President Clinton was finally signing the Everglades bill, America's effort to restore Graham's boyhood playground, to re-create the watery wonderland that sheltered millions of wading birds before pioneers like his father began trying to tame it. For Graham, this was bigger than *Bush v. Gore.*

Graham liked to say that when Hamilton Disston first saw the panoramic sawgrass marshes of the Everglades, he must have thought: This doesn't look anything like Philadelphia. It looked strange and unique, and the young industrialist had been determined to convert it into something familiar and productive. But Graham liked strange and unique, as one might expect of a politician who recorded his breakfast choices every morning for posterity. Yes, restoring the Everglades would preserve aquifers and promote ecotourism, but Graham really wanted to restore the Everglades because it was singular, because it distinguished south Florida from other sprawling concatenations of tract homes, strip malls, CVS, and KFC. Marjory Stoneman Douglas had made that point with the first sentence of her book: "There are no other Everglades in the world." It was an American original, it was dying, and now it would receive open-heart surgery. Graham's only concern was that as years passed, billions of dollars were spent, and the patient remained critical, enthusiasm would wane, money would be diverted elsewhere, and the Everglades would be abandoned mid-operation.

Everyone at the bizarre bipartisan White House ceremony knew the Everglades still faced a multitude of threats. There were still 50,000 tons of phosphorus sitting at the bottom of Lake Okeechobee, and 2 million acres of exotic vegetation marching across the Everglades. Red tides were massacring dolphins, manatees, and sea turtles in the estuaries, while plagues and other diseases were killing off the coral reefs. Secretary Babbitt was concerned about the runaway sprawl that continued to chew up the edges of the ecosystem, forcing the Army Corps to paint its restoration masterpiece on an ever-shrinking canvas. Senator Smith threw his arm around Michael Davis, who was moving to Florida to oversee the restoration project, and gestured toward Governor Bush and his aides. "You've got to watch those guys," he whispered to Davis. "They're going to try to grab all that water." President Clinton, shooting the breeze with two legislative aides after the ceremony, flagged another dire threat: rising sea levels. "If we don't do something about climate change," he said, "your Everglades is going to be underwater."

But this was a day to imagine a better future, to reclaim the Everglades in a new way. It was now as unifying a force as it had been during the drainage era, except that the new consensus called for undraining it. While Florida was roiling over "undervotes" and "overvotes," everyone was holding hands over the swamp. "I'd be happy to speculate about the Supreme Court!" Graham told the swarm of reporters gathered outside the West Wing. He then grinned and returned to his preferred subject: "This is a very happy day for the Everglades, and a signal day for the movement around the world to try to repair damaged environmental systems." Smith pointed out that there were no alligators in New Hampshire; the Everglades seemed to transcend state lines, just as it seemed to transcend party lines. It had become a symbol of America's responsibility to make amends to Mother Nature. "We worked together to save a national treasure," Smith said. "It didn't get a lot of ink in what's going on today, but it's very, very important."

The power of the Everglades lay in its example. The twentieth century had been an era of mess-making; the twenty-first century could be a time to clean up the messes. And not just the toxic petrochemical messes that had set rivers on fire and thinned the shells of bald eagles during the sixties, but the ordinary messes created by man's routine dominion over nature. Man's efforts to tame the Everglades had taken a toll—the death tolls of the 1926 and 1928 hurricanes, the near-extinction of panthers and sparrows and gourds, the soil losses and water shortages and traffic jams on the Palmetto Expressway—but they had created homes and vacation destinations for millions of people, and more were on the way. Everglades restoration could set an international standard for sustainable development. It could prove that man and nature could coexist in peace.

After Governor Bush dodged more questions about his brother—"Marvin? He's doing well. That's very kind of you to think about him."—Babbitt stepped forward to talk about the Everglades as a model, a paradigm for thinking on a landscape scale. He suggested a partial list of endangered

American ecosystems that could follow south Florida's example: the Great Lakes, San Francisco Bay, New York Harbor, and the Missouri and Mississippi River basins.

Babbitt also mentioned the Louisiana coast, where—due largely to the enduring battle between the Corps and the Mississippi River—wetlands were disappearing at the astonishing rate of twenty-five square miles per year, decimating fish and wildlife while exposing New Orleans to storm surges. Governor Bush predicted that the ripples from the Everglades would extend even further than that: "This is a model—not just for our country, but for projects around the world."

THE UNANSWERED QUESTION WAS whether it would turn out to be a new model. Would it be a true restoration project, revamping man's approach to the Everglades, or just another dirt-moving Corps water project, "environmental infrastructure" with better press? Would it inaugurate a new relationship between the human and natural environments in south Florida, encouraging man to limit his footprint and live in harmony with the ecosystem, or would it just facilitate additional growth and sprawl, luring millions more people into the path of the next hurricane? Would politicians and engineers begin to consider the needs of birds, bears, and bays in addition to the needs of man, or would water continue to flow uphill toward money?

On the same day that President Clinton celebrated the new politics of the Everglades at the White House, while the Supreme Court prepared to choose his successor, a meeting in West Palm Beach suggested that the old politics of the Everglades was not quite dead. South Florida was suffering through one of its worst droughts ever, and Lake Okeechobee was so low that the water management district's guidelines prohibited releases for irrigation. But a consultant for the sugar industry had demanded to see the district's engineers, warning in an e-mail that "users will never sit still for zero water-supply releases." He got his meeting on December 11, and with no public input, the engineers agreed to tweak their guidelines so that growers could receive half their usual releases. That winter, the lake plunged below nine feet for the first time in recorded history. A third of the lake disappeared, along with most of its bass, and the region was battered so badly that Governor Bush declared an economic state of emergency. But the sugar industry enjoyed its fourth-largest harvest ever. "Thanks for all your work and for continuing to improve the process," the consultant wrote to the district's engineers.

That money-talks process has damaged the Everglades for more than a century, and it has damaged ecosystems around the world. CERP is supposed to change that, by making sure there is enough water for nature and the public as well as special interests. At a time when fresh water is emerging as the oil of the twenty-first century, Everglades restoration will be a crucial test of man's ability to stave off the bloody water wars that some analysts expect to erupt in the coming decades. If south Florida can't solve problems limited to one state in the wealthiest nation on earth, with billions of dollars to spend and fifty-five annual inches of rain to distribute, it's hard to imagine solving cross-border water disputes in poorer and drier regions. And south Florida has a trump card—the Everglades, the most beloved wetland on the planet, and the most intensely studied. If man can't save the Everglades, what can he save?

Senator Graham is probably the starkest example of the Everglades dilemma. His father was a sugar grower, a cattle rancher, and a real estate developer who dreamed of draining the Everglades. Graham launched the movement to restore the Everglades. But he also continued to support sugar farming, cattle ranching, and real estate development. He wouldn't have been the most popular politician in Florida if he hadn't. In fact, Graham's notes reveal that on the afternoon of December 11, after he watched fifteen minutes of MSNBC commentary about the Supreme Court hearing but before he bought a half-gallon of low-fat milk, the senator spoke to the Cuban-American leader Jorge Mas about the Homestead airport. Graham had declared that he would remain neutral and respect the Clinton administration's final decision, but everyone knew he was still pushing for the airport.

Graham still wanted to save the Everglades; he had started Save Our Everglades. But every politician had to strike a balance between nature and people.

ON JANUARY 16, 2001, four days before Clinton left office, the administration announced its decision on Homestead. It rejected the airport. It was too late to change the Nader votes of south Florida's environmentalists, but the Everglades had dodged another bullet. It was a reminder that money doesn't always talk. People talk, too. "This is a victory for common sense and public input over special interests," one activist said. It was also a victory for hard-line Ivory Soap environmentalists over the moderates who had considered the airport war a lost cause—although the greenest vice president in history turned out to be a casualty of that war.

As the new millennium dawned, the Everglades was not yet saved. But it was not yet doomed, either. Millions of acres of the ecosystem remained in public ownership. Water quality was improving. And America was now formally committed to restoring the Everglades, with billions of dollars and the prestige of a nation on the line. That didn't mean it would happen, but it meant there was a chance.

EPILOGUE

The Future of the Everglades

"It's Not Restoration!!!"

I N THE FIRST FEW YEARS of the twenty-first century, man has already proven that he can restore nature. The evidence is just above Lake Okeechobee, where engineers have filled in seven miles of the ruler-straight C-38 Canal, almost instantly re-creating fourteen miles of the zig-zagging Kissimmee River. Art Marshall's Alamogordo has been a grand success. The long-imprisoned river is once again overflowing its banks and watering its floodplain, transforming eleven thousand acres of drained pastures back to tangled marshes. Dominated by cattle and cattle egrets for three decades, the Kissimmee basin is again attracting waterfowl and wading birds. The river's bass fishery is recovering, its sandbars are reemerging, and gators are sunning themselves on its banks. "It's natural again," said Lou Toth, the South Florida Water Management District biologist who designed the project. "All we had to do was get out of its way."

Officials from Japan, England, Brazil, Italy, and Hungary have visited the Kissimmee to learn how to bring a river back from the dead. The secret, they have discovered, is to undo man's manipulations and let it flow. Man created the C-38 by digging a huge ditch and building five huge dams; he has begun to dismantle the C-38 by backfilling some of the ditch, blowing up one of the dams, and buying back 85,000 acres of pastures. Ultimately, dechannelizing one-third of the Kissimmee will cost at least ten times more than channelizing the entire river. But the project has already demonstrated that Mother Nature can be resurrected. Art Marshall was right: The river was still there.

*   *   *

THE LEADERS OF THE EVERGLADES restoration have hailed the Kissimmee restoration as their model, an inspiring exhibit of man's ability to atone for his ecological sins. But in 2002, shortly after he was named the district's employee of the year, Lou Toth told the *Washington Post* that the Everglades project would never replicate the Kissimmee's success—because it's a multi-purpose water project instead of a restoration project, led by engineers instead of scientists, tightening human control of nature instead of remov-ing barriers and letting nature heal itself. "They just don't get it," Toth said of his Corps counterparts. "I hate to say it, but these guys haven't learned anything about restoring an ecosystem."

Toth's bosses have demoted him and cut his pay. But as this book went to press in the fall of 2005, his warnings seemed prescient. The Shark Slough restoration in Everglades National Park was still paralyzed after sixteen years, and its price tag had quadrupled. The related effort to restore Taylor Slough was also stalled. And CERP was stumbling badly out of the gate. Many Americans think the Everglades has already been saved. In fact, while some of the Kissimmee is on the mend, the rest of the ecosystem is still in trouble.

WATER QUALITY IN THE EVERGLADES is better than it used to be, but it needs to be pristine, or else the marsh will continue to deteriorate. It's not pristine, and thanks to Big Sugar, Governor Bush, and the Florida legisla-ture, it's not clear when it will get pristine.

By 2003, the Everglades Forever Act had reduced phosphorus levels from 200 to 30 ppb, and the spread of cattails from six to two acres per day. But cattails were still spreading like a tumor, and phosphorus levels were still above 10 ppb, so the Everglades was still dying—just a bit slower than before. The cleanup's leaders had converted 40,000 acres of sugar fields into filter marshes in Phase One, but they had no plan for a Phase Two to achieve 10 ppb. Sugar growers were afraid that once Florida missed its December 2006 cleanup deadline, there would be more law-suits, and they would be forced to give up more land and cough up more money.

So Big Sugar turned again to the Republican-controlled legislature, unleashing a phalanx of forty-six lobbyists—including two former House speakers and two former gubernatorial chiefs of staff—on Tallahassee to amend Everglades Forever, seeking to extend the final deadline to 2026 and increase the final limit to 15 ppb. After an outcry from environmentalists and the Miccosukee Indians, Governor Jeb Bush's aides helped fashion a compromise bill that pushed the deadline back to 2016 while keeping the limit at 10 ppb—but with obvious enforcement loopholes. The legislature overwhelmingly passed it.

Senator Graham, Representative Shaw, and other key congressmen in both parties warned that the so-called Everglades Whenever Act would reinforce perceptions that Big Sugar controlled Florida, jeopardizing federal funding for the Everglades. Officials at Justice, Interior, and EPA were also appalled, but none said so publicly, because the president was Governor Bush's brother, and the White House had ordered federal agencies to defer to the state on Everglades issues. Even Judge Hoeveler denounced the bill—and after serving as an Everglades watchdog for a decade and a half, the eighty-one-year-old Hoeveler was removed from the case after the sugar industry complained that his criticism demonstrated bias. "Score one for Big Sugar and Governor Jeb Bush," the *Herald* wrote.

Judge Hoeveler, a Democrat, was replaced by Judge Federico Moreno, a Republican, and some activists thought the cleanup was doomed. But in June 2005, Judge Moreno ruled that Florida has violated the 1992 consent decree by allowing too much phosphorus in the Loxahatchee refuge, and chastised Governor Bush's aides for "stating 'all is well' and nothing more needs to be done except further meetings and studies." Moreno vowed that his legal remedy will include "<u>specific acts</u> to be performed and <u>specific dates</u> by when those acts must be completed."

At press time, Florida still had no plan to get 10 ppb. But Govenor Bush was holding meetings with officials in his brother's administration, trying to persuade them to adopt a joint strategy to get the consent decree dropped.

At least Judge Moreno is trying to move the cleanup forward. But there's no judge to jump-start the restoration of water flows in the Everglades. In March 2005, the Corps planner overseeing CERP in Washington warned in an internal memo that the project was already dramatically over budget, behind schedule, and off track: "It's different from what we told Congress we would do—and it's not restoration!"

The planner, Gary Hardesty, noted that "we haven't built a single project in the first five years of CERP." In fact, his memo continued, the Corps has not even built the "critical" pilot projects that were supposed to test the plan's uncertain technologies.

"I'm hearing statements like 'CERP is dead,'" Hardesty wrote.

The main problem has been money. The federal budget surplus of 2000 has given way to yawning deficits, thanks to sluggish economic growth, President Bush's tax cuts, the wars in Afghanistan and Iraq, and now Hurricane Katrina. The environment has not been a GOP priority, and as Hardesty wrote, soaring costs have been a "huge issue" for CERP, with $1 billion worth of overruns on the first four components alone. Overall, there are sixty-eight components, and the official cost estimate has climbed to $10.5 billion. One former Corps leader has predicted that restoration will eventually cost $80 billion, which would have made it by far the most expensive public works project in American history—before the rebuilding of New Orleans.

The Everglades is still a popular cause. Even President Bush, who will never be confused with John Muir, has hailed its restoration as the model for a "new environmentalism for the twenty-first century." But the enthusiasm of 2000 has faded, as the leading Everglades advocates have left Congress. Senators Mack and Graham retired. Senator Smith lost his seat to a Republican primary challenger; he moved to Florida and is now a professional Everglades activist. He was replaced as chairman of the Environment and Public Works Committee by James Inhofe, the project's sole opponent in the Senate. Under Inhofe, the committee has shown zero interest in the Everglades. "There have been no hearings, no requests for briefings, and no general inquiries about CERP," Hardesty wrote. And President Bush's budget aides have held up one of the few CERP projects popular with environmentalists and economic interests, a $ 1 billion effort to restore the St. Lucie estuary.

At the state level, Governor Bush has kept his funding promises to the Everglades. But he has also practically seized control of the restoration project, which has only intensified its emphasis on Florida residents and businesses over national parks and refuges. In October 2004, he unveiled a $1.5 billion plan to accelerate eight CERP components, most of them water-supply reservoirs. His "Acceler8" plan did include a few environmental components—Florida is already filling in the canals the

Rosen brothers once dredged in Golden Gate Estates—but the Audubon Society and the Nature Conservancy were the only prominent environmental groups to endorse it. After the Everglades Whenever dispute, Govenor Bush made it clear he doesn't care what environmentalists think. "We don't need their permission to save the Everglades," he told aides.

CERP is not dead, but it still relies on the technical plan that was savaged by Everglades National Park, the plan that only aimed to provide 70 percent of the park's historic flows by 2036—and now its water-supply components are more front-loaded than ever. It still depends on aquifer storage wells to store billions of gallons of water, even though its leaders are losing faith in that technology. And it still proposes only token efforts to restore natural flow, even though a water management district scientist named Christopher McVoy has demonstrated that natural flow was vital to the natural Everglades. McVoy's work was dismissed at first—in part because he's a quirky peace activist who performs in a Chilean dance troupe and teaches yoga on the side—but it has been endorsed by the National Academy of Sciences. Now even CERP managers say they plan to focus on water flow as well as quality, quantity, timing, and distribution.

So far, though, saying has not translated into doing. CERP's great innovation was supposed to be its tremendous flexibility, but for the most part, the original plan is still the plan. The 1999 modeling that helped persuade many environmentalists that more top-to-bottom flow could provide real environmental benefits has been shelved. The extra 79 billion gallons that were supposed to go to the park are still "under study." Many of the environmentalists who helped push the plan through Congress are feeling pangs of buyer's remorse. "We were so focused on passing the bill, we glossed over a lot of ugly details," says David Guggenheim, the cochair of the Everglades Coalition in 2000. The Sierra Club has withdrawn its support for CERP, and even Nathaniel Reed, who testified that year that CERP would "unequivocally" restore the Everglades, warned in a 2005 e-mail that "we are witnessing the potential end of a great experiment in restoration."

Reed usually trusts the good intentions of public servants, and at seventy-two, he is desperate to see the Everglades restored in his lifetime. But he sometimes wonders if he was gullible to support CERP. "I get so depressed," he says. "I'm not getting any younger."

A Threatened Landscape

IF LAKE OKEECHOBEE can still be considered the heart of the Everglades, then as this book went to press, the ecosystem was having a massive coronary. In the summer of 2004, four hurricanes blasted through Florida, churning up giant globs of phosphorus that had accumulated on the lake's bottom, wiping out its native vegetation, slathering its surface with coffee-colored crud. Scientists began warning that the 730-square-mile lake was becoming a dead zone, and water managers began flushing even more of the turbid lake into the fragile estuaries, which were clobbered by red tides so toxic that beachgoers had to wear surgical masks. Environmentalists sued, and the *Palm Beach Post* asked: "R.I.P. Lake Okeechobee?" In the fall of 2005, Hurricane Wilma ripped across the Everglades and pummeled the lake once again, taking several sizable chunks out of the Hoover Dike. There was no breach, but a few more hours of pounding could have created a catastrophe. And the lake's coffee hue has darkened from latte to espresso.

The hurricanes were natural disasters, but the collapse of Lake Okeechobee is a man-made disaster. Ever since the lake was imprisoned by the Hoover Dike after the killer storm of 1928, it has been used as a reservoir and a sewer for farms, dairies, and cities. The control of the lake has enabled people to live and prosper in south Florida, but the slimy gunk that is suffocating the lake is a legacy of those people. So are the red tides that have increased 1,500 percent in the Caloosahatchee estuary, and the invasive Old World climbing fern that is spreading through the Everglades like kudzu, and the water shortages plaguing a region that leads the nation in per capita water use. Man's impact has pervaded the ecosystem, from Disney World to Key West, from the herbicide-resistant hydrilla clogging the Kissimmee chain of lakes to the sixty-mile-wide cloud of "black water" that wiped out half the coral in western Florida Bay in 2002. Nature is clearly out of whack in south Florida. The watershed has dozens of parks and wildlife refuges, not to mention highway underpasses to protect panthers from cars and no-wake zones to protect manatees from boats, but sixty-eight of its species are still on the endangered list.

The Everglades has dodged the bullets of the Big Cypress and Homestead airports, but the ecosystem still faces a variety of threats. The Collier family has retained the right to drill for oil in Big Cypress. Rock miners are still shredding wetlands in the eastern Everglades. As President Clinton observed, rising seas could inundate the entire ecosystem. Invasive species are proliferating, from bugs that suck the sap out of native trees to seaweed that blankets native coral to pythons that entertain tourists by attacking native alligators. The problem was demonstrated recently in gruesome allegorical form when an exotic twelve-foot python devoured a native six-foot gator, then died when its stomach ruptured after the scaly meal. Meanwhile, new attention is being paid to scientific theories that the collapse of Florida Bay, once attributed to "hypersalinity" caused by the reduction of freshwater flow through the Everglades, was actually triggered by nitrogen pollution. If those theories are correct, much of the entire restoration project may have to be revamped.

THE MOST DAUNTING THREAT to the Everglades is the runaway development that is still wiping out its wetlands and stressing its aquifers. The Miami–Fort Lauderdale–West Palm Beach conurbation has become America's sixth-largest metropolitan area, obliterating almost every patch of green space between the Atlantic and the perimeter levee. Postwar Everglades suburbs such as Coral Springs, Hialeah, Miami Gardens, Miramar, Pembroke Pines, and Sunrise have all attracted 100,000 residents, and are approaching build-out. Southeast Florida's office sprawl is just as intense; one study declared the region "the most centerless large office market in the U.S.," the ultimate "Edgeless City."

Westward sprawl has become the area's hottest political issue. Miami-Dade County has already approved two developments outside its "urban service boundary"—one built by Governor Bush's former business partner—and is now embroiled in a battle over proposals to shift the entire boundary west and south. Broward County's western frontier is almost completely paved. And in Palm Beach County, a war is raging over the Scripps Research Institute's plan for a 2,000-acre biotechnology campus at the edge of the Everglades, bolstered by $369 million in subsidies from Governor Bush and the legislature. It is easy to see the allure of biotech as an economic engine that could wean the region from its dependence on tourism and real estate, while providing life-saving medical research near God's Waiting Room. But it is hard to see why the engine has to be located along the Loxahatchee River's headwaters, on pristine wetlands and rural farmland with no access to existing roads or sewers. Environmental critics—led by Nathaniel Reed, with funding from Paul Tudor Jones—have denounced the current plan as a billion-dollar development play disguised as a scientific venture, and have sued to try to force Scripps to choose a less destructive site closer to the coast.

Then again, one proposed Scripps site could have spawned even more sprawl and ecological damage: The Fanjul brothers wanted the campus built in the Everglades Agricultural Area. Their offer was rejected, but it has launched a debate about the 700,000-acre farm empire, a debate that could determine the fate of the Everglades. Predictions that soil subsidence would eliminate farming in the upper Glades by 2000 have been proven wrong, but the soil is still shrinking, land prices are still soaring, and sugar growers are besieged by low-carb diets, artificial sweeteners, and free-trade agreements. They have warned that when it is no longer profitable to grow sugar in the Everglades, they will grow condos instead. And sugar farming is a relatively benign use of the muck, compared to hundreds of thousands of homes.

Everglades activists still dream of converting the sugar fields into reservoirs, and perhaps even flowways reconnecting Lake Okeechobee to the River of Grass. But in the coming years, their top priority will be preventing the conversion of sugar fields into bedroom communities. There are already 30,000 residents in eager-to-expand Everglades Agricultural Area communities such as Belle Glade, South Bay, and Pahokee, and the fast-growing horse town of Wellington—where Tommy Lee Jones plays polo, and Bruce Springsteen and Michael Bloomberg take their daughters to equestrian competitions—is also maneuvering to expand west. U.S. Sugar and the Fanjul interests are developing plans for new subdivisions and rock mines, and Governor Bush is convening a commission to study the future of development in the agricultural area. As strange as it sounds, environmentalists may come to yearn for the days when Big Sugar ruled the upper Glades.

THE OTHER MAJOR BATTLEGROUND in the Everglades sprawl wars is southwest Florida, which is rapidly expanding eastward into the peninsula's watery interior. Environmental agencies have been helpless in the face of the intense development pressure—and equally intense political pressure. "We are permitting in SW Florida as fast as we can the same types of development and associated environmental degradation we are spending billions of dollars trying to fix on the SE coast," the EPA's south Florida director e-mailed the top Army Corps regulator in Florida. "Haven't we learned our lessons? Apparently not!"

History repeats itself daily in the Naples–Fort Myers–Cape Coral area, where regulators who raise red flags about impacts to water quality, water flows, and endangered species are routinely overwhelmed by the political clout of developers such as recently retired WCI Communities CEO Al Hoffman, who also served as President Bush's campaign cochairman and Governor Bush's finance chairman, and is now U.S. ambassador to Portugal. The Fish and Wildlife Service recently admitted that the science it has used to rubber-stamp thousands of homes in panther habitat was flawed. An environmental impact study of the region's growth has languished in the bowels of the Army Corps bureaucracy for almost a decade. And the EPA's top regulator in southwest Florida quit after President Bush's appointees began pushing a developer-funded study claiming that natural wetlands caused pollution.

On the east coast, the perimeter levee has served as a final limit to westward development into the Everglades, but on the west coast, there is no levee to stop the eastward surge of driveways, highways, and fairways. Misnamed subdivisions keep steamrolling wetlands and farmland: Old Cypress, Winding Cypress, Naples Lakes, Collier Lakes. Southwest Florida is already getting a taste of southeast Florida's traffic jams, lawn-watering restrictions, polluted beaches, overstuffed schools, and a vanishing sense of place. Even Governor Bush's developer-friendly Growth Management Commission, chaired by Mel Martinez, an Orlando Republican who is now a U.S. senator, warned that sprawl was out of control throughout south Florida. "The developers are very, very powerful, but obviously something has got to change," says Martinez, who also served as President Bush's housing secretary. "We're going to lose the Everglades. We're already losing quality of life." In coming decades, as sprawl marches east into the Everglades and west into the Everglades Agricultural Area, south Florida could become an uninterrupted asphalt megalopolis stretching from Naples to Palm Beach. Perhaps it could be called Napalm Beach.

This is a constant refrain in south Florida, especially from newcomers who believe their paradise is being spoiled by additional newcomers. When the *Herald* ran a series of articles on Broward County's sprawl, the reader reaction was furious. "It's time to stop the growth: The concreting and asphalting over of everything green," a Pembroke Pines resident wrote. "My belief is, enough is enough. I live in Weston and we have gone far enough into the Glades," another reader agreed. "To the builders and developers and to our county, state and federal legislators: No, a thousand times no, on moving westward," said a Sunrise man. "We need green. We need the Everglades. We don't need more buildings," added a Davie woman.

Of course, those antisprawl letter-writers all lived in sprawling suburbs in the former Everglades. Now that they were settled in their gated communities, they wanted to slam the gate behind them. It is easy to fulminate about the costs of south Florida's growth—its gridlock, environmental degradation, inadequate municipal services, and cookie-cutter landscape—but there is no denying the allure of its 75-degree January afternoons. Even if south Florida fails to manage its growth or preserve its natural beauty, it will still be more attractive than Cleveland or Buffalo in the winter. And even if it fails to diversify its economy or protect its aquifers, it will still look like paradise to residents of Havana or Caracas. Some observers warn that Florida real estate is as overvalued now as it was before the 1926 hurricane, but the bubble didn't burst after the four Florida hurricanes of 2004. People are still flocking to the sunshine, and the land rush is expected to accelerate as heat-seeking baby boomers reach retirement age. The Hoover Dike leaks when Lake Okeechobee gets high, and the Corps say it could fail if lake levels rise seven feet above normal, unleashing the monstropolous beast on millions of people. Hurricane forecasting has improved dramatically since 1928, but it's not perfect, and the unexpected failure of Corps floodwalls during Katrina was a reminder that federal engineering isn't, either.

CERP is designed to feed south Florida's growth addiction, not to cure it. The project aims to supply enough water to help the region double its population, which will increase the demands on aquifers and wetlands that prompted the project in the first place. In fact, one little-noticed CERP provision will launch a $12 million study of a future CERP-style restoration project for southwest Florida. So while public officials are spending billions to repair the damage of past development in southeast Florida, they are already preparing to spend billions more to mitigate future development on the other side of the peninsula. They know they are in a hole, but they seem intent to keep digging.

Nature is resilient, and the Kissimmee restoration shows that drainage can be reversible. But development is harder to undo. And if the last century of human meddling with the Everglades has proven anything, it's that ecological damage is easier and cheaper to prevent than it is to reverse. CERP's leaders have been commissioned to paint a restoration masterpiece, but their canvas is shrinking. Lake Okeechobee may have reached its tipping point, that "snowballing degeneration of major resources" that Art Marshall predicted years ago. No one knows where that point lies for the Everglades.

*   *   *

Eden Again

AT THE EVERGLADES COALITION CONFERENCE in Miami Beach in 2004, an engineer named Azzam Alwash told a gripping story of ecological destruction. He described a shallow-water marsh that had once seemed endless, supporting hundreds of species of flora and fauna, as well as native people who had thrived on its abundant fish. The marsh had been desiccated by ditches and dikes—Alwash called this the "environmental crime of the century"—and a unique native culture was destroyed along with it. But Alwash vowed that the marsh would be restored, along with the region's water supply. "It is parched today, but it can be wet tomorrow," he said. "Nature is wonderful, isn't it?"

Alwash was not talking about the Everglades. He was describing the "Garden of Eden" marshes between the Tigris and Euphrates Rivers in Iraq, a wetland twice the size of the Everglades. Some scholars believe these Mesopotamian marshes supported the original Garden of Eden; they later became the cradle of western civilization, and eventually the homeland for 250,000 "Marsh Arabs" who plied their waters in kayaks. After the first Gulf War in 1991, Saddam Hussein punished the Marsh Arabs for a failed uprising by draining their swamp, building massive canals with names like Mother of All Battles, Loyalty of the Leader, and Saddam, converting 6,000 square miles of wetlands into desert. But after the fall of Baghdad in 2003, Alwash and other Iraqi exiles began pushing to reflood Eden. Their model, Alwash said, is the effort to restore the Everglades. "We want to do what you're doing," said Alwash, whose father was once the region's irrigation engineer. He wants to turn back time, and bring back Eden.

Today, with some help from the Army Corps, the Eden Again project has begun. In fact, as Governor Bush and Senator Graham predicted after the historic bill-signing in 2000, the Everglades is becoming a restoration model for damaged ecosystems around the globe, including the Danube and Nile Rivers, the Black, Baltic and Aral Seas, the Pantanal wetlands of Brazil, and the Okavango Delta of Botswana. The World Bank has cited CERP as a paradigm for sustainable development, a worldwide guide for resolving the water conflicts that could dominate twenty-first-century geopolitics. And as Secretary Babbitt suggested that chilly December day in 2000, CERP is already the restoration blueprint for America, inspiring multibillion-dollar megaprojects for Chesapeake Bay, San Francisco Bay, the upper Mississippi River, and the

Great Lakes. "This will be modeled after the landmark Florida project to save the Everglades," Virginia governor Mark Warner said of the Chesapeake initiative. The Corps now spends 20 percent of its budget on environmental work, a figure likely to increase.

The most ambitious Everglades knockoff is a $14 billion plan to restore Louisiana's tattered web of coastal marshes, which a bipartisan coalition of Louisiana interest groups has rebranded as "America's Wetland," a pitch modeled on "America's Everglades." The Corps helped ravage those marshes, by straightjacketing the Mississippi and choking off the natural delta-building process that carried its silt to the coast. So the coalition has argued that the Corps has a duty to fix its mistakes in southern Louisiana, just as it is doing in the Everglades. The Louisianans have also noted that the marshes provided natural hurricane protection for much of America's oil and gas infrastructure, and for the low-lying city of New Orleans. President Bush's aides have pressured Louisiana to scale back its ambitions, but that was before August 29, 2005, when Hurricane Katrina roared up the Gulf, buckled several Corps floodwalls, and inundated New Orleans, the most expensive natural disaster in U.S. history.

Katrina refocused attention on the Louisiana coast and its disappearing natural buffer. But the storm also refocused attention on the Corps and its priorities. For four decades, scientists and local critics urged the Corps to close its little-used Mississippi River Gulf Outlet, the so-called hurricane highway that had eroded 20,000 acres of nearby marshes. But the Corps insisted the outlet was economically justified, even though cargo ships avoided it like an iceberg. Now experts have calculated that the outlet amplified Katrina's surge by as much as 20 percent, and that the floodwall failures were concentrated in areas where marshes had eroded. America's skewed water priorities were even starker along the nearby New Orleans Industrial Canal, where underdesigned and underfunded Corps floodwalls collapsed during Katrina, a stone's throw from a $750 million navigation boondoggle the Corps was building for port interests. Before Katrina, the Corps already spent more in Louisiana than any other state, most of it on work that had nothing to do with hurricane protection. Louisiana also had the second-largest Corps construction budget—just behind Florida, thanks to Everglades restoration.

The Corps will probably receive even more money and power in the aftermath of its Katrina failures, because it's still America's flood-control agency, and New Orleans clearly needs new protection. Katrina may also provide the impetus for the $14 billion coastal restoration, which would give the Corps a new platform even larger than the Everglades. But the Corps has not yet proved it's up to the challenge. It's now America's environmental restoration agency, but it's struggling to develop an environmental culture. It's learned to talk green, but it's still the Corps of Engineers.

PRESIDENT BUSH may be right: Ecosystem restoration may be the new environmentalism of the twenty-first century. In the twentieth century, conservationists tried to stop the destruction of nature—first by protecting beloved species, such as wading birds, and beloved places, such as Paradise Key; later by cracking down on air and water pollution, while offering new protections to wetlands and endangered species. Today, America's air and water are much cleaner. Its rivers no longer catch fire; its bald eagles are no longer endangered. The national wildlife refuge system that began with five acres at Pelican Island now protects 96 million acres. Suburban sprawl and invasive species still pose serious threats to nature in America, and global warming is a twentieth-century-style pollution problem that the country cannot ignore much longer. But reviving entire ecosystems is the challenge of the future. It will require Americans to think on a landscape scale, to clean up their own messes, to gore someone's ox now and then. It will require the Corps to embrace its environmental mission as more than a new way to move dirt, to change its culture as well as its rhetoric, to surrender some of its historical battlefield to Mother Nature.

On a landscape scale, restoring the Everglades would help people as well as panthers and periphyton. But the Everglades is more than a test of our ability to help ourselves. You don't have to worship Gaia or God to sense that we have done something wrong to the earth in south Florida; you just need to drive through the region's strip-mall hellscapes. There is only one Everglades, and we have just about destroyed it. It is our ability to recognize this, and to make amends, that sets us apart from other species.

Before the war in Iraq, Secretary of State Colin Powell invoked the "Pottery Barn rule" for invading sovereign nations: You break it, you own it. The same rule should apply to ecosystems. We broke the Everglades, so we ought to fix it. "The Everglades is a test," the environmentalists say. "If we pass, we may get to keep the planet." It is a test of our scientific knowledge, our engineering prowess, and our political will. It is a test of the concept of sustainable development. But most of all, the Everglades is a moral test. It will be a test of our willingness to restrain ourselves, to share the earth's resources with the other living things that moveth upon it, to live in harmony with nature. If we pass, we may deserve to keep the planet.

Notes

A Note on Sources

All the facts in this book came from letters, diaries, pamphlets, reports, and other original documents; interviews with more than 1,000 contemporary sources; newspaper, magazine, and journal articles; and books by other authors. I did not attempt to guess what anyone must have said or thought; the quotations all come from documents or interviews. The only sections of the text that have necessarily involved a bit of speculation are my descriptions of the natural Everglades. I consulted scores of descriptions of the ecosystem before drainage, mostly from the nineteenth century, but there is no way to be sure precisely what it looked like in prehistoric times.

These notes provide specific sources for almost all the information in the book. But for the sections that rely heavily on interviews—especially the description of the Everglades in the first chapter, and the modern history in the last few chapters—I listed my most important sources on general topics in the notes, rather than attribute every specific fact to everyone who mentioned it. In any event, I am indebted to everyone who spoke to me, as well as the journalists and authors who tackled south Florida before me. I do want to draw attention to four unfortunately obscure works that helped me immeasurably: Julius Dovell's unpublished 1947 dissertation *A History of the Everglades of Florida,* Alfred Jackson Hanna and Kathryn Abbey Hanna's 1948 *Lake Okeechobee: Wellspring of the Everglades,* Nelson Blake's 1980 *Land into Water, Water into Land: A History of Water Management in Florida,* and David McCally's 1998 *The Everglades: An Environmental History.*

Some of my reporting about the modern plan to restore the Everglades appeared in a four-day series in *The Washington Post* in June 2002, "The Swamp." I also drew on some of my *Post* stories from 2000 about problems at the U.S. Army Corps of Engineers and from 2005 about Hurricane Katrina. I have written two essays for *The New Republic* about south Florida, "Water World" (2/24/2004) and "Swamp Things" (11/8/2004), as well as an article about the sugar industry, "Sugar Plum" (4/24/2003). I also wrote a piece for *Slate* about Everglades restoration, "Swamp Thing" (6/15/05).

Abbreviations for Archives:

ABER   Amos Beebe Eaton Papers, Archives and Special Collections, Otto G. Richter Library, University of Miami, Coral Gables, FL.

ACEJD   Army Corps of Engineers Archives, Jacksonville District, Jacksonville, FL.

ACEHQ   Army Corps of Engineers Archives, Andrew A. Humphreys Engineer Center, Alexandria, VA.

ARMP   Arthur R. Marshall, Jr., Papers, Special and Area Studies Collections, George A. Smathers Libraries, University of Florida, Gainesville, FL.

AMP   Arthur E. Morgan Papers, Special and Area Studies Collections, George A. Smathers Libraries, University of Florida, Gainesville, FL.

ECP   Ernest Coe Papers, Everglades National Park, Homestead, FL.

EGP   Ernest Graham Papers, Special and Area Studies Collections, George A. Smathers Libraries, University of Florida, Gainesville, FL.

EL   Everglades Litigation Collection, University of Miami, digitized documents at http://exchange.law.miami.edu/everglades/

FEF   Fred C. Elliot file, Title and Land Records Section, Division of State Lands, Florida Department of Environmental Protection, Tallahassee, FL.

FHC   Florida Heritage Collection. Digitized documents at http://susdl.fcla.edu/fh/.

FSA   Florida State Archives, R. A. Gray Building, Tallahassee, FL.

GNBP   Governor Napoleon B. Broward Papers, R. A. Gray Building, Florida State Archives, Tallahassee, FL.

GRAP   Governor Reubin O. Askew Papers, R. A. Gray Building, Florida State Archives, Tallahassee, FL.

GSHC   Governor Spessard L. Holland Papers, R. A. Gray Building, Florida State Archives, Tallahassee, FL.

HFP   Henry Flagler Papers, Henry Flagler Museum, Palm Beach, FL.

HHP   Herbert Hoover Papers, Herbert Hoover Presidential Library, West Branch, IA.

HMSF   Historical Museum of South Florida archives, Miami, FL.

HSPBC   Historical Society of Palm Beach County archives, Palm Beach, FL.

JPP   John Pennekamp Papers, Special and Area Studies Collections, George A. Smathers Libraries, University of Florida, Gainesville, FL.

MSDP   Marjory Stoneman Douglas Papers, Archives and Special Collections, Otto G. Richter Library, University of Miami, Coral Gables, FL.

NA   National Archives, Washington, DC.

NBP   Napoleon B. Broward Papers, Special and Area Studies Collections, George A. Smathers Libraries, University of Florida, Gainesville, FL.

RTE   Reclaiming The Everglades: South Florida's Natural History, 1884–1934. Digitized documents at http://everglades.fiu.edu/reclaim.

SFWMD   South Florida Water Management District archives, West Palm Beach, FL.

SHP   Spessard L. Holland Papers, Special and Area Studies Collections, George A. Smathers Libraries, University of Florida, Gainesville, FL.

TWP   Thomas E. Will Papers, Special and Area Studies Collections, George A. Smathers Libraries, University of Florida, Gainesville, FL.

WJP   William S. Jennings Papers, Special and Area Studies Collections, George A. Smathers Libraries, University of Florida, Gainesville, FL.

ZTP   Zachary Taylor Papers, Library of Congress Manuscript Collection, Washington, DC.

I also received modern documents from dozens of sources, but I would especially like to thank Joe Browder, who allowed me to peruse his extensive collection of Everglades records, and Robert Mooney, who constantly posts valuable information on the Everglades Commons. Stephanie Daigle, C. K. Lee, Joette Lorion, Catherine Ransom, Tim Searchinger, and Brad Sewell also provided useful documents.

Introduction   "A Treasure for Our Country"

PAGE

1   *Senator Robert Smith of New Hampshire:* Robert Smith recalled his thoughts and actions on December 11, 2000, in interviews and e-mails. He says he sat about fifteen rows from the bench, facing Justice Clarence Thomas.

1   *a former small-town civics teacher:* Philip D. Duncan and Brian Nutting, *CQ's Politics in America 2000,* (Washington: Congressional Quarterly, 1999), p. 838. Interviews with Smith.

2   *an unabashed ideologue:* Senator Smith was named the most conservative senator by the right-wing John Birch Society, ahead of North Carolina's Jesse Helms. He was also named the most fiscally responsible senator by the National Taxpayers Union. He received a 100 percent rating from Citizens Against Government Waste.

2   *"just a can of Coke per citizen per day":* Senate Environment and Public Works Committee Hearing, 5/11/2000, Federal News Service transcript.

2   *some of the Democratic Party's top environmentalists:* I reconstructed the scene at the White House by interviewing almost everyone there, including Tom Adams, Secretary Babbitt, Mary Barley, Michael Collins, Michael Davis, Mary Doyle, Frampton, Carrie Meek, Senator Bob Graham, Gary Guzy, Bill Leary, John Podesta, Senator Smith, David Struhs, and Joseph Westphal.

2   *"The last time I was here":* The lobbyist was the Audubon Society's Tom Adams.

2   *One Clinton appointee:* That was Joseph Westphal, the assistant Army secretary who oversaw the Army Corps of Engineers.

2   *Jeb even said hi to a Miami congresswoman:* That was Carrie Meek, who is the daughter of a sharecropper and the granddaughter of a slave. "I was polite to the governor," she recalled, "but my rage ran deep."

2   *Jeb's top environmental aide:* That was Florida Department of Environmental Protection chief David Struhs.

16 *South Florida ended up:* As any barefoot beachgoer knows, land gets hotter than water under the summer sun; in south Florida, this disparity creates sea breezes that carry moist air inland, where it rises and forms clouds through a process called convection. This is the source of most of the region's summer showers. There is less convection over the cooler waters of Lake Okeechobee, so much less that satellite photos sometimes show clouds blanketing all of south Florida except for a round gap over the lake. South Florida's weather is also influenced by the warm currents of the Gulf Stream, which skirts the peninsula like a river within the sea. James A. Henry, Kenneth M. Portier, and Jan Coyne, *The Climate and Weather of Florida* (Sarasota: Pineapple Press, 1994), pp. 7–8, 31–34, 101–5.

17 *in its natural state, 70 percent of the region:* In south Florida, there is practically no distinction between surface water and groundwater.

18 *It is a common nutrient in nature:* Phosphorus is found in colas, fine chinas, baking powders, and matches, as well as most commercial fertilizers. Neil Santiello, "Glades Element at Center of Dispute, *Fort Lauderdale Sun-Sentinel,* 5/25/2003.

18 *"The water is pure and limpid":* Smith, "The Report of Buckingham Smith," p. 28.

19 *"They do not attempt to fly":* Archie Williams, "Across South Central Florida in 1882: The Account of the First New Orleans Times-Democrat Exploring Expedition, Part I," edited by Morgan Dewey Peoples and Edwin Adams Davis, *Tequesta* 10 (1950), p. 73.

19 *"Feasted sumptuously on wild turkey":* George Henry Preble, "A Canoe Expedition into the Everglades in 1842," reprinted in *Tequesta* 15 (1945), p. 40.

20 *"Their number and variety are simply marvelous":* Willoughby, *Across the Everglades,* p. 67.

20 *In fact, the southwest edge of the Everglades:* Interviews with Michael Russo and Margo Schwadron; *see also:* Robin C. Brown, *Florida's First People: 12,000 Years of Human History* (Sarasota: Pineapple Press, 1994), pp. 31–38; Michael Russo, et al., "Final Report on Horr's Island: The Archaeology of Archaic and Glades Settlement and Subsistence Patterns," submitted to Key Marco Developments, 1991; Michael Russo, "Why We Don't Believe in Archaic Ceremonial Mounds and Why We Should: The Case from Florida," *Southeastern Archaeology* 13, no. 2 (winter 1994), pp. 93–109.

20 *Horr's Island, a squiggle-shaped clump of mangroves:* In the 1990s, image-conscious developers incorporated Horr's Island into the upscale community of Key Marco. Apparently, prospective buyers were getting the wrong idea when they heard the name "Horr's," unaware it was named for a nineteenth-century settler.

20 *where a twentieth-century entomologist:* Raymond F. Dasmann, *No Further Retreat: The Fight to Save Florida* (New York: Macmillan, 1971), p. 140.

21 *Calusa Indians controlled southwest Florida:* I relied on the following books and articles for general information about the Calusa and other native Florida tribes: John W. Griffin, *The Archaeology of the Everglades,* Jerald T. Milanich and James J. Miller, eds. (Gainesville: University Press of Florida, 2002); William E. McGoun, *Prehistoric Peoples of South Florida* (Tuscaloosa, AL: University of Alabama Press, 1993); Randolph J. Widmer, *The Evolution of the Calusa: A Nonagricultural Chiefdom on the Southwest Florida Coast* (Tuscaloosa: University of Alabama Press, 1988); Robert S. Carr and John G. Beriault, "Prehistoric Man in South Florida," in Gleason, ed., *Environments of South Florida;* Jerald T. Milanich, "Original Inhabitants," in Gannon, ed., *The New History of Florida;* Charlton W. Tebeau, *Man in the Everglades: 2000 Years of Human History in Everglades National Park* (Coral Gables: University of Miami Press, 1968); McCally, *The Everglades.*

21 *"men of strength":* Hernando d'Escalante Fontaneda, *Memoir of d'Escalante Fontaneda Respecting Florida, Written in Spain, about the year 1575* (reprint of the 1854 translation by Buckingham Smith, Miami: University of Miami, 1944), pp. 20–21.

21 *Fontaneda catalogued the marine cuisine:* Ibid., pp. 12–13.

22 *One archaeologist was amazed:* Frank Hamilton Cushing, *Exploration of Ancient Key-Dweller Remains on the Gulf Coast of Florida* (originally published in 1896; Gainesville: University Press of Florida, 2000), pp. 64, 101–2. Cushing's excavation of Marco Island is still considered one of the great digs on U.S. soil.

22 *"The men onely use deere skins":* John Sparke, *The Voyage Made by M. John Hawkins Esquire, 1565* (originally reprinted in *Early English and French Voyages;* New York: Charles Scribner's Sons, 1906; published electronically by Wisconsin Historical Society, www.americanjourneys.org).

23 *Calusa chiefs performed human sacrifices:* Rene Goulaine de Laudonniere, *History of the First Attempt of the French (The Huguenots) to Colonize the Newly Discovered Country of Florida.* Reprinted in *Historical Collections of Louisiana and Florida,* B. F. French, ed. (New York: J. Sabin & Sons, 1869; published electronically by the Wisconsin Historical Society, www.americanjourneys.org); McCally, *The Everglades,* pp. 46–47.

23   *"In view of the fact that they lived there"*: Charlton W. Tebeau, *Florida's Last Frontier: The History of Collier County* (Coral Gables: University of Miami Press, Coral Gables, 1966), p. 28.

2   The Intruders

24   *"We appeal to the Great Father"*: John T. Sprague, *The Origins, Progress and Conclusion of the Florida War* (originally published in 1848; Gainesville: University Press of Florida, 1964), p. 57.
24   *But the most illuminating parts of Sparke's account:* Sparke, *The Voyage,* pp. 120–128.
24   *"islands surrounded by swamplands"*: Bartolome Barrientos, *Pedro Menéndez de Avilés,* translated from the 1567 ed. by Anthony Kerrigan (Gainesville: University Press of Florida, 1965), pp. 24–26.
25   *the first white intruder in Florida:* The hostile reaction of the Calusa suggests that they might have received an earlier visit from whites, but none has been documented. Frederick T. Davis, "History of Juan Ponce de León's Voyages to Florida: Source Records," *Florida Historical Society Quarterly* 14, no. 1 (July 1935), pp. 3–70; Widmer, *The Evolution of the Calusa;* Michael Gannon, "First European Contacts," in Gannon, ed., *The New History of Florida,* pp. 16–21; Charlton W. Tebeau, *A History of Florida* (Coral Gables: University of Miami Press, 1971), pp. 20–22.
25   *"gold and other metals"*: Davis, "Source Records," pp. 9–14.
25   *After landing near Cape Canaveral:* Ponce initially headed south down the Atlantic coast to the islands of the Keys, which he named Los Martires because he thought they looked like suffering men. He then headed west to the Dry Tortugas, which he named for their loggerhead turtles. He then turned north up the Gulf coast.
25   *"The natives of the land"*: Davis, "Source Records," p. 60.
26   *The next conquistador:* The rapacious Panfilo de Narvaez landed in Tampa Bay in 1528, and the even more despicable Hernando de Soto followed suit in 1539, but both headed north in search of gold. Narvaez was lost at sea; de Soto died of a fever after discovering the Mississippi River, which at the time was known as the Rio Grande de la Florida.
26   *Pedro Menéndez de Avilés:* I relied on the following books and articles about Menéndez: the eyewitness account of Gonzalo Solis de Meras, *Pedro Menéndez de Avilés: Memorial,* translated from the 1567 ed. by Jeannette Thurber Connor (Gainesville: University of Florida Press, 1964); the contemporary account by Barrientos, *Pedro Menéndez de Avilés;* a twentieth-century biography by Albert Mauncy, *Florida's Menéndez: Captain General of the Ocean Sea* (St. Augustine: The St. Augustine Historical Society, 1965); original documents edited by Eugene Lyon, *Pedro Menéndez de Avilés* (New York: Garland Publishing, 1995); Eugene Lyon, "Settlement and Survival," in Gannon, ed., *The New History of Florida,* pp. 40–46; Tebeau, *A History of Florida,* pp. 32–38; Douglas, *The Everglades: River of Grass,* pp. 148–167.
26   *America's oldest settlement:* St. Augustine residents like to point out that their city was ready for urban renewal by the time the Pilgrims arrived at Plymouth Rock.
26   *He was a muscular, strange-looking man:* Mauncy, *Florida's Menéndez,* p. 14.
26   *In an early example of Florida boosterism:* Douglas, *The Everglades: River of Grass,* p. 149.
26   *supplies that included 3,182 hundredweight of biscuits:* Lyon, *Pedro Menéndez de Avilés,* pp. 105–6.
26   *"At them!"*: Solis de Meras, *Memorial,* p. 100.
26   *a war of fire and blood:* Ibid., p. 112. Legend has it that Menéndez posted a sign over the bodies of some of his victims: "I do this not unto Frenchmen, but unto Lutherans." Two years later, a French mercenary took revenge by slaughtering scores of Spaniards, and supposedly left a sign of his own: "I do this not unto Spaniards, but unto Traitors, Robbers and Murderers." Frank Parker Stockbridge and John Holliday Perry, *Florida in the Making* (New York: De Bower Publishing, 1926), pp. 10–11; Charlton W. Tebeau and Ruby Leach Carson, *Florida: From Indian Trail to Space Age, Vol. 1* (Delray Beach, FL: The Southern Publishing Co., 1965), p. 30; Mark Derr, *Some Kind of Paradise: A Chronicle of Man and the Land in Florida* (Gainesville: University Press of Florida, 1998), pp. 239–40.
26   *"They came and surrendered"*: Marjory Stoneman Douglas, *Florida: The Long Frontier* (New York: Harper & Row, 1967), p. 75.
27   *"by every right he could have burnt them alive"*: Barrientos, *Pedro Menéndez de Avilés,* p. 69.
27   *Florida's new* adelantado: Spanish adelantados were explorer-governors enshrined by royal contracts.

27 *But Carlos had foolishly surrendered:* McGoun, *Prehistoric Peoples of South Florida,* p. 9.

27 *He invited Menéndez to celebrate:* Solis de Meras, *Memorial,* pp. 148–9.

27 *"The Adelantado showed much [desire]":* Ibid., p. 150.

28 *"She told him she wished that God might kill her":* Ibid., p. 190.

28 *"She was very sorrowful":* Ibid., p. 192.

28 *Menéndez decided that Carlos:* Ibid., p. 220.

28 *"When I showed them clearly and to their face":* John H. Hann, editor and translator, *Missions to the Calusa* (Gainesville: University Press of Florida, 1991), p. 239.

29 *You have two hearts:* Solis de Meras, *Memorial,* p. 229.

29 *"After the salvation of my soul":* Mauncy, *Florida's Menéndez,* p. 94.

29 *"liable to be overflowed":* Jeannette Thurber Connor, *Colonial Records of Spanish Florida: Letters and Reports of Governors and Secular Persons* (Deland, FL: Florida State Historical Society, 1925), p. 41.

29 *"a blood lust for killing Christians":* Ibid., p. 39.

29 *Florida became a pawn:* Charles W. Arnade, "Raids, Sieges and International Wars," in Gannon, *The New History of Florida,* pp. 100–15; Tebeau, *A History of Florida,* pp. 57–72.

30 *"soldiers and savages excepted":* William Stork, *An Account of East-Florida, with Remarks on Its Future Importance to Trade and Commerce* (London: printed for G. Woodfall, 1766), p. 67.

30 *the backwater of the backwater:* McCally, *The Everglades,* p. 57. McCally compares the Spanish experience in south Florida to a bad horror movie: The climactic scenes come in the first fifteen minutes.

30 *When Spain ceded Florida:* Bernard Romans, *A Concise Natural History of East and West Florida* (facsimile reproduction of the 1775 edition, Gainesville: University of Florida Press, 1962). Some ethnologists argue that Romans could not have been sure that every last Calusa was gone. In any event, the Calusa way of life was gone.

30 *the Seminole Indians:* The Seminoles included Creeks, Apalachicolas, Yamassees, Uchees, Talahassees and Miccosukees, before whites began using the word "Seminoles" to describe all Florida Indians. The term has become widely accepted, but I don't mean to diminish the importance of the individual bands by using it. The following books and articles were my guides to the Seminoles and their history: James W. Covington, *The Seminoles of Florida* (Gainesville: University Press of Florida, 1993); Joe Knetsch, *Florida's Seminole Wars, 1817–1858* (Charleston, SC: Arcadia Publishing, 2003); John Mahon, *History of the Second Seminole War, 1835–1842* (Gainesville, University Press of Florida, 1987); Robert V. Remini, *Andrew Jackson and His Indian Wars* (New York: Penguin, 2001); Brent Richards Weisman, *Unconquered People: Florida's Seminole and Miccosukee Indians* (Gainesville: University Press of Florida, 1999); John K. Mahon and Brent R. Weisman, "Florida's Seminole and Miccosukee Peoples," in Gannon, ed., *The New History of Florida,* pp. 183–206; Joe Knetsch, Willard Steele, and Buffalo Tiger also helped educate me about the Seminoles in interviews.

30 *They were known as* cimmarones: Some Seminoles object to the translation "wild ones," preferring "breakaways," "frontiersmen," or "free people."

30 *by 1800, their permanent villages only stretched:* Covington, *The Seminoles,* p. 26. In an interview, Buffalo Tiger said that the Miccosukees were first directed toward "the pointed land" by a tree limb that pointed south.

30 *"Here our naval strings were first cut":* Mahon, *History of the Second Seminole War,* p. 2.

31 *"as blithe and free as the birds of the air":* William Bartram, *Travels,* edited by Francis Harper from the 1791 edition (New Haven: Yale University Press, 1958).

31 *The Seminoles clashed constantly:* Mahon, *History of the Second Seminole War,* pp. 19–20; Weisman, *Unconquered People,* pp. 43–45. After the failure of the Patriot Army, a brigade of Tennessee volunteers made a foray into Spanish Florida in 1812, burning 386 Seminole homes and driving off Seminole cattle.

31 *"uncommon cruelty and barbarism":* Remini, *Andrew Jackson and His Indian Wars,* p. 120.

31 *"I view the Possession of the Floridas":* Ibid., p. 134. Jackson told Monroe that Spanish Cuba was strategically vital as well, and promised that with a few more men and a frigate, he could take it within a few days. Monroe never took him up on the offer.

32 *"It is a land of swamps":* The congressman was John Randolph of Virginia, who was arguing against Florida's admission to the Union.

32 *"a natural and necessary part of our empire":* Annals of Congress, 7th Cong., 2nd sess., col. 204, quoted in Albert K. Weinberg, *Manifest Destiny: A Study of Nationalist Expansionism in American History* (Baltimore: Johns Hopkins Press, 1935), p. 47.

32 *A Kentucky journal declared: Frankfort Commentator,* 5/28/1819, quoted in Weinberg, *Manifest Destiny,* p. 50.

32 *Secretary of State John Quincy Adams soon negotiated a treaty:* In exchange for Florida, the United States agreed to assume $5 million worth of debts that Spain supposedly owed American settlers for failing to protect them from Seminole raids. Spain also gave up its claim to Oregon, while the United States gave up its claim to Texas—but not for long, as it turned out. Incidentally, Secretary Calhoun argued during a cabinet meeting that Jackson should be censured for his Florida adventure—which Jackson only found out while Calhoun was his vice president, accelerating the rupture between the two men.

32 *"This rendered it still more unavoidable":* Norman Graebner, ed., *Manifest Destiny* (Indianapolis: Bobbs Merrill, 1968), p. xxiv.

32 *"God hath consumed the natives":* Anders Stephanson, *Manifest Destiny: American Expansion and the Empire of Right* (New York: Hill and Wang, 1995), p. 11.

32 *"the design of Providence":* Ibid., p. 11.

33 *"To have acquired a territory":* Meyer M. Cohen, *Notices of Florida and the Campaigns* (Charleston, S.C.: Burges & Honour, 1836), p. 49.

33 *"the progress of mankind is arrested":* Annals of Congress, 15th Cong., 2nd sess., col. 838, quoted in Weinberg, *Manifest Destiny,* p. 80.

33 *"The hatchet is buried":* Mahon, *History of the Second Seminole War,* pp. 44–45. For more than a century before Mahon's history appeared in 1965, the definitive account of the conflict was Lt. John Sprague's *The Origin, Progress and Conclusion of the Florida War.* Other contemporary accounts include Cohen, *Notices of Florida;* Jacob Rhett Motte, *Journey Into Wilderness: An Army Surgeon's Account of Life in Camp and Field during the Creek and Seminole Wars, 1836–1838,* edited by James F. Sunderman (Gainesville: University of Florida Press, 1953); Woodburne Potter, *The War in Florida: Being an Exposition of Its Causes, and an Accurate History of the Campaigns of Generals Clinch, Gaines and Scott* (reproduction of the 1837 edition, Ann Arbor: University Microfilms, 1966). I also consulted the letters and diaries of many U.S. soldiers, including Amos Beebe Eaton, James Elderkin, George McCall, and George Preble, as well as Generals Thomas Jesup, Zachary Taylor, and Alexander Webb.

33 *"We rely on your justice and humanity":* Mahon, *History of the Second Seminole War,* p. 45.

34 *less than a penny an acre:* Ibid., p. 47. The Seminoles gave up more than 28 million acres of land in Florida, in exchange for about $221,000 in cash and other considerations.

34 *"It is not necessary to disguise the fact":* Ibid., pp. 48–49.

34 *"The best of the Indian lands":* Covington, *The Seminoles of Florida,* p. 60. The official was William Duval, Jackson's successor as territorial governor.

34 *"I was in several of their houses":* Knetsch, *Florida's Seminole Wars,* pp. 53–54.

34 *"We were promised justice":* Sprague, *The Origin, Progress and Conclusion of the Florida War,* pp. 50–51.

35 *"The Treaty of 1823 deprived them":* Mahon, *History of the Second Seminole War,* pp. 73–74.

35 *Gadsden persuaded several chiefs:* Gadsden threatened to stop feeding the Seminoles if they did not agree to removal. According to an army officer named Ethan Allen Hitchcock, Gadsden also secured the tribe's consent by bribing its black interpreter, Abraham, who was specifically granted $200 in the treaty. John Phagan, the government agent for the Seminoles, then escorted the tribal delegation to Fort Gibson, and allegedly threatened to leave them there if they did not agree to sign another treaty. Phagan, who was later fired for stealing Seminole annuities, apparently altered a few words of the treaty as well, eliminating a requirement that it would have to be ratified by the Seminole people. Ibid., pp. 75–86; Covington, *The Seminoles of Florida,* pp. 63–66.

35 *"My people cannot say they will go":* Covington, *The Seminoles of Florida,* p. 74.

35 *Osceola, a mixed-blood Alabama Creek:* It was widely believed that Osceola's father was an English trader named William Powell, and he was often called Powell. His mother had some white blood, too.

36 *"indomitable firmness":* Potter, *The War in Florida,* pp. 234–35.

36 *"I will make the white man red with blood":* Sprague, *The Origin, Progress and Conclusion of the Florida War,* p. 86.

36 *"My children: I have never deceived":* Ibid., pp. 78–81.

36 *At first, Osceola tore his hair:* Patricia R. Wickham, *Osceola's Legacy* (Tuscaloosa: University of Alabama Press, 1991), p. 34.

37 *"The Seminole of the present day":* Knetsch, *Florida's Seminole Wars,* p. 69.

37 *"I cannot see that any danger":* Mahon, *History of the Second Seminole War,* p. 67.

37 *"I have no doubt that the object"*: Sprague, *The Origin, Progress and Conclusion of the Florida War*, p. 88.

37 *"The passions of a people"*: Ibid., p. 93. The Dade massacre was the U.S. Army's worst defeat at the hands of Indians before Custer's Last Stand at the Little Bighorn in 1876.

38 *"You have guns, and so have we"*: Potter, *The War in Florida*, p. 126.

38 *40,000 federal regulars and state militiamen*: Mahon, *History of the Second Seminole War*, p. 325.

38 *"so officers may make additions"*: One copy of this 1837 map, titled "Theatre of Military Operations," hangs in the archives of the Big Cypress Seminole Indian Reservation.

38 *One army engineer scoffed*: John LeConte, *Observations on the Soil and Climate of East Florida* (reproduction of the 1822 edition, edited by Richard Adicks, Orlando: University Press of Florida, 1978), p. 21.

38 *"beyond the ultimate limits"*: *Niles Weekly Register*, 3/18/1826, quoted in Nelson M. Blake, *Land Into Water—Water Into Land: A History of Water Management in Florida* (Tallahassee: University Press of Florida, 1980), p. 15.

38 *"extensive inundated Region covered with Pine and Hummock Islands"*: Charles Vignoles, *Observations Upon the Floridas* (facsimile reproduction of the 1823 edition, Gainesville: University Press of Florida).

38 *"I found it impracticable to navigate"*: George E. Buker, *Swamp Sailors in the Second Seminole War* (originally published in 1975; Gainesville: University Press of Florida, 1997), p. 55. Powell's only battle with the Seminoles was a lopsided defeat that he blamed on his "lame—blind—deaf and idiotic" men: "I could have taught them to make watches as easily as to learn the one to handle an oar and the other a musket."

39 *"the everglades may be impenetrable"*: Zachary Taylor letters to Thomas Jesup, 2/10/1838, 2/13/1838, 2/22/1838, ZTP, Reel 1, Series 2.

39 *"I would not trade one foot"*: Willard Steele, *The Battle of Okeechobee* (Miami: Florida Heritage Press, 1987), p. 22.

3 Quagmire

40 *"Florida is certainly"*: Motte, *Journey into Wilderness*, p. 199.

40 *"To surround what?"*: Remini, *Andrew Jackson and His Indian Wars*, p. 275–76.

40 *"swampy, hammocky, low"*: Mahon, *History of the Second Seminole War*, p. 179.

41 *Just when he thought he had his finger on them*: Cohen, *Notices of Florida and the Campaigns*, p. 205.

41 *"We are not inaptly compared to a prize-ox"*: Ibid., p. 222.

41 *"The white man . . . wants to catch water"*: William Hartley and Ellen Hartley, *Osceola: The Unconquered Indian* (New York: Hawthorn Books, 1973), p. 179.

41 *The Second Seminole War was America's first Vietnam*: Norman J. Jones, "America's First Vietnam: The Seminole Wars," *Command* 41 (January 1997), p. 66. There was a burst of Seminole War scholarship during the Vietnam era. Jones noted that "the tactics, geography and climate of Florida in many ways gave the conflict an eerie similarity to the war that would later be fought in Indochina."

41 *"Our troops generally fought with great bravery"*: Theodore Roosevelt, *Life of Thomas Benton* (Boston and New York, 1886), p. 187, quoted in Hartley and Hartley, *Osceola*.

41 *a disgruntled lieutenant*: George A. McCall, *Letters from the Frontiers* (reproduction of the 1868 edition, Gainesville: University Press of Florida, 1974; republished electronically by Hillsdale College, www.hillsdale.edu/dept/history/war/American/Indian).

41 *"Millions of money"*: Amos Beebe Eaton, 11/18/1837 diary entry, ABEP.

41 *"How vastly wide"*: Ibid., 11/13/1837 diary entry.

42 *"Campaigning in Florida"*: Motte, *Journey into Wilderness*, pp. 143–45.

42 *"It was intolerable—excruciating!"*: Ibid., p. 232.

42 *"It is in fact a most hideous region"*: Ibid., p. 199.

42 *"Before us and on either side of us"*: Ibid., pp. 186–87.

43 *"The doctors at one time thought"*: Preble, "A Canoe Expedition into the Everglades in 1842," p. 49.

43 *One officer suggested that anyone*: Mahon, *History of the Second Seminole War*, p. 318.

43 *"a thick crop of sharply pointed knives"*: Motte, *Journey Into Wilderness*, p. 231.

43 *"Every rod of the way"*: James D. Elderkin, *Biographical Sketches and Anecdotes of a Soldier of Three Wars, as Written by Himself* (Detroit: published electronically by Hillsdale College: www.hillsdale.edu/academics/history/War/America/Indian/1841), p. 26.

43   *"Their everlasting hum never ceases":* Benjamin Strobel, "Dr. Strobel Reports on Southeast Florida, 1836," edited by E. A. Hammond, *Tequesta* 21 (1961), pp. 69–70.

43   *General Alexander Webb's war diaries:* Alexander S. Webb, "Campaigning in Florida in 1855," *The Journal of Military Service Institution* 45 (November–December 1909), pp. 397–489.

44   *"general sinking of the system":* Buker, *Swamp Sailors in the Second Seminole War,* p. 125.

44   *At one point, five battalions:* Mahon, *History of the Second Seminole War,* pp. 209–210; Blake, *Land into Water—Water into Land,* p. 17.

44   *"My company now left":* Andrew Humphreys letter to Samuel Humphreys, 8/10/1836, Andrew A. Humphreys Papers, Pennsylvania Historical Society. Humphreys and his epic hubris were immortalized in John Barry's terrific history of the 1927 Mississippi River flood. John Barry, *Rising Tide: The Mississippi Flood of 1927 and How It Changed America* (New York: Simon & Schuster, 1997).

44   *One regiment attributed:* Steele, *The Battle of Okeechobee,* p. 17.

44   *"Oh!" Motte wailed:* Motte, *Journey into Wilderness,* p. 209.

44   *In the frenzy of finger-pointing:* Mahon, *History of the Second Seminole War,* pp. 161–66; Cohen, *Notices of Florida and the Campaigns,* p. 222.

44   *while Jackson growled that Floridians:* Remini, *Andrew Jackson and His Indian Wars,* pp. 275–76. Jackson was not the only politician who got passionate about the war. A New Hampshire congressman was killed by a Kentucky congressman in a duel after mocking a colleague's sympathy for the Seminoles on the House floor.

44   *"the impervious swamps and hammocks":* Sprague, *The Origin, Progress and Conclusion of the Florida War,* p. 158.

44   *"The most prominent cause of failure":* Cohen, *Notices of Florida and the Campaigns,* p. 222.

45   *This retardation continued:* Knetsch, *Florida's Seminole Wars,* pp. 100–101.

45   *"This is the true secret":* Motte, *Journey into Wilderness,* pp. 143–44.

45   *"I may be permitted to say":* Steele, *The Battle of Okeechobee,* p. 17.

46   *"Their whole object is to avoid":* Taylor letter to Jesup, 4/3/1838, ZTP.

46   *"They had no difficulty finding plenty of food":* Richard J. Procyk, *Guns Across the Loxahatchee* (Melbourne: Florida Historical Society Press, 1999), p. 94. Other future Civil War generals who fought in the Florida war included Joseph Johnston, George Meade, and Jubal Early.

46   *the Seminole names of Everglades plants:* The Seminoles also collected plants named "to make axe handle," "tobacco seasoner," and—years before Viagra—"to harden penis." William C. Sturtevant, *The Mikasuki Seminoles: Medical Practices and Beliefs* (Yale University dissertation, 1955; reprinted by University Microfilms), p. 453.

46   *"There is no country in the world":* John T. Sprague, "Macomb's Mission to the Siminoles," edited by Frank F. White, Jr., from Sprague's 1839 journal, *Florida Historical Quarterly* (October 1956), p. 178.

47   *"God-abandoned" hellscape:* Milton Meltzer, *Hunted Like a Wolf: The Story of the Seminole War* (New York: Farrar, Straus and Giroux, 1972), p. 178.

47   *Sherman saw the peninsula:* Procyk, *Guns Across the Loxahatchee,* p. 94.

47   *South Florida, General Jesup concluded:* Sprague, *The Origin, Progress and Conclusion of the Florida War,* p. 200.

47   *He described it as "a Negro war":* Mahon, *History of the Second Seminole War,* pp. 201–3.

47   *"The Indians are a persecuted race":* Samuel Forry, "The Letters of Samuel Forry," *Florida History Quarterly* 6 (January 1928), p. 135.

47   *The very existence of the Seminoles:* Knetsch, *Florida's Seminole Wars,* p. 104.

47   *Secretary of War Joel Poinsett:* Mahon, *History of the Second Seminole War,* p. 207.

47   *"We disclaim all participation":* Meltzer, *Hunted Like a Wolf,* p. 140.

48   *"the strong & oppressive hand of the white people":* Wickham, *Osceola's Legacy,* p. 100.

48   *In a scene immortalized in verse:* Walt Whitman, "Osceola," reprinted in Walt Whitman, *Complete Poetry and Selected Prose.,* edited by James E. Miller, Jr. (Boston: Houghton Mifflin Co., 1959), p. 379.

48   *"Thus has a great savage":* Wickham, *Osceola's Legacy,* p. 100.

48   *Twenty-two towns:* Mahon, *History of the Second Seminole War,* pp. 213–14.

48   *about 1,500 bedraggled American troops:* Procyk, *Guns Across the Loxahatchee,* p. 78.

48   *"The Indians yelled and shrieked":* Motte, *Journey into Wilderness,* p. 194.

49   *"plunged into the swift torrent":* Ibid.

49   *"met, beat and dispersed":* Thomas Jesup letter to Secretary of War Joel Poinsett, 2/18/38, NA, Adjutant General's Office, Thomas Jesup Papers, Record Group 94, Microfilm 565.

49 *"There before us lay death"*: Motte, *Journey into Wilderness,* p. 195.
49 *"In regards to the Seminoles"*: Sprague, *The Origin, Progress and Conclusion of the Florida War,* p. 200.
50 *William Harney was another Jackson protégé:* I consulted two biographies of Harney: George Rollie Adams, *William S. Harney: Prince of Dragoons* (Lincoln, NE: University of Nebraska Press, 2001); L.U. Reavis, *The Life and Military Services of Gen. William Selby Harney* (St. Louis: Bryan, Brand & Co., 1878).
50 *"physically the finest specimen of man"*: Charles M. Brookfield and Oliver Griswold, *They All Called It Tropical: True Tales of the Romantic Everglades, Cape Sable and the Florida Keys* (originally published in 1949; Miami: Historical Association of Southern Florida, 1985), p. 36.
50 *And he had some anger issues:* Adams, *William S. Harney,* pp. 36, 47.
50 *On a stop for provisions:* Hester Perrine Walker, "Massacre at Indian Key, August 7, 1840, and the Death of Dr. Henry Perrine," *Florida Historical Quarterly* 5, no. 1 (July 1926), pp. 21–22.
50 *160 Indians led by the hulking warrior Chakaika:* Some ethnologists have suggested that Chakaika's band of Spanish-speaking Indians with ties to Cuba may have included descendants of the Calusa.
50 *"There must be no more talking"*: Adams, *William S. Harney,* p. 73. Harney was also furious at Secretary Poinsett, who had told white Floridians that the Indian reservation in South Florida should only be considered a temporary measure, further inflaming the Seminoles.
51 *"After their repast was over"*: Walker, "Massacre at Indian Key," pp. 27–32.
51 *Harney led ninety men:* Harney had a guide named John, a captured black man who had taken refuge with the Seminoles. Lieutenant John McLaughlin had asked to use John in 1839, and later claimed he could have prevented the Indian Key massacre if permission had been granted.
51 *"expressly intended as a retreat"*: Anonymous, "Notes on the Passage Across the Everglades," p. 59.
51 *"We have now crossed the long fabled and unknown Everglades"*: Ibid., pp. 62–64.
51 *Harney killed or captured only:* Harney later claimed that he was personally responsible for ending the war, even though it didn't end for another two years. He complained that *The Origin, Progress and Conclusion of the Florida War* did not give his company enough credit, ripping Lieutenant Sprague as "one of the most contemptible liars and puppies in the whole army." William Harney, letter to Theodore Rodenbough, 12/24/1875, unfiled original in the Seminole Tribe of Florida's archives.
52 *"The commands in canoes"*: Sprague, *The Origin, Progress and Conclusion of the Florida War,* p. 353. The Mosquito Fleet's prized schooner was the *Flirt,* which is how Lake Flirt at the headwaters of the Caloosahatchee River got its name.
52 *"If our labors have not been rewarded"*: John T. McLaughlin letter to Navy Secretary A. P. Upshur, 11/25/1841, reprinted in "Report of Buckingham Smith," p. 112.
52 *"At night as we lay down"*: Abner Doubleday, *My Life in the Old Army: The Reminiscences of Abner Doubleday,* edited by Joseph E. Chance from the collections of the New York Historical Society (Fort Worth: Texas Christian University Press, 1948), p. 189.
52 *He was enchanted:* Motte, *Journey into Wilderness,* pp. 156–59, 161, 177.
53 *"Nothing, however, can be imagined"*: Ibid., pp. 191–92.
53 *"The further pursuit"*: Sprague, *The Origin, Progress and Conclusion of the Florida War,* p. 493.
53 *mercurial chief Billy Bowlegs:* Bowlegs was a notoriously unsteady leader. The "King of the Everglades" had regaled reporters with his drunken monologues during his 1852 tour of America, in which he marveled at Wall Street's banks, enjoyed a minstrel show and a ballet, and met with President Millard Fillmore, drinking copious amounts of "fire-water" at every stop. *New York Times,* 9/21/1852, p. 3; 9/24/1852, p. 8; 9/25/1852, p. 8; 9/27/1852, p. 8; *Harper's Weekly,* 6/12/1858.
53 *more accurate maps and descriptions:* The best was a map drawn by Lieutenant J. C. Ives, directed by then Captain Humphreys, under orders from Secretary of War Jefferson Davis. It was accompanied by a dull but thorough narrative that may be the most detailed description of the "comparatively unknown" south Florida environment before drainage. J. C. Ives, *Memoir to Accompany a Military Map of the Peninsula of Florida, South of Tampa Bay* (New York: M.B. Wynkoop, Book and Job Printer, 1856).
53 *"This country should be preserved for the Indians"*: Webb, "Campaigning in Florida in 1855," p. 423.

4   A New Vision

PAGE

54   *"Its being made susceptible of cultivation":* General William S. Harney letter to Buckingham Smith, 1/23/1848; reprinted in "Report of Buckingham Smith," p. 45.

54   *He was an unhealthy and unlucky:* Edward Jelks, "Dr. Henry Perrine," *Journal of the Florida Medical Association* (April 1934), in Henry Perrine file, HMSF.

54   *He knew it was considered:* Henry Perrine letter to Secretary of State Louis McLane, 10/23/1834, quoted in Sarah R.W. Palmer, "Henry Perrine, Pioneer Botanist and Horticulturist." *Florida Historical Quarterly* 5, no. 2 (October 1926).

55   *"How many years have I fruitlessly labored":* Jerry Wilkinson, *Dr. Henry Perrine* (Tavernier, FL: Wilkinson Publishing, 1995), p. 30.

55   *"sheltered seashore of an ever-verdant prairie":* Brookfield and Griswold, *They All Called It Tropical,* p. 46.

55   *South Florida's white population:* "Report of Buckingham Smith," p. 25.

56   *"rich as dung":* Stork, *An Account of East-Florida,* p. iii–iv.

56   *Edward Judson:* Cooper Kirk, "Edward Zane Carroll Judson," *Broward Legacy* (fall 1979), p. 16. As Ned Buntline, Judson also instigated the catastrophic Astor House Opera riots in New York City in 1849. Michael Grunwald, "Shakespeare in Hate: 150 Years Ago, 23 People Died in a Riot Over 'Macbeth,'" *Washington Post,* 3/28/1999.

56   *"the richest land I have ever seen":* E.Z.C. Judson, "Sketches of the Florida War," November 1894 to April 1894, reprinted in Kirk, "Edward Zane Carroll Judson" (fall 1979), pp. 21–27.

56   *"was like talking of limiting the stars":* Norman Graebner, ed., *Manifest Destiny,* p. xxi.

56   *"America is a land of wonders":* Alexis de Toqueville, *Democracy in America* (originally published 1834; New York: Vintage Books, 1945), vol. 1, p. 443.

56   *Meanwhile, the U.S. Army Corps of Engineers:* U.S. Army Corps of Engineers Office of History, *The History of the US Army Corps of Engineers,* Washington, 1998; Martin Reuss and Paul K. Walker, "Financing Water Resources: A Brief History," U.S. Army Corps of Engineers Office of History; Michael Grunwald, "A River in the Red," *Washington Post,* 1/9/2000; Grunwald, "An Agency of Unchecked Clout," *Washington Post,* 9/10/2000.

57   *John Quincy Adams asked the House clerk: Congressional Globe,* 29th Cong., 1st sess, 2/9/1846; quoted in Graebner, *Manifest Destiny,* pp. 340–42.

57   *"Could it be drained":* Williams, *The Territory of Florida,* p. 151.

57   *"survey the Everglades":* Resolution in relation to the Ever Glades of Florida, 12/10/1845, reprinted in "Report of Buckingham Smith," p. 74; "The Everglades of Florida," Senate Document 89, p. 34.

58   *"I entertain no doubt of the practicability":* Thomas Jesup letter to Senator James Westcott, 2/12/1848, reprinted in "Report of Buckingham Smith," p. 42.

58   *"The results of such a work":* L. M. Powell letter to Senator James Westcott, 3/1/1848, reprinted in "Report of Buckingham Smith," p. 49.

58   *"I do not know of a project":* Harney letter, reprinted in "Report of Buckingham Smith," p. 44.

58   *One man who had tried:* S. R. Mallory letter to Buckingham Smith, September 1847, reprinted in "Report of Buckingham Smith," p. 55.

58   *Still, Senator Westcott pestered:* Senator James Westcott letter to Treasury Secretary R. J. Walker, 5/11/1847, reprinted in "Report of Buckingham Smith," p. 66.

58   *He was not an engineer:* Dumas Malone, ed., *Dictionary of American Biography,* vol. 17 (New York: Charles Scribner's Sons, 1935), p. 243. Smith was included in the dictionary for his historical work; his entry does not even mention the Everglades. In fact, Smith could not resist the temptation to indulge his obsession with Florida history in his report, peppering his analysis of the Everglades with dozens of irrelevant anecdotes about everything from buccaneers to the Calusa trade in ambergris to a British plan to build mega-forts called "Pharuses."

59   *"Imagine a vast lake of fresh water":* "Report of Buckingham Smith," p. 28.

59   *"The first and most abiding impression":* Ibid., p. 29.

59   *"The statesman whose exertions":* Ibid., p. 34.

59   *Smith made a meticulous case:* Ibid., pp. 10–38.

60   *"The elevation of the Ever Glades":* Gen. James Gadsden letter to Treasury Secretary R. J. Walker, 5/4/1847, reprinted in "Report of Buckingham Smith," p. 42.

60   *"A bountiful Providence":* Powell letter, reprinted in "Report of Buckingham Smith," pp. 49–50.

61 *"too considerable to have been undertaken"*: "Report of Buckingham Smith," p. 12.
61 *"That such work would reclaim"*: Harney letter, reprinted in "Report of Buckingham Smith," p. 44.
61 *"beyond question, defray all outlay"*: "Report of Buckingham Smith," p. 17.
61 *Harney was even more optimistic*: Tebeau, *Man in the Everglades*, p. 129. He wrote a letter asking U.S. Senator David Levy Yulee to pull some strings: "Can you do me the favor to get an order to have it surveyed at once, so that I can employ men to commence in clearing and planting at once?"
61 *"the best sugar land in the south"*: Harney letter, reprinted in "Report of Buckingham Smith," p. 45.
61 *"as valuable as any in the world"*: Jesup letter, reprinted in "Report of Buckingham Smith," p. 43.
61 *"That the results must be"*: "Report of Buckingham Smith," pp. 33–34.
61 *"the making of salt by solar evaporation"*: Ibid., p. 32.
61 *"In less than five years"*: Harney letter, reprinted in "Report of Buckingham Smith," p. 45.
62 *"To be identified"*: "Report of Buckingham Smith," p. 31.
62 *Stephen Mallory, a customs official*: Mallory letter, reprinted in "Report of Buckingham Smith," pp. 53–55.

5   Drainage Gets Railroaded

PAGE
64 *"Draining of the Everglades"*: Joe Knetsch, "John Darling, Indian Removal, and Internal Improvements in South Florida, 1848–1856," *Tampa Bay History* (fall/winter 1995), p. 13. Darling had a particularly ingenious idea for draining the Everglades. He suggested dredging a canal that would divert the Kissimmee River west into Peace Creek and out to sea, so it would no longer spill into Lake Okeechobee, and the Lake would no longer overflow into the Everglades. Fortunately, no one ever tried it.
64 *"the progress of Florida"*: "Eight Years of Progress," p. 3.
64 *His bill imposed strict conditions*: *Congressional Globe*, 30th Cong., 2nd sess. 12/20/1848, pp. 69–70; 12/22/1848, pp. 87–91.
64 *"would make the grant utterly valueless"*: *Congressional Globe*, 30th Cong., 2nd sess., 12/20/1848, p. 69, Dovell, "The Everglades Before Reclamation," p. 37.
65 *His ancestors were driven out of Spain*: Much of my information about David Levy Yulee came from two doctoral dissertations: Joseph Gary Adler, *The Public Career of Senator David Levy Yulee* (Ann Arbor, MI: University Microfilms, 1973, Case Western Reserve University dissertation); Arthur W. Thompson, *David Yulee: A Study of Nineteenth Century Thought and Enterprise* (Ann Arbor, MI: University Microfilms, 1971, Columbia University dissertation); Alfred Jackson Hanna and Kathryn Abbey Hanna also wrote about Yulee in *Florida's Golden Sands* (Indianapolis: The Bobbs-Merrill Company, 1950).
65 *increasingly estranged from his father*: Moses became most distressed about David's drift away from Judaism, but he also rejected his son's legal career and political activities. David urged Moses not to dwell on their philosophical differences, but refused to change his own to suit his father. "We will continue to cherish the affectionate interest in you which the relation you bear to us and the excellence of your character induces," David wrote to Moses on July 21, 1849. "But in respect to our religious views and conduct you must consent to leave us unquestioned and responsible only to God." Thompson, *David Yulee*, pp. 220, 224, 229.
65 *"So far from the practice of cruelty"*: *Congressional Globe*, 27th Cong., 2nd sess., 6/13/1842, p. 499.
65 *Levy saved his highest dudgeon*: Yulee letter, 9/7/1840, reprinted in Thompson, *David Yulee*, pp. 247–248.
66 *Levy was the father of Florida statehood*: Thompson, *David Yulee*, pp. 23, 55–56.
66 *At the time, the state's only railroad*: Blake, *Land into Water*, p. 39; George E. Buker, *Sun, Sand and Water: A History of the Jacksonville District of the U.S. Army Corps of Engineers*, 1821–1975 (Washington, D.C.: U.S. Army Corps of Engineers, Office of History), p. 39.
66 *In an era when a railroad lawyer*: David H. Donald, *Lincoln* (New York: Simon & Schuster, 1995), p. 210; Stephen E. Ambrose, *Nothing Like It in the World: The Men Who Built the Transcontinental Railroad, 1863–1869* (New York: Simon & Schuster, 2000), p. 28.
66 *"impositions and exactions"*: Thompson, *David Yulee*, p. 56.
66 *The day statehood was approved*: Junius Elmore Dovell, *A History of the Everglades of Florida* (Unpublished dissertation, University of North Carolina, 1947), p. 285. Courtesy Joe Knetsch.
66 *"spread a belt of civilization"*: Thompson, *David Yulee*, p. 307.

66   *Yulee instructed one Army engineer:* Yulee letter to Captain Smith, 10/11/54, reprinted in Thompson, *David Yulee,* p. 388.

66   *"You suffer the penalty":* Thompson, *David Yulee,* p. 124.

67   *He told his six-year-old son:* Hanna and Hanna, *Florida's Golden Sands,* p. 128.

67   *The swamplands act eventually granted:* The swampland grant to Florida was the largest land grant from the federal government to a state in American history. Knetsch, "John Darling," p. 14.

67   *"the product of the brain":* Dovell, *A History of the Everglades of Florida,* p. 102.

67   *One clause did authorize:* Blake, *Land into Water,* pp. 39–41. Blake properly describes the clause as a mere "sop" to drainage advocates. The law was so railroad-focused that it even established a prohibition on driving trains while drunk.

67   *More than a year after the law was enacted: Minutes of the Board of Trustees of the Internal Improvement Fund of Florida,* vol. 1, 8/5/1856, p. 32.

67   *"The rapid enhancement of the general wealth":* George W. Pettengill, Jr., *The Story of the Florida Railroads* (Boston: The Railway and Locomotive Historical Society, 1952), p. 20.

68   *And Florida finally enjoyed a mild railroad boom:* Florida added 368 miles of track in the first eight years of the internal improvement law. Blake, *Land into Water,* p. 39.

68   *The senator and his partners:* Yulee's only political defeat came in 1851, when he lost his Senate seat in a stunning upset to fellow Democrat Stephen Mallory of Key West, thanks to an alliance between Key West boosters (who opposed any cross-peninsula railroad that would draw traffic away from the Florida Straits) and Tampa boosters (who wanted a cross-peninsula railroad with the Gulf terminus in their town). So Yulee later authored a state report recommending Tampa as the Gulf terminus, which helped ensure his reelection in 1855. But that was just a feint. He had already bought all the land in Cedar Key, and once he returned to the Senate, he once again listed Cedar Key as the line's Gulf terminus. Adler, *The Public Career of Senator David Levy Yulee,* pp. 104–5; Arthur W. Thompson, "The Railroad Background of the Florida Senatorial Election of 1851," *Florida Historical Quarterly* 1 (1953), pp. 181–95.

68   *The* American Railroad Journal *predicted:* Thompson, *David Yulee,* p. 95.

68   *Yulee used his chairmanship:* It didn't hurt that the postmaster general was Yulee's brother-in-law, and a secret investor in Yulee's operations. Adler, *The Public Career of Senator David Levy Yulee,* p. 131.

68   *"Railroads are useful": Minutes,* vol. 1, 6/1/1859, p. 135. Perry's pique was somewhat less than high-minded—Yulee had refused to divert the railroad through his plantation—but his charges were on the mark. Joe Knetsch, "Madison Starke Perry vs. David Levy Yulee: The Fight for the Tampa Bay Route," *The Sunniland Tribune, Journal of the Tampa Historical Society* 23 (November 1997), pp. 13–23.

68   *"I remember him in the House":* Thompson, *David Yulee,* p. 146. The senator was Andrew Johnson of Tennessee, the future president.

68   *He pleaded with the Confederate commander, Robert E. Lee:* Yulee letter, 12/11/1862, reprinted in Thompson, *David Yulee,* p. 528.

69   *"I humbly trust I may not be wanting":* Yulee letter, 6/4/1863, reprinted in Thompson, *David Yulee,* p. 546.

69   *"a wild run":* Dovell *A History of the Everglades of Florida,* p. 112.

69   *Florida's antebellum Democratic power brokers:* After his release, Yulee still argued that the South needed "some form of compulsory labor," so that it wouldn't be "Africanized and ruined." He didn't quite catch the point of the war. John T. Foster, Jr., and Sarah Whitmer Foster, *Beechers, Stowes, and Yankee Strangers: The Transformation of Florida* (Gainesville: University Press of Florida, 1999), p. 32.

69   *If the Wright brothers:* D. Graham Copeland, "A Report of the Board of Commissioners of the Everglades Drainage District," 9/15/1930.

69   *"The manipulations have, in all cases":* The land agent was Samuel Swann, who later went to work for Yulee. Thompson, *David Yulee,* p. 188.

70   *a pair of crooks:* The flim-flammers were Milton Littlefield, a former Union general known as the Prince of the Carpetbaggers, and George Swepson; they once gave a state official $5,000 to accept a bad $472,000 check on behalf of the fund. Gregg Turner, *A Short History of Florida Railroads* (Charleston, S.C.: Arcadia Publishing, 2003); Foster and Foster, *Beechers, Stowes and Yankee Strangers,* pp. 82–83.

70   *The trustees eagerly accepted:* Gleason was authorized to buy one square mile of land for every 50,000 cubic feet he dug, the equivalent of a ditch just one mile long, three feet

wide, and three feet deep. *Minutes,* vol. 1, pp. 270–71, 351–58; Blake, *Land into Water,* pp. 43–46.

70 *But Gleason was Florida's archetypal carpetbagger:* The best source on Gleason is: Lewis Hoffman Cresse, Jr., *A Study of William Henry Gleason: Carpetbagger, Politician, Land Developer* (Ann Arbor: University Microfilms International, 1979, University of South Carolina dissertation).

70 *After the war:* Ibid., pp. 8–11.

71 *And when Reed exhibited a few unscripted flashes:* Ibid., pp. 37–41.

71 *"Gleason wore a fine beaver hat":* John Wallace, *Carpetbag Rule in Florida: The Inside Workings of the Reconstruction of Civil Government in Florida after the Close of the Civil War* (reproduction of the 1888 edition, Gainesville: University of Florida Press, 1964).

71 *ruled Gleason ineligible for high office:* Gleason was disqualified because he had not been a Florida citizen for three years. The new residency and citizenship requirements had been designed to keep blacks out of high office, but Governor Reed used them against his fellow carpetbagger.

71 *Gleason also seized control:* Thelma Peters, *Biscayne Country, 1870–1926* (Miami: Banyan Books, 1981), p. 21. Gleason tried to steal another election from Pig Brown in 1876, but his shenanigans happened to coincide with the national controversy over that year's presidential race. In a bizarre foreshadowing of the chaos of 2000, the 1876 election depended on Florida's electoral votes, and the Florida results remained incomplete for weeks because Gleason had sequestered Dade County's ballots. "Where in the hell is Dade?" one election official asked. Gleason's skullduggery was ultimately foiled, because Democrats regained control of the legislature and refused to seat him. Arva Moore Parks, "Miami in 1876," *Tequesta* 35 (1975), pp. 89–139.

71 *continued to float new proposals: Minutes,* vol. 1, pp. 293, 296–7, 319–20, 355–358, 364–9, 403–4, 420–1, 456–9; vol. 2, pp. 8, 61–3, 95, 298.

71 *One angry legislator compared Gleason:* Wallace, *Carpetbag Rule,* pp. 156–58.

71 *"Far better for Gleason if he had remained":* Parks, "Miami in 1876," p. 106.

72 *the great giveaway was shut down:* Blake, *Land into Water,* p. 49.

72 *"a little grasping fellow":* Douglas, *The Everglades,* p. 269.

72 *David Yulee then persuaded Governor Reed:* Canter Brown, Jr, "Carpetbagger Intrigues, Black Leadership and a Southern Loyalist Triumph: Florida's Gubernatorial Election of 1872," *Florida Historical Quarterly* 72, no. 3 (January 1994), p. 282; Foster and Foster, *Beechers, Stowes and Yankee Strangers,* p. 83.

72 *was even denied 3,840 acres:* Gleason's company dredged 300,000 cubic feet, which entitled him to buy six sections of land for just $240, but the fund refused to release the parcels because of the Vose court order. *Minutes,* vol. 2, 5/25/1875. p. 94.

72 *Soon the fund was so broke:* Blake, *Land into Water,* pp. 52–53, 58–60.

72 *"Our development from internal improvements was stagnant":* William Bloxham, "The Disston Sale and the State Finances. A Speech Delivered at the Park Theatre, Jacksonville, Fla," 8/26/1884, FSA.

73 *By 1880: Statistical Abstract of the United States, 1882,* Washington: Government Printing Office, 1883.

73 *north Florida at least showed signs of life:* In 1870, 14,000 tourists visited the Jacksonville area; by 1875, the figure was 50,000. Foster and Foster, *Beechers, Stowes and Yankee Strangers,* pp. xviii, 1–2, 102–03, 111–12. Ex-President Ulysses S. Grant visited in 1880. Shofner, Jerrell H. "Reconstruction and Renewal, 1865–1877," in Gannon, ed., *The New History of Florida,* pp. 257–59.

73 *Yulee built a luxury hotel:* Hanna and Hanna, *Florida's Golden Sands,* p. 260.

73 *Even Stowe began to resent:* Harriet Beecher Stowe, "Protect the Birds," *The Semi-Tropical,* December 1877, reprinted in Foster and Foster, *Beechers, Stowes and Yankee Strangers,* p. xx.

73 *the census reported just 257 white residents:* 1880 U.S. Census, Florida. By contrast, the same geographical area is now home to more than five million people, about 2 percent of the U.S. population.

74 *"How still it is here!":* Iza Duffus Hardy, *Oranges and Alligators: Sketches of South Florida Life* (London: Ward and Downey, 1887). The aptly named Iza Duffus summed up the region she incorrectly described as "South Florida" with a racist ditty: "The land of the possum, mosquito and jigger. Where the rattlesnake crawls in the burning hot sand. And the red-bug he bites both the white man and nigger!"

74 *A few enterprising cattlemen:* Alfred Jackson Hanna and Kathryn Abbey Hanna, *Lake Okeechobee: Wellspring of the Everglades* (Indianapolis: The Bobbs-Merrill Company, 1948), pp. 84–89. Hendry named LaBelle for his daughters, Laura and Belle. Summerlin, reportedly the first

American born in Florida after it became a U.S. territory, once financed a courthouse to make sure the Orange County seat remained in Orlando.

74 *grazed wild herds of wiry "scrubs":* The artist Frederic Remington, best known for his portraits of the West, described Florida's cattle as "scrawny creatures not fit for a pointer-dog to mess on." The whips, according to most accounts, gave the crackers their nickname. Derr, *Some Kind of Paradise,* p. 98,

74 *"We call this God's country":* Loren G. Brown, *Totch: A Life in the Everglades* (Gainesville: University Press of Florida, 1993), p. 10. The pioneer was Charles McKinney, Brown's grandfather.

74 *When Fort Lauderdale's lighthouse keeper:* Stuart McIver, *Death in the Everglades: The Murder of Guy Bradley, America's First Martyr to Environmentalism* (Gainesville: University Press of Florida, 2003), p. 14.

74 *In 1879, a visitor:* James Henshall, *Camping and Cruising in Florida* (Cincinnati: Robert Clarke & Co., 1884; reprinted by Florida Classics Library, Port Salerno, 1991), pp. 72–73.

74 *Even the area's most exuberant booster:* Cresse, *A Study of William Henry Gleason,* p. 113.

74 *"They are unbearable by anyone":* James Buck "Biscayne Sketches at the Far South," An 1878 diary, with an introduction by Arva Moore Parks, *Tequesta* 29 (1979), p. 73.

75 *Boss Tweed passed through:* "William M. Tweed: Romance of His Flight and Exile," *Harper's Weekly,* 4/14/1877.

75 *the notorious Edgar "Bloody" Watson:* Tebeau, *Man in the Everglades,* pp. 86–89. The Watson story is still bathed in myth, but Peter Matthiessen did the best research on the topic for his brilliant (although fictional) Watson trilogy, especially *Killing Mister Watson* (New York: Vintage Books, 1990).

75 *The surviving Seminoles:* Clay MacCauley, *The Seminole Indians of Florida* (Reproduction of the 1887 report by the Smithsonian Institution, Gainesville: University Press of Florida, 2000); Covington, *The Seminoles of Florida,* pp. 146–149; Buck, "Biscayne Sketches at the Far South," pp. 82–83.

75 *"The Seminole, living in a perennial summer":* MacCauley, *The Seminole Indians of Florida,* p. 504.

75 *"They are the producers of perhaps the finest sugar cane":* Ibid., p. 511.

75 *They taught their children:* Covington, *The Seminoles of Florida,* pp. 146–50.

75 *"When informed that the negroes were free":* "The Times-Democrat's Expedition to the Everglades," *Weekly Floridian,* 9/25/1883.

76 *except when their "wild natures":* Buck, "Biscayne Sketches at the Far South," p. 83.

76 *"If the native Floridian does not extend":* Frederick A. Ober, "Ten Days with the Seminoles," *Appleton's Journal,* 8/7/1875, p. 173. Another white visitor could not understand why some Seminoles had failed to understand him—even though he had "tried some Indian words out of Longfellow's Hiawatha." Harry Bullock, *Journey Through the Everglades: The Log of the Minnehaha,* edited by Pat Dodson from the 1891 journal (Tampa: Trend Publications, 1973).

76 *Lake Okeechobee was still so inaccessible:* Dovell, *A History of the Everglades of Florida,* p. 117.

76 *"It has slept":* Maurice Thompson, *The Witchery of Archery* (New York: Charles Scribner's Sons, 1878), quoted in Dovell, "The Everglades Before Reclamation," p. 42.

76 *"Am in despair":* Kirk Munroe, "A Lost Psyche: Kirk Munroe's Log of a 1,600 Mile Canoe Cruise in South Florida Waters, 1881–1882," edited by Irving A. Leonard, *Tequesta* 28 (1968), pp. 80–83.

76 *"so keenly appreciated":* Angelo Heilprin, *Exploration on the West Coast of Florida and in the Okeechobee Wilderness, with Special Reference to the Geology and Zoology of the Floridan Peninsula* (Philadelphia: Wagner Free Institute of Science, 1887), p. 45.

77 *The Everglades also remained a mystery:* New Orleans *Times-Democrat,* 12/3/1883, quoted in Hanna and Hanna, *Lake Okeechobee,* pp. 107–108.

77 *"The singular and wonderful region":* Henshall, *Camping and Cruising, in Florida,* p. 72

77 *"Bear, deer, otter, mink":* Elizabeth Ogren Rothra, *Florida's Pioneer Naturalist: The Life of Charles Torrey Simpson* (Gainesville: University Press of Florida, 1995), p. 29.

77 *In the late nineteenth century:* William Cronon, *Nature's Metropolis: Chicago and the Great West* (New York: W.W. Norton & Co., 1991), pp. xv–xvi; Sean Dennis Cashman, *America in the Gilded Age: From the Death of Lincoln to the Rise of Theodore Roosevelt* (New York: New York University Press, 1988), p. 131.

77 *But a* Harper's *writer who visited:* "Along the Florida Reef," *Harper's New Monthly Magazine,* 1871, reprinted in Oppel and Meisel, eds., *Tales of New Florida,* pp. 265–309.

78 *"The Everglades will always retain its present state":* Henshall, *Camping and Cruising in Florida,* p. 72..

78 *sponsored an expedition to the Everglades:* Williams, "North to South Across the Glades," p. 33.

6   The Reclamation of a Kingly Domain

PAGE

81   *"A radical and recent change": Minutes,* vol. 3, p. 319.

81   *In 1881, Bloxham found his man:* "Florida's Governor. He Talks of the Enterprise of Philadelphia Capitalists," *Philadelphia Press,* 5/13/1881.

81   *Today, Disston is often recalled:* There are no biographies of Hamilton Disston; I hope Joe Knetsch, my invaluable guide to Disston's work, will write one someday. For now, the best published source is a chapter in *Lake Okeechobee* by Alfred Jackson Hanna and Kathryn Abbey. Otherwise, Disston has gotten a rough ride. "One can only marvel at Florida's choice of Disston as its last and greatest hope," John Rothchild wrote in *Up for Grabs: A Trip Through Time and Space in the Sunshine State* (New York: Viking Penguin, 1985), p. 29. To Rothchild, whose book is otherwise spot-on as well as hilarious, the choice of Disston was damning evidence of Florida's "self-perceived inferiority, its hat-in-hand attitude, its inability to distinguish substance from pose, its susceptibility to bluster." Similarly, Charles E. Harner described Disston as "one of history's great wheeler-dealers" in his book *Florida's Promoters: The Men Who Made It Big* (Tampa: Trend House, 1973). Harney portrays Disston as a hard-drinking rogue who embarrassed his family.

83   *Hamilton Disston's ancestors:* Joe Knetsch, "Hamilton Disston and the Development of Florida," *Sunniland Tribune,* 1/24/1998, pp. 1–2; Jacob S. Disston, Jr., "Henry Disston: Pioneer, Industrialist, Inventor and Good Citizen," Newcomen Publications, 1950. Courtesy Joe Knetsch.

83   *He endured countless setbacks:* William D. Disston, Henry Disston and William Smith, "The Disston History," company history, May 1920, pp. 12–14. Courtesy Joe Knetsch.

83   *he had to warn customers:* Ibid., p. 25.

83   *He built a paternalistic company town:* Rival enterprises were banned from Tacony, which is now part of Philadelphia. Disston, Disston, and Smith, "The Disston History," p. 7; Louis M. Iatarola, "The Life and Influence of Hamilton Disston," The Historical Society of Tacony, 2001; Harry C. Silcox, "Henry Disston's Model Industrial Community: 19th Century Paternalism in Tacony," *The Pennsylvania Magazine of History and Biography* CXIV, no. 4 (October 1990). Henry Disston had occasional tensions with his laborers, but for the most part, he was a popular employer. One worker returned to his job the day after Henry had fired him. "If you don't know when you've got a good man, I know when I've got a good boss," the worker told Henry.

83   *Henry threatened to sack him:* The next time the fire whistle blew, Hamilton supposedly slipped out a factory window, and Henry yelled that he should never come back again. The next day, Hamilton simply showed up to work as if nothing had happened. Hanna and Hanna, *Lake Okeechobee,* p. 93.

83   *He twice ran off to enlist:* According to the Civil War Service Records at the National Archives, Hamilton enlisted as a private in the Pennsylvania Militia's infantry in June 1863. He was mustered out of the militia in August 1863. The records suggest that Hamilton's company was assigned to guard the bridges and roads leading into Philadelphia in case of a Confederate advance, and never left the state.

83   *serving as a ward leader:* Knetsch, "Hamilton Disston and the Development of Florida," pp. 2–4.

83   *lent his yacht:* During Quay's fishing excursion with some Republican national committeemen on Disston's yacht, one of the guests fell down the hatchway and broke his leg. "The Political Campaign: Curious Increase in the Colored Vote," *New York Times,* 10/17/1888; "Mr. Fessenden Severely Injured," *New York Times,* 8/9/1889; "Republican Fishermen Ashore," *New York Times,* 8/13/1889.

84   *He founded the Protective Tariff Club:* "Philadelphia's Loyalty Shown," *New York Times,* 9/26/1880.

84   *"He can drink plenty of champagne":* "Mr. Disston's Plaything: Millions of Acres and a Sugar Mill," *New York Daily Tribune,* 3/27/1892.

84   *The firm's 2,000 workers:* Hanna and Hanna, *Lake Okeechobee,* p. 94.

84   *President Rutherford B. Hayes:* "The President and Mrs. Hayes," *New York Times,* 4/27/1878; *Frank Leslie's Illustrated Newspaper,* 5/18/1878; Disston, Disston, and Smith, "The Disston History."

84   *He invested in a chemical firm:* Henry had also speculated in Atlantic City real estate, but otherwise Hamilton was striking out on his own. Knetsch, "Hamilton Disston and the Development of Florida," p. 4; Hanna and Hanna, *Lake Okeechobee,* p. 94.

85   *"If this remarkable enterprise":* "Draining the Everglades," *Manufacturer and Builder* 13, no. 3 (March 1881), p. 53.

85 *Pig Iron Kelley hailed:* Hanna and Hanna, *Lake Okeechobee,* p. 95.

85 *The* New York Times *decreed:* "Draining the Everglades: The Scheme by Which a Philadelphia Company Expects to Reclaim Twelve Million Acres of Land," *New York Times,* 2/18/1881, p. 2.

85 *"All know the value of the lands":* "Contract for Drainage," *Weekly Floridian,* 2/1/1881, p. 1.

85 *his heirs and other creditors:* "The Contract for the Four Million Sale," *The Weekly Floridian,* 2/28/1881, p. 1. The Floridian later provided a long history of the fund before the Disston sale. "The Internal Improvement Fund and the Disston Sale," The *Weekly Floridian,* 9/5/1882, p. 1.

86 *"This growing cancer":* Bloxham, "The Disston Sale and the State Finances," p. 19.

86 *"the largest purchase of land ever made":* The *Times* mangled the name of the buyer, calling him Hamilton Desson. "Buying Four Million Acres. An Immense Sale of Land by the State of Florida," *New York Times,* 6/17/1881, p. 5.

86 *the* Floridian *exulted:* "Sale of Four Million Acres," *Weekly Floridian,* 6/21/1881.

86 *A Fort Myers telegraph operator:* Karl H. Grismer, *The Story of Fort Myers* (Reproduction of the 1949 edition, Fort Myers Beach: Island Press Publishers, 1982), p. 103.

86 *"Both Democrats and Republicans":* "Press Comments on the Great Land Sale," *Weekly Floridian,* 7/12/1881.

86 *Not everyone was ecstatic:* Homesteaders and squatters also feared that Disston would evict them, but he treated them generously. Railroads were concerned that Disston would try to keep them off his land, but he welcomed them, and even invested in a few. And Tampa boosters were worried that Disston would squelch their development by hogging all the area's good land, but his advertisements only helped promote the city. Knetsch, "Hamilton Disston and the Development of Florida," pp. 8–9; Hanna and Hanna, *Lake Okeechobee,* p. 96; Edward C. Williamson, *Florida Politics in the Gilded Age, 1877–1893* (Gainesville: University Press of Florida, 1976), p. 78.

86 *it awarded one man 98,000 acres:* That man was named John Henderson; his heirs later sold the land to the Seattle developer James Moore, who founded the town of Moore Haven on Lake Okeechobee.

86 *It was true that other potential buyers:* Syndicates based in New York, Boston, England, and Germany all made runs at the fund during this period, as did Ambassador Sanford. The land agent Samuel Swann offered an extra penny per acre for four million acres after the Disston deal was done, but it was never clear how he would raise the cash. Blake, *Land into Water,* pp. 76–77.

86 *There was more griping:* Agreement between Edward J. Reed and Hamilton Disston, 1/18/82, Title and Land Records Section, Division of State Lands, Department of Environmental Protection. Over the years, the Reed deal has been reported as a $400,000 sale and a $500,000 sale, but the agreement was for $600,000. Reed was required to pay Disston $50,000 up front, another $50,000 in twenty days, and cover the rest of the $500,000 he owed the fund. It is possible that Disston never collected the $100,000 due directly to him, but he should have. Incidentally, it took Reed an extra year to complete his $500,000 in payments to the fund, but he did complete them.

86 *"when this great incubus of incumbrance":* Bloxham, "The Disston Sale and the State Finances," p. 20.

87 *In the four years following the sale:* Bloxham, "The Disston Sale and the State Finances," p. 21; Arnold Marc Pavlovsky, *We Busted Because We Failed: Florida Politics, 1880–1908* (Unpublished Princeton University dissertation, 1973); Joe M. Richardson, "The Florida Excursion of President Chester A. Arthur," *Tequesta* 24 (1964), p. 41.

87 *"Would that Florida had a thousand Disstons":* "The Disston Company," *Weekly Floridian,* 11/8/1881.

87 *"The scheme has outgrown":* Edward N. Akin, *Flagler: Rockefeller Partner and Florida Baron* (originally published in 1988; Gainesville: University Press of Florida, 1991), p. 118. The original Ponce de Leon Hotel is now Flagler College; the Tampa Bay Hotel is now the University of Tampa.

87 *He imported 250 New Yorkers:* Hanna and Hanna, *Lake Okeechobee,* p. 101. As a humanitarian gesture, Disston offered forty acres to Jewish émigrés who fled to Philadelphia after pogroms in Russia.

87 *He later founded the coastal resorts:* Jack W. McClellan, "Hamilton Disston in Florida" (Unpublished University of North Florida thesis, 1987); Tebeau, *Florida's Last Frontier,* p. 169.

88   *"there is only one Fort Myers":* Michael Grunwald, "Growing Pains in Southwest Fla.," *Washington Post,* 6/5/2002. The quotation is featured at the Thomas Edison estate, and online at www.edison-ford-estate.com.

88   *"little Eden in the wilderness":* Williams, "North to South Through the Glades," Part One, p. 42.

88   *In 1883, Disston arranged a visit to Kissimmee:* Richardson, "The Florida Excursion of President Chester A. Arthur"; "The President's Vacation," *New York Times,* 4/11/1883; "Fishing in South Florida," *New York Times,* 4/12/1883; "The President's Holiday," *New York Times,* 4/22/1883.

88   *One devout Shaker:* Russell Anderson, "The Shaker Community in Florida," *Florida Historical Quarterly* 38 (July 1959), p. 29.

89   *opening real estate offices in England: Minutes,* vol. 3, pp. 80–81, 11/8/1881 letter.

89   *"lands of inexhaustible fertility without fertilizing":* "The 'Disston' Okeechobee Land and Drainage Company of Florida," undated advertisement, UFA, Miscellaneous Manuscripts, Box 6, Disston Land Co.

89   *"The immigrant from Europe":* "Descriptive List Catalogue, Disston Lands in Florida, Owned by the Florida Land and Improvement Co., Atlantic and Gulf Coast Canal and Okeechobee Land Co., and Kissimmee Land Co," 1885 catalogue, FSA, p. 2.

89   *"You secure a home":* "The Wonderful Country—Where Farming Pays," undated Disston Land Co. pamphlet, UFA, Miscellaneous Manuscripts, Box 6, Disston Land Co., p. 7.

89   *"ATTENTION!! FARMERS!!":* "Kissimmee Land Company, 200,000 Acres Best Land in Florida," 1884 brochure, UFA, Miscellaneous Manuscripts, Box 6, Disston Land Co.

90   *"as dry as a bone":* Grismer, *The Story of Fort Myers,* p. 105.

90   *"not only a sure and safe investment":* James Kreamer, "The Atlantic and Gulf Coast Canal and Okeechobee Land Company (of Florida)," 1881 prospectus, FSA, p. 5.

90   *Disston's drainage strategy:* "Draining Lake Okeechobee," *Weekly Floridian,* 2/8/1881; "The Okeechobee Drainage Scheme," *Weekly Floridian,* 11/29/1881.

90   *Disston's chief engineer, James Kreamer:* Kreamer, "The A&GCC&OLC (of)" Atlantic and Gulf prospectus, pp. 14–16. Kreamer predicted that the St. Lucie Canal would flow at 2.63 miles per hour. The shorter Caloosahatchee Canal would have much less slope, and therefore much less velocity.

91   *"The groundwork is laid":* "The Okeechobee Drainage Scheme," *Weekly Floridian,* 11/29/1881.

91   *"The arrival of the dredge":* David Von Drehle, "Bury My Heart at Southwest 392nd Terrace," *The Miami Herald Tropic Magazine,* 10/8/1989, p. 14.

91   *The dredge itself was a lumbering hunk:* Conrad Menge, "Early Dredging in the Lake Okeechobee Region" (Unpublished 1947 manuscript at the Clewiston Museum); Will Wallace Harney, "The Drainage of the Everglades," *Harper's Magazine* 68, no. 406 (March 1884), pp. 598–605. Lawrence Will did not work for Disston, but his memoir of dredging the "soup-doodle muck" of the Everglades, *A Dredgeman of Cape Sable* (Belle Glade, FL: Glades Historical Society, 1984), is the most vivid account of drainage life. Christopher McVoy, Stuart Appelbaum, and James Vearil all helped teach me Drainage for Dummies.

92   *"The huge crane swings":* Harney, "The Drainage of the Everglades," p. 601.

92   *"We had to drag our boats":* Menge, "Early Dredging in the Lake Okeechobee Region," pp. 11–12.

92   *James Dancy reported on his progress: Minutes,* vol. 3, pp. 243–47. Dancy later returned to make an even more exuberant report. Ibid., pp. 332–39.

93   *A year later, state engineer H. S. Duval:* Ibid., pp. 314–23.

93   *Mr. Frazier, who seemed to materialize:* For example, Captain Hendry took the *Times-Democrat's* correspondents to visit Mr. Frazier's homestead, where they saw "tomatoes and okra in abundance." Williams, "North to South Through the Glades," Part I, p. 50.

93   *"During an experience of 12 years": Minutes,* vol. 3, p. 338.

93   *Duval and Dancy certified:* Dancy's first report certified that 535,285 acres had been reclaimed. Duval then signed off on an additional 2,182,412 acres. Dancy's follow-up then added 234,401 more, for a total of 2,952,098 acres.

93   *His sugar plantation produced U.S.-record yields:* "The Disston Sugar Plantation: Its Success and Its Failure," Annual Report of the State Chemist of Florida, 1919, FSA; Pat Dodson, "Hamilton Disston's St. Cloud Sugar Plantation, 1887–1901," *Florida Historical Quarterly* (1971), pp. 356–69.

93   *the sugar king of Hawaii:* His name was Claus Spreckels. "Report of the Trustees of the Internal Improvement Fund on the Operations of the Atlantic and Gulf Coast Canal and Okeechobee Land Company," 5/24/1893, p. 12, RTE.

94   *The federal government's chief chemist:* His name was H. W. Wiley. Ibid., p. 11.
94   *"There is not one of our little party":* Williams, "Across South Central Florida," Part I, p. 60.
94   *"None but those who are fault-finders":* "The Disston Sale," *Weekly Floridian,* 2/26/1884, p. 1.
94   *The commission did acknowledge:* "Report of the Committee Appointed by the Governor on the Work of the Atlantic and Gulf Coast Canal and Okeechobee Land Company," *Weekly Floridian,* 2/17/1887.
95   *This revisionism has shaped Disston's image:* For example, one historian wrote that Disston had "nearly total ignorance of the region he moved to remake." Derr, *Some Kind of Paradise,* p. 90.
95   *"That style of tribunal":* H. S. Duval, "A Reply to the Okeechobee Commission Report," *Weekly Floridian,* 4/28/1887. Duval sniped that the commission's three members must have been hydrophobic: "Without being impertinent, I would like to ask . . . if one of them wouldn't feel more at home in a law library, and the other two feel less strained . . . in another branch of engineering."
96   *"While the company has not progressed":* "Report of the Committee," *Weekly Floridian,* 2/17/1887.
96   *Disston and the fund's trustees eventually reached:* Minutes, vol. 3, pp. 501–5. Disston agreed to spend $125,000 in order to keep the 1,174,943 acres he had already received from the fund. He would then receive one acre for every additional 25 cents he spent up to $206,264, for a maximum of two million acres.
96   *Disston ultimately dug more than eighty miles:* "Report of the Board of Trustees." This report cites rainfall data from Jacksonville to "completely dispose" of the 1887 report's suggestion that the early 1880s had been drought years. The trustees were probably correct about the early 1880s, but this didn't prove their case; rainfall in Jacksonville suggests very little about rainfall further south.
96   *he never did dig canals south and east:* Disston's focus had shifted so far away from the heart of the Everglades that when residents along the Caloosahatchee continued to complain about flooding, he offered to sever the river's connection to Lake Okeechobee. It never happened, though.
96   *there's little evidence to support it:* The indefatigable Florida historian Joe Knetsch deserves the credit for debunking this long-standing myth. Knetsch, "Hamilton Disston and the Development of Florida."
96   *he later rescinded the cuts:* "Advanced the Wages of Its Men," *New York Times,* 5/23/1895.
96   *Disston's estate was valued:* "Hamilton Disston's Will," *New York Times,* 5/9/1896; "Hamilton Disston's Life Insurance," *New York Times,* 5/15/1896; "Insured for a Million Dollars," *New York Daily Tribune,* 5/2/1896.
96   *All but one of Disston's obituaries:* "Hamilton Disston," *New York Daily Tribune,* 5/1/1896; "Hamilton Disston Found Dead," *New York Times,* 5/1/1896; "Sudden Death of Mr. Disston," *New York Herald,* 5/1/1896; "He Died Without Warning," *Washington Post,* 5/1/1896. The less-than-reputable *Philadelphia Press,* a Democratic paper, included the only account of Disston's purported ruin and suicide.
97   *"These are nature's silent witnesses": Minutes,* vol. 3, p. 317.

7   The Father of South Florida

PAGE
98   *"Think of pouring all that money out":* Ron Chernow, *Titan: The Life of John D. Rockefeller, Sr.* (New York: Random House, 1998), p. 345.
98   *"Oh, Lord! Oh, God!":* Church, "A Dash Across the Everglades," p. 23.
98   *the owner of Ingraham's railroad, Henry Plant, had asked him to survey a line:* Plant, a Connecticut Yankee, did not seem to recognize the logistical challenge of his request. "Mr. Plant, that is right across the Everglades of Florida!" Ingraham said. "What of it?" Plant replied. "So far as I know, only two white men ever made that trip!" Ingraham said. Dovell, *A History of the Everglades of Florida,* p. 150.
98   *"Locomotion is extremely difficult":* Wallace R. Moses, "The Journal of the Everglades Exploring Expedition," edited from the 1892 log by Watt P. Marchman, *Tequesta* 7 (1947), p. 19.
98   *"I was so tired I had lost interest in everything":* A friendly Seminole eventually found the men floundering in the Everglades and helped lead them to Miami. Church, "A Dash Across the Everglades," pp. 28, 32.
99   *he believed its Everglades backcountry could be drained:* "Where Nature Smiles," p. 5. Ingraham proposed his own scheme to drain 500,000 acres to the internal improvement board, but it was rejected. *Minutes,* vol. 4, pp. 198, 206.

99   *"a great tract of land":* James Ingraham, "Draining of the Everglades," *Success Magazine,* quoted in "Where Nature Smiles," Florida Everglades Land Company brochure, 1909, p. 5; "Statement of J. E. Ingraham before the Joint Committee of the Senate and the House in re Everglades Drainage Matters," 4/17/1917, JIP, Box 1.

99   *Henry Morrison Flagler was born poor:* I consulted three biographies of Henry Flagler: Edward Akin's *Flagler;* David Leon Chandler, *Henry Flagler: The Astonishing Life and Times of the Visionary Robber Baron Who Founded Florida* (New York: Macmillan, 1986); and Sidney Walter Martin, *Henry Flagler: Visionary of the Gilded Age* (originally published in 1949; Lake Buena Vista, FL: Tailored Tours Publications, 1998). I also referred to Ron Chernow's brilliant biography of John D. Rockefeller Sr., *Titan,* and Les Staniford's engaging narrative of the Key West extension, *Last Train to Paradise: Henry Flagler and the Spectacular Rise and Fall of the Railroad That Crossed an Ocean* (New York: Crown Publishers, 2002). All of those books rely on the archives at the Henry Flagler Museum; Micki Blakely directed me to additional documents there.

99   *kept one of them all his life:* Flagler later explained that the coin reminded him of the New Testament parable (Matthew 25:14–30) of the man with one talent. Flagler knew exactly what his own talent was.

99   *six decades later:* Akin, *Flagler,* p. 196.

100   *"I had scruples about the business":* Chernow, *Titan,* p. 107.

100   *he invented a horseshoe:* Chandler, *Henry Flagler,* p. 51.

100   *"I trained myself in the school":* Ibid., p. 260.

100   *Rockefeller agreed to make Flagler a partner:* At first, Flagler was chosen more for Mary's family money than his talent; he persuaded her cousin Stephen Harkness to provide seed money for Rockefeller, but as a condition of his investment, Harkness insisted that Rockefeller take on Flagler as a partner.

100   *Rockefeller once remarked that in thirty-five years:* Akin, *Flagler,* p. 255.

100   *Flagler kept a quotation on his desk:* The quotation was from the popular novel *David Harum.* Chandler, *Henry Flagler,* p. 82.

101   *"If you think the perspiration":* Akin, *Flagler,* p. 67,

101   *"No, sir, I wish I'd had the brains":* Martin, *Henry Flagler,* p. 48.

101   *"He was a man of great force":* Chernow, *Titan,* p. 109.

101   *"It suits me to go elsewhere for advice":* Akin, *Flagler,* p. 91; Martin, *Henry Flagler,* p. 67.

102   *He once compared himself:* Martin, *Henry Flagler,* p. 91.

102   *Flagler considered his Florida projects:* "I am convinced that he did not regard his Florida properties in the same light as he would have looked on a commercial enterprise," James Ingraham said. Thomas Graham, "Henry Flagler's St. Augustine," *El Escribano: The St. Augustine Journal of History* 40 (2003), pp. 1–9.

102   *"I see that you are wheeling the muck":* Akin, *Flagler,* p. 128.

102   *If he was going to build hotels:* Joe Knetsch, "Flagler's Business System," *El Escribano: The St. Augustine Journal of History* 40 (2003), pp. 75–76.

103   *"I comfort myself with the reflection":* Akin, *Flagler,* p. 121.

103   *"Hire another cook":* Edwin Lefevre, "Flagler and Florida," *Everybody's Magazine* 20, no. 2 (February 1910), p. 178.

103   *In Florida, Flagler wanted to create:* Les Staniford, *Last Train to Paradise,* p. 50.

103   *"Permanence appeals to him":* Lefevre, "Flagler and Florida," p. 174.

103   *Rockefeller was so appalled:* Chernow speculates that Rockefeller and his wife came to see the Flaglers as the kind of "gaudy arrivistes [they] had always abhorred." Chernow, *Titan,* pp. 344–45.

103   *"Not a day passes but that I call myself to account":* Chandler, *Henry Flagler,* p. 113.

103   *He had already spent ten times more:* Martin, *Henry Flagler,* p. 108.

104   *"I have found a veritable Paradise!":* Chandler, *Henry Flagler,* p. 135.

104   *"In a few years, there will be a town over there":* Akin, *Flagler,* p. 144.

104   *His steel ribbon soon unspooled:* Martin, *Henry Flagler,* pp. 111–12. Flagler wanted a link to Juno, the Dade County seat, but local landowners who thought they had Flagler over a barrel demanded such exorbitant prices that he bypassed the town completely. That was the beginning of Juno's decline. William Gleason, the former Miami booster, helped persuade Flagler to stop in his new hometown of Eau Gallie. Cresse, *Gleason,* p. 174.

104   *requiring 2,400 gallons of paint:* "The Story of a Pioneer," Florida East Coast Railway pamphlet.

104   *"Yesterday a swamp was here":* Lefevre, "Flagler and Florida," p. 171.

104 *The Royal Poinciana soon became:* Martin, *Henry Flagler,* pp. 117–19; Derr, *Some Kind of Paradise,* pp. 43–44; *The Palm Beach Post: Our Century,* edited by Jan Tuckwood (New York: Mega-Books/Progressive Publishing, 2000), pp. 20–23.

105 *"more wonderful than any palace":* "Artists of the World Hardworked to Furnish Whitehall," *New York Herald,* 3/30/1902.

105 *"I feel that these people are wards of mine":* Henry Flagler letter to Rev. Charles Stevens, 9/4/1901, HSPBC.

105 *temperatures dipped to fourteen degrees:* Larry Wiggins, "The Birth of the City of Miami," *Tequesta* 55 (1995).

105 *Florida's yearly citrus production:* Helen Muir, Miami U.S.A. (originally published in 1953; Gainesville: University of Florida Press, 2000).

105 *"It is the dream of my life":* Arva Moore Parks, *Miami: The Magic City* (Tulsa, OK: Continental Heritage Press, 1981), p. 63.

107 *But now Ingraham returned:* Legend has it that Turtle sent the orange blossoms to Flagler, but Ingraham recounted in a 1920 speech to the Miami Women's Club that he carried them himself. It is not clear whether Tuttle gave them to him. Chandler, *Henry Flagler,* pp. 168–69; "The Story of a Pioneer," p. 18.

107 *he had one of his Standard Oil lobbyists secure $300,000:* Akin, *Flagler,* p. 170.

107 *"If I owned both Miami and Hell":* Parks, *Miami,* p. 76. Sometimes the "follow the crowd" story is told in the reverse, as Plant's advice to Flagler on how to find Tampa.

107 *"What we want for some little time to come":* Akin, *Flagler,* p. 189.

108 *"Most Productive Soil in Existence":* Florida East Coast *Homeseeker,* April 1910, Everglades Special, p. 115.

108 *Japanese immigrants who started a now-defunct farm colony:* "Dade County's Japanese Colony," *Miami Metropolis,* 3/17/1905.

108 *Years later, a reporter for* Everybody's Magazine: Lefevre, "Flagler and Florida," p. 181.

108 *"My domain begins in Jacksonville":* Akin, *Flagler,* p. 190.

109 *"being rapidly destroyed":* Simpson, *In Lower Florida Wilds,* p. 49.

110 *"There was a most magnificent":* Parks, *Miami,* p. 76.

110 *"It may seem strange":* Willoughby, *Across the Everglades,* p. 13.

110 *"Some men believe the Everglades should be drained":* Samuel Proctor, *Napoleon Bonaparte Broward: Florida's Fighting Democrat* (originally published in 1950; Gainesville: University Press of Florida, 1993), p. 250.

110 *"With the money spent on hotels":* Church, "A Dash Through the Everglades," pp. 20–21.

111 *"land development scheme":* "Florida's Rich Rivals," *New York Times,* 3/18/1896.

111 *In 1898, Ingraham and Rufus Rose: Minutes,* vol. 4, pp. 433–50. The company's name was patterned on Flagler's Florida East Coast Railroad Company and Florida East Coast Hotel Company.

111 *For Flagler, Captain Rose designed a purely local plan: Minutes,* vol. 4, pp. 456–57; *Minutes,* vol. 5, pp. 31–32.

111 *Drainage would be "a simple process":* "The Everglades of Florida: Prospectus of the Florida East Coast Drainage and Sugar Company," UFA, 1902, p. 4.

112 *Governor Bloxham proclaimed:* Dovell, *A History of the Everglades of Florida,* p. 192.

112 *"It may be taken as assured fact":* "Draining Glades; The Project Perfectly Feasible and Practical," *Tropical Sun,* 3/21/1902, PBCHS.

112 *"As the bottom of this basin is above tide water":* "Draining the Everglades; A Remarkable Work That Has Been Undertaken. Object of the Operations," *Tropical Sun,* 12/17/1902; Irvine Mather, "Draining the Everglades," *Florida Magazine* 4, no. 5 (May 1902), pp. 259–63.

112 *The United States was now the richest nation on earth:* Steven J. Diner, *A Very Different Age: Americans of the Progressive Era* (New York: Hill & Wang, 1998); Edmund Morris, *Theodore Rex* (New York: Modern Library, 2002), p. 20; David von Drehle, "Origin of the Species," *Washington Post Magazine,* 7/25/2004.

113 *"I have no command of the English language":* Akin, *Flagler,* p. 206. Flagler also expressed his fond hopes that Roosevelt would be eaten by a lion during his post-presidency journey through Africa.

113 *He found that Florida had given away:* As of August 6, 1904, the fund had received 20,113,837.42 acres from the federal government, and had only 3,076,904.68 acres left. "Message of the Governor," Napoleon Broward message to the legislature, 5/3/1905, *Florida Senate Journal,* pp. 378–413.

113 *When Flagler's railroad tried to claim 156,000 acres:* James Ingraham letter, 8/23/1906, HFP.
114 *The governor had just traveled to California:* "The Story of the Everglades," p. 122.
114 *"So far as I am personally concerned":* Henry Flagler letter, 1/28/01, to James Ingraham, HFP.
115 *One of Flagler's closest friends gasped:* George Morgan Ward, "In Memoriam: Henry M. Flagler." Eulogy presented in Palm Beach, 3/15/1914, reprinted by Matthews-Northrup Works, Buffalo.
115 *"The financiers considered the project":* "The Story of a Pioneer."
115 *"It was very strange, at first":* LeFevre, "Flagler and Florida," p. 174.
115 *Krome and his crew:* William J. Krome, "Railway Location in the Florida Everglades," with an introduction by Jean C. Taylor, *Tequesta* 39 (1979), pp. 5–16.
115 *"I found a most God-forsaken region":* Standiford, *Last Train to Paradise,* p. 87.
116 *"The muck with proper drainage":* Krome, "Railway Location," p. 16.

8   Protect the Birds

PAGE
117 *"Florida has been considered":* Harriet Beecher Stowe, "Protect the Birds," reprinted in Foster and Foster, *Beechers, Stowes and Yankee Strangers,* p. xx.
118 *Wading birds are extraordinarily demanding creatures:* Ed Carlson at Corkscrew Swamp gave me a tour of his sanctuary's wood stork rookeries. Lodge, *The Everglades Handbook,* 1st ed. pp. 153–154.
118 *"Here I felt I had reached the high-water mark":* Herbert K. Job, *Wild Wings: Adventures of a Camera-Hunter Among the Largest Wild Birds of North America on Sea and Land* (New York: Houghton Mifflin & Co., 1905), p. 54.
119 *As many as 2.5 million wading birds:* William B. Robertson Jr. and James A. Kushlan, "The Southern Florida Avifauna," in Gleason, ed., *Environments of South Florida,* p. 230.
119 *"It was truly a wonderful sight":* McIver, *Death in the Everglades,* pp. 36–37.
119 *Snowy egrets with bright yellow feet:* Henri Dauge, "Mr. Wegg's Party on the Kissimmee," *Harper's New Monthly Magazine,* 1886, in Oppel and Meisel, eds., *Tales of Old Florida,* p. 320.
119 *"When do they sleep?":* Willoughby, *Across the Everglades,* pp. 116.
119 *During an 1832 visit:* Audubon, "Ornithological Biography," quoted in Proby, *Audubon in Florida,* pp. 327–37.
119 *"The flocks of birds that covered the shelly beaches":* Ibid., p. 332.
119 *"Our first fire among a crowd of the Great Godwits":* Ibid., p. 332.
119 *Florida's first environmental broadside:* Stowe, "Protect the Birds," in Foster and Foster, *Beechers, Stowes and Yankee Strangers,* pp. xx–xxii.
120 *In February 1886, a birdwatcher named Frank Chapman:* McIver, *Death in the Everglades,* pp. 1–2.
120 *At the height of the fad:* Pierce, "The Cruise of the Bonton," p. 23; McIver, *Death in the Everglades,* pp. 40–41, 46–53; Harry A. Kersey, Jr., *Plumes, Pelts and Hides: White Traders Among the Seminole Indians, 1870–1930* (Gainesville: University Press of Florida), 1982.
120 *"What do you hunt?":* McIver, *Death in the Everglades,* p. 41.
121 *Florida's most notorious plumer, Jean Chevelier:* McIver, *Death in the Everglades,* pp. 16–17.
121 *The logkeeper for one Chevelier expedition:* Pierce, "The Cruise of the Bonton," pp. 26, 55; William B. Robertson, Jr., "Ornithology of 'The Cruise of the Bonton,'" *Tequesta* 22 (1962), p. 70.
121 *They used quiet weapons:* McIver, *Death in the Everglades,* pp. 2–3.
121 *"Hundreds of broken eggs":* Ibid., p. 40.
122 *"The Indian leaves enough of the old birds":* A. W. Dimock and Julian A. Dimock, *Florida Enchantments* (originally published in 1908; Detroit: Gale Research Company, 1975), p. 299.
122 *This kill-them-all strategy took its toll:* William B. Robertson Jr. and James A. Kushlan, "The Southern Florida Avifauna," in Gleason, ed., *Environments of South Florida,* pp. 230–31; Derr, *Some Kind of Paradise,* p. 140.
122 *"I don't think in my reincarnation":* McIver, *Death in the Everglades,* p. 96.
122 *the author of* The Territory of Florida: Williams, *The Territory of Florida,* pp. 62, 65, 76.
122 *Occasionally, a writer like Buckingham Smith:* "Report of Buckingham Smith," p. 29.
123 *The industrialization and deforestation:* Robert L. Dorman, *A Word for Nature: Four Pioneering Environmental Advocates, 1845–1913* (Chapel Hill: University of North Carolina Press, 1998). This book was my most important guide to the origins of the American conservation movement.

123   *"We need the tonic of wilderness"*: Henry David Thoreau, *Walden* (originally published in 1854; Ware, England: Wordsworth American Library, 1995), pp. 190, 215.

123   *"I love Nature partly because she is not a man"*: Dorman, *A Word for Nature*, p. 70.

123   *He wanted to preserve the nastiest rattlesnakes*: Ibid., p. 119.

123   *"All nature is linked together by invisible bonds"*: Ibid., p. 33.

124   *"The conservation of natural resources"*: Char Miller, *Gifford Pinchot and the Making of Modern Environmentalism* (Washington, DC: Island Press, 2001), p. 229. Miller argues that Pinchot had a bit of Thoreau and Muir in him, too. He waxed especially lyrical about south Florida's fish.

124   *T. R. began his career*: Edmund Morris, *The Rise of Theodore Roosevelt* (New York: Modern Library, 2001), pp. 17–20, 36–38, 65–67, 109.

125   *"This is bully!"*: Morris, *Theodore Rex*, p. 231.

125   *"Conservation means development as much as it does protection"*: Roosevelt said this in a speech in Kansas in August 1910. The quotation is now on the wall near the entrance to the Museum of Natural History in New York, and at www.theodoreroosevelt.org.

125   *"This is the last pitiful remnant"*: Job, *Wild Wings*, p. 54.

125   *By 1900, "Audubon societies"*: McIver, *Death in the Everglades*, pp. 97–101. A history of the movement can be found on the National Audubon Society's website, www.audubon.org.

125   *The president turned to his aides*: "Pelican Island: Restoring a Legacy," U.S. Fish and Wildlife Service publication, November 1999; Morris, *Theodore Rex*, p. 519.

125   *"Birds should be saved for utilitarian reasons"*: Theodore Roosevelt, *A Book-Lover's Holidays in the Open*, 1916, Quotation published at www.theodoreroosevelt.org

126   *Even a hardened Everglades pioneer*: McKinney's grandson, Totch Brown, became a renowned Everglades gator hunter, marijuana smuggler, and author. Brown, *Totch*, pp. 10–11, 249–65.

126   *"a sturdy, fearless fellow"*: McIver, *Death in the Everglades*, p. 9. I am indebted to McIver's excellent account of Guy Bradley's story in particular and plume hunting in general.

126   *But he later renounced bird slaughter*: Ibid., p. 115.

126   *He also gave tours to visiting ornithologists*: Ibid., p. 142.

127   *"You ever arrest one of my boys again"*: Ibid., p. 144.

127   *On the morning of July 8*: Ibid., pp. 152–53.

127   *"There is no community sufficiently law-abiding"*: Ibid., p. 163.

128   *"Though we saw birds everywhere"*: Grey, *Tales of Southern Rivers*, p. 56.

128   *South Florida's leading conservationist*: John C. Gifford, *The Everglades and Other Essays Relating to Southern Florida* (Kansas City: 1911), pp. 101–102; John C. Gifford, *Living by the Land* (Coral Gables, FL: Park Art Printing Association, 1945); Henry Troetschel, Jr., "John Clayton Gifford: An Appreciation," *Tequesta* 10 (1950), pp. 35–42.

128   *Do not think that conservation*: Gifford, *Living by the Land*, p. 8.

128   *"In southern California"*: Gifford, *The Everglades and Other Essays*, p. 102.

129   *"It is a natural swamp tree"*: Gifford, *Living by the Land*, p. 80.

129   *Thomas Will, a self-made man*: Junius Elmore Dovell, "Thomas Elmer Will, Twentieth Century Pioneer," *Tequesta* 8 (1948), pp. 21–55. TWP.

129   *"He was capable of extreme exertion"*: Lawrence Will quoted in ibid.

129   *"the underdog against vested interests"*: John Newhouse quoted in ibid.

129   *"Remember, I'm on the job"*: Thomas Will letter to G. P. Alliston, 2/25/1932, TWP, quoted in Dovell, *The History of the Everglades of Florida*, p. 45.

9   "Water Will Run Downhill!"

PAGE

130   *"Yes, the Everglades is a swamp"*: Broward, "The Call of the Everglades," *Florida East Coast Homeseeker* 12, no. 4.

130   *"It would indeed be a sad commentary"*: Napoleon Bonaparte Broward, "Open Letter of Governor N.B. Broward to the People of Florida," 11/25/1905, NBP, Box 11.

130   *Awestruck journalists gushed*: "Broward Is Dubbed the Father of the Everglades," *Miami Metropolis*, 7/23/1910; "Correcting an Error," *Miami Metropolis*, 3/13/1911; Joe Hugh Reese, "The Everglades," Part One, *The Hollywood Magazine* I, no. 6–7 (April-May 1925), p. 5.

131   *"It might be said of me"*: Proctor, *Napoleon Bonaparte Broward*, pp. 259–60. The late Samuel Proctor's authorized biography is a bit hagiographic, but it is lively and informative, and it is

the only source for a great deal of personal information that Broward's family entrusted to Proctor. I also relied heavily on the governor's public papers at the University of Florida and the state archives in Tallahassee.

131   *"Had it not been for Broward":* Thomas E. Watson, "Governor Broward and the Everglades," *Watson's Jeffersonian Magazine* 2, no. 5 (May 1908), p. 263.

131   *"the man who makes two blades of grass":* Gifford, *The Everglades,* p. 2.

131   *"the seductive and enslaving power of corporate interests":* Proctor, *Napoleon Bonaparte Broward,* p. 190.

131   *"draining the people instead of the swamps":* Florida Times-Union, 2/28/1904; Proctor, *Napoleon Bonaparte Broward,* p. 190.

131   *the richest one percent of Americans:* Diner, *A Very Different Age,* pp. 4, 28.

131   *Napoleon Bonaparte Broward was born:* Proctor, *Napoleon Bonaparte Broward;* Napoleon Bonaparte Broward, "Napoleon B. Broward, Candidate for Governor of Florida: Autobiography, Platform, Letter and Short Story of the Steamer 'Three Friends,' and a Filibustering Trip to Cuba," 1900 pamphlet.

132   *"We were not discouraged, but immediately went to work":* Broward, "Autobiography," p. 2.

132   *one of Broward's letters:* Napoleon Broward letter to Pulaski Broward, 10/9/1907, Box 5A, NBP. This letter casts rather significant doubt on Proctor's claim that Broward's mother "failed gradually." Proctor, *Napoleon Bonaparte Broward,* p. 20.

132   *Nature, he liked to say:* "Governor Broward Visits Fair," *Miami Metropolis,* 3/8/1906.

132   *By the time he was thirty:* Proctor, *Napoleon Bonaparte Broward,* p. 35.

133   *When he remarried, a local paper:* Proctor, *Napoleon Bonaparte Broward,* p. 38.

133   *In 1888, Broward's influence and friends:* Ibid., pp. 39–41.

133   *He cemented his reputation:* Ibid., pp. 73–76.

133   *"He is not one of the high-falutin":* Ibid., p. 180.

134   *Teddy Roosevelt once needled Broward:* Ibid., pp. 264–65.

134   *"I'm going to . . . talk to the farmers":* Ibid., p. x.

134   *Instead, he proposed to reclaim the Everglades:* Blake, *Land into Water,* p. 95.

134   *Broward's drainage dreams were ridiculed:* Proctor, *Napoleon Bonaparte Broward,* p. 200.

134   *"If a graveyard has been despoiled":* Broward, "Open Letter," p. 2.

135   *"Is it for legal services":* Broward, "Autobiography," p. 14.

135   *"Laugh if you like":* Proctor, *Napoleon Bonaparte Broward,* p. 204.

135   *a bold progressive agenda:* Broward promised equal treatment for all white Floridians. As for blacks, he believed that America should buy their property and relocate them to a sovereign nation of their own, which was actually a relatively enlightened stand for a southern politician in those days.

135   *"tap the wealth of the fabulous muck":* Proctor, *Napoleon Bonaparte Broward,* p. 191.

135   *"The Everglades of Florida should be saved":* Ibid., p. 210.

135   *Broward sent the legislature a special message on drainage:* "Message of the Governor," pp. 396–97.

136   *Broward had run dredges for years:* Transcript of the 1907 Commission for the Investigation of the Acts and Doings of the Trustees of the Internal Improvement Fund, p. 434, Series 654, FSA.

136   *He grumbled that by the time the studies were done:* Gifford, *The Everglades,* p. 99.

136   *"If my friends will hold the knockers in check":* Proctor, *Napoleon Bonaparte Broward,* pp. 247–48.

136   *"I consider the launching of a dredge":* William Jennings letter to Governor Napoleon Broward, 1/21/1905, reprinted in "The Everglades of Florida in Acts, Reports and Other Papers, State and National, Relating to the Everglades of the State of Florida and Their Reclamation," Senate Document 89, 62nd Cong., 1st sess., 1911, p. 54.

136   *"Shall the sovereign people of Florida":* Broward, "Open Letter," p. 1.

136   *"He desires to own all lands":* Proctor, *Napoleon Bonaparte Broward,* p. 260.

136   *the state would need dipper dredges:* Joe Knetsch, "Governor Broward and the Details of Dredging," *Broward Legacy* 14, no. 1–2 (winter–spring 1991), p. 41.

137   *One day, the governor wrote to tell a contractor:* Broward letter to Marion Steam Shovel Company, 9/10/1908, Box: Governor Correspondence, 1908, Folder: Correspondence—Marion Steam Shovel Co,. NBP.

137   *He wrote another letter asking why a bulkhead had been thickened:* Broward letter to Tampa Foundry & Machine Co., 8/27/1908, Box: Governor Correspondence, 1908, Folder: Tampa Foundry & Machine Co., NBP.

137   *His letters to the project's chief engineer:* Broward letter to Newman, 5/13/07 and 7/8/07, General Correspondence—Broward, Series 32, vol. 75, NBP.

137   *"The governor just naturally sweats dope"*: *Florida Times-Union,* 8/20/1906, quoted in Proctor, Napoleon Bonaparte Broward, NBP, pp. 245–46.

137   *Like Disston:* Nelson Blake points out that Broward was much more popular around the New River than the St. Lucie area. Captain Rose, who had avoided the St. Lucie while working for Disston as well, wrote the governor in 1905 to point out that while a St. Lucie canal would help lower the lake, it would not drain local wetlands; the New River Canal would achieve both objectives. Blake, *Land into Water,* p. 97; Rufus Rose letter to Broward, November 1905, Box 11, GNBP.

137   *the governor was making monthly visits:* The Internal Improvement Fund's minutes record Broward's reimbursements for at least eight trips to Fort Lauderdale during late 1905 and early 1906.

137   *The populist rabble-rouser Thomas Watson:* Watson, "Governor Broward and the Everglades," p. 266.

138   *An engineering magazine predicted:* A. B. Clark, "To Drain the Florida Everglades," *The Technical World Magazine* (May 1907), p. 253.

138   *"It has been said that man can never improve on nature":* R. V. Blackman, "First Farm in the Everglades," *Florida East Coast Homeseeker* (April 1910), p. 138.

138   *Swampland the state had sold to settlers:* "Where Nature Smiles," p. 22.

138   *At a time when farmers were struggling:* Walter Waldin, *Truck Farming in the Everglades* (Kansas City: Florida Everglades Land Sales Company pamphlet, 1910), FHC, p. 6.

138   *"My prophecy is that this great Everglades district":* Ibid, p. 139.

138   *The knockers, however, kept knocking:* Proctor, *Napoleon Bonaparte Broward,* pp. 241–43; "A Plain Answer to Governor Broward's Open Letter to the People of Florida," 1905 pamphlet; NBP, Miscellaneous Material on the Everglades; Dovell, *A History of the Everglades of Florida,* p. 250.

139   *Frank Stoneman, the editor of a newspaper:* Christopher F. Meindl, "Frank Stoneman and the Florida Everglades During the Early 20th Century," *The Florida Geographer* 29 (1998), pp. 50–51.

139   *As usual, Broward dismissed his critics:* Proctor, *Napoleon Bonaparte Broward,* p. 247; Marjory Stoneman Douglas, *Voice of the River: An Autobiography with John Rothchild* (Sarasota: Pineapple Press, 1987), p. 99.

139   *the state's land grant corporations:* Farmers and other residents owned 185,020 acres in the district, while corporations and other absentee owners owned 4,044,500 acres. Blake, *Land into Water,* p. 98.

139   *"This rich, fertile land":* Proctor, *Napoleon Bonaparte Broward,* p. 247.

139   *He mused that if he were to change his mind:* Proctor, Ibid., p. 245.

139   *"Is there any steal in this":* James Ingraham letter to *Florida Times-Union,* 8/23/06, HFA.

140   *"because he is the man who is draining the Everglades of Florida":* Proctor, *Napoleon Bonaparte Broward,* p. 289.

140   *His administration touted:* "New Glades Lands to Be Platted," *Miami Metropolis,* 11/29/1907; *Minutes,* vol. 7, p. 122; "The Everglades of Florida," p. 110.

140   *Broward now wanted at least six dredges:* Transcript of the 1907 Commission, p. 428; Stephen S. Light and J. Walter Dineen, "Water Control in the Everglades: A Historical Perspective," in Davis and Ogden, eds., *The Everglades,* pp. 47–84.

140   *He eventually resolved their claims:* McCally, *The Everglades,* p. 93.

141   *"Money will assuage almost every other grief":* George W. Hallam, *Bolles: The Standard Bearer* (Jacksonville, FL: The Bolles School, 1983), p. 16.

141   *The son of a New York doctor:* Hanna and Hanna, *Lake Okeechobee,* pp. 137–39; Hallam, *Bolles,* pp. 14–16.

141   *Some of his buyers later discovered:* "The Leslie J. Lyons Hearings," Hearings before the Committee on the Judiciary on House Resolution 488, 62nd Cong., 2nd sess., 1912.

141   *"The people of our country are land-hungry":* "The Story of the Everglades," *Florida East Coast Homeseeker* (April 1910), p. 122.

142   *By the convention's end:* The developer R. P. Davie, also of Colorado Springs, held an option to buy the 108,000 acres from the Southern States Land and Timber Co., a corporation whose major investors included New York governor Herbert Lehman and his brother. But Davie decided not to exercise the option, so Broward and Jennings asked the company's agent to offer the land to Bolles. "Land Deal as Told by Broward," *Miami Metropolis,* 4/14/1910, p. 1.

142   *Bolles declared the project: Miami Metropolis,* 8/16/1909; Napoleon Bonaparte Broward, "Draining the Everglades," in *Independent Magazine,* 6/25/1908, reprinted in "The Florida Everglades Land Co."

142   *Frank Stoneman continued to sound alarms:* Meindl, "Frank Stoneman and the Florida Everglades," pp. 51–52.

143   *Charles Elliott, chief of the U.S. Department of Agriculture's new drainage bureau:* Charles G. Elliott letter to Governor Napoleon Broward, quoted in "Everglades of Florida Hearings," U.S. House Committee on Expenditures in the Department of Agriculture, 62nd Cong., 1st sess., 1912, p. 1259; Dovell, *A History of the Everglades of Florida,* pp. 218–19; Aaron D. Purcell, "Plumb Lines, Politics and Projections: The Florida Everglades and the Wright Report Controversy," *Florida Historical Quarterly* 80, no. 2 (Fall 2001), p. 168.

143   *Broward had welcomed the bureau's assistance:* Governor Napoleon Broward letter to Agriculture Secretary James Wilson, 1/16/1906; Wilson letter to Broward, 1/26/1906, quoted in "Everglades of Florida Hearings," pp. 208–09.

143   *Broward did not really want an investigation:* Broward privately explained his reasoning with blunt candor: "I could hire an engineer; but there is a fight on down here, a political one. If I should hire personally an engineer for making this survey these people would say, 'It is Broward's engineer and Broward's report,' and for that reason I wanted to get the department, which I know is interested in the matter, to send a man down here to make an investigation." "The Everglades of Florida" Hearings, p. 77.

143   *Elliot was a fastidious, apolitical engineer:* Arthur E. Morgan, "The Florida Everglades Incident." Unpublished Section of Autobiographical Writings, Yellow Springs, Ohio, 1954, AMP, Box 1; Arthur E. Morgan, *Dams and Other Disasters: A Century of the Army Corps of Engineers in Civil Works* (Boston: Porter Sargent Publisher, 1971), p. 372.

143   *Wright's main talent was speechmaking:* Christopher F. Meindl, et al., "On the Importance of Environmental Claims-Making: The Role of James O. Wright in Promoting the Drainage of Florida's Everglades in the Early Twentieth Century," *Annals of the Association of American Geographers* 92, no. 4, p. 695.

143   *he routinely accepted gratuities:* There is strong circumstantial evidence, and some direct evidence, that Wright was on the take in Florida as well. For one example, see: "Charges Filed by R. F. Ensey," FEP. Alfred Jackson Hanna and Kathryn Abbey Hanna claimed that a confidential report by the House committee investigating the Everglades in 1912 further implicated Wright. *Lake Okeechobee,* p. 159.

143   *"Mr. Wright is afraid":* Dapray letter to Broward, 11/8/1907, Box 5A, NBP, p. 2 Dapray represented the National Drainage Association when Broward was its president.

144   *"With Mr. W. at the head":* Ibid., p. 4.

144   *"I feel sure he can be trusted":* Ibid., p. 4.

144   *With Broward's help:* The governor personally lobbied Elliott to put Wright in charge of the project during a meeting of the National Drainage Conference in Baltimore. Elliott testified that Broward told him: "I wish you would send that old man Wright down there to continue the work." "Everglades of Florida Hearings," p. 1098.

144   *"no engineering difficulties to overcome":* Christopher F. Meindl, "On the Eve of Destruction: People and Florida's Everglades from the Late 1800s to 1908," *Tequesta* 63 (2003), p. 26.

144   *Everglades land companies quickly began citing the Wright report:* Hanna and Hanna, *Lake Okeechobee,* p. 159; "The Richest Land Not Under Cultivation Today," *Florida Fruit Lands Review,* January 1909, UFA, Miscellaneous Materials, Box 9.

144   *Bolles paid a Kansas city firm $400,000:* Cooper Kirk, "The Abortive Attempt to Create Broward County in 1913," *Broward Legacy* 12, nos. 1–2 (winter–spring 1989), p. 5.

145   *His ads even quoted Secretary Wilson:* "Everglades of Florida Hearings," p. 1323.

145   *"the tentative experiments that have been made":* Meindl, "Frank Stoneman and the Florida Everglades," p. 52.

145   *"one of the greatest enterprises on record":* Hanna and Hanna, *Lake Okeechobee,* p. 140.

145   *"The drainage of the Everglades is well under way":* Gifford, *The Everglades,* pp. 1–2.

145   *major Everglades landowners finally agreed:* As part of the deal, Flagler got the state to agree to remove 60,000 acres of his land west of Homestead from the drainage district. Bolles, on the other hand, agreed to speed up his payments to the Internal Improvement Fund.

145   *"We do not believe":* Journal of the State Senate of Florida of the Session of 1911, p. 1761; Dovell, *A History of the Everglades of Florida,* p. 273; Joe Hugh Reese, "To Dig 235 Miles of Drainage Canals in Florida," *Manufacturer's Record,* 5/5/1910.

145   *"There is no 'if' nor 'but'":* "The Pledge of a State," *Florida East Coast Homeseeker* (April 1910), p. 140.

145   *The land syndicates dispatched propagandists:* "Extracts from Reports of Experts," The Florida Everglades Land Company pamphlet, 1910, UFA, Box 9, Miscellaneous Materials on Everglades;

"Says Everglades Is Talk of Town, Even in Chicago," *Miami Metropolis,* 7/28/1910; "To Drain 10,000 Square Miles of Florida Land," *Manufacturer's Record,* 6/25/1910; "Reclaiming the Everglades," *Cassiers Magazine,* March 1911.

146   *The* Homeseeker *predicted that within a decade:* "America's Winter Garden," *Florida East Coast Homeseeker,* April 1910, p. 130.

146   *Fletcher also distributed brochures:* "What Has Broward Done for the People?" 1908 brochure, Duncan Fletcher for U.S. Senate, NBP.

146   *"United States Official Indorsement":* Everglades Land Company advertisement, *Washington Star,* 2/5/1912, quoted in Dovell, *A History of the Everglades of Florida,* p. 259.

146   *"It is a peach!":* Vance Helm to Thomas Will, 12/11/911, TWP; Dovell, "Thomas Elmer Will," p. 32.

147   *A natural booster with a background in Florida real estate:* Ric A. Kabat, *Albert W. Gilchrist: Florida's Progressive Governor.* Unpublished master's thesis, Florida State University, 1987, p. 56.

147   *"Opposition is rapidly disappearing":* "Where Nature Smiles," p. 25.

147   *"It is a question only in the minds":* Gifford, *The Everglades,* p. 42.

147   *the mimeographed circular warned:* The Everglades of Florida Wright Hearings, pp. 140–41.

147   *"I believe this Company is thoroughly responsible":* Broward letter, 5/1/1909, Box 9, NBP; Broward letter to W. R. Marion, 7/9/1909, Box 9, NBP; Broward letter to Philip Delaney, 4/30/1909, Box 9, NBP.

148   *Broward and Jennings had received lucrative payoffs:* "How Broward as Governor Came to Own 27,000 Acres of Land," *Miami Metropolis,* 4/4/1910, p. 1; "L'Engle Says He's Not the Only Well Fed Candidate in the Race for Senator," *Miami Metropolis,* 4/20/1910, p. 1.

148   *Broward claimed the tracts:* "Did Best He Could—Has Nothing to Apologize for, Says Broward," and "Land Deal as Told by Broward," *Miami Metropolis,* 4/14/1910, p. 1. R. P. Davie, who introduced Bolles to Broward, also wrote the *Metropolis* to back up the governor, saying he had refused to take an interest in any deal involving state lands. But Davie admitted that Broward had expressed interest in private land deals, and had even mentioned that "if he could make a commission he would appreciate it very much." "Says Broward's Dealing Strictly on the Square While in Office," *Miami Metropolis,* 4/22/1910, p. 1.

148   *a new allegation of a $24,500 cash kickback:* "Gov. Broward's Land Transactions Exposed by Congressman Lamar," 5/26/1910, Box 11, NBP.

149   *"If Broward had the good fortune to be located":* Broward Is Dubbed the Father of the Everglades," *Miami Metropolis,* 7/23/1910, reprinted from *Everglades Magazine.*

149   *"The stunning and hardly comprehensible announcement":* "Napoleon Bonaparte Broward," *Miami Metropolis,* 10/1/1910, p. 4.

149   *he was mourned across the state:* "Comment of the State Press on the Life and Death of Hon. Napoleon Bonaparte Broward," *Miami Metropolis,* 10/4/1910 and 10/5/1910.

149   *his widow later netted a $167,500 profit:* Kirk, "The Abortive Attempt to Create Borward County in 1913," p. 7; Samuel Proctor, Broward's sympathetic biographer, concluded that the governor's most enduring monument was the fact that he died poor. *Napoleon Bonaparte Broward,* p. xiv.

149   *A month after Broward's death: Minutes,* vol. 8, p. 587.

149   *"placed the fund on Easy Street":* Rufus E. Rose, *The Swamp and Overflowed Lands of Florida: A Reminiscence* (Tallahassee: T. J. Appleyard, 1916), p. 13; *Minutes,* vol. 8, p. 567.

149   *"The Florida Everglades will be dry in two years":* "Draining the Everglades," *Florida East Coast Homeseeker* 12, no. 4 (April 1910), p. 121.

150   *"Your committee is of the opinion":* Journal of the State Senate of Florida of the Session of 1911, p. 1763.

150   *Meanwhile, Flagler spent his twilight years:* Les Staniford's *Last Train to Paradise* is an engaging narrative of the Overseas Railroad. Flagler felt so strongly about the project that he amended his will to make sure it continued if he died before its completion. Akin, *Flagler,* pp. 212–22.

150   *a dozen dredges:* Martin, *Henry Flagler,* p. 166.

150   *Flagler was eighty-two when he rode the first train:* Akin, *Flagler,* pp. 222–24; Staniford, *Last Train to Paradise,* pp. 201–06.

150   *"I can hear the children":* Staniford, *Last Train to Paradise,* pp. 204–05.

10   Land by the Gallon

PAGE
151   *"The real estate propaganda said"*: John Newhouse, *History of Okeelanta*. Unpublished manuscript, 1932, FSA, p. 2.

151   *"The Village of Yesterday Today a Seething Mass of Bustling Humanity"*: J. H. Reese, *Miami Metropolis*, 3/11/1911.

152   *"The air of expectancy pervading this place"*: Reese, "The Village of Yesterday."

152   *But ex-Governor Jennings, still on the Bolles payroll*: Ibid.

152   *The visitors heard from all the leading drainage advocates*: "Investigation of the Everglades: As Seen by the Brightest Minds of Today," Chambers Land Company pamphlet, 1912.

152   *James Wright, who assured them*: "Everglades Will Never Overflow Again After This Year, Says Chief Engineer Wright," *Miami Metropolis*, 4/5/1912.

152   *A Grand Rapids News correspondent*: "Investigation of the Everglades," p. 6.

153   *"The most superlative adjectives"*: Ibid., p. 19.

153   *"Seeing is believing"*: Ibid., p. 8.

153   *"I had read with the proverbial grain of salt"*: Ibid., p. 10.

153   *"I can think of no sufficient expressive adjectives"*: Ibid., p. 20.

154   *"I have bought land by the acre"*: Hanna and Hanna, *Lake Okeechobee*, p. 134.

154   *As one Illinois schoolteacher discovered*: Ibid.

154   *When a Chicago man*: R. H. Little, *Pioneering in the Everglades* (Jacksonville, FL: Works Progress Administration State Office, Historical Records Survey, 1938), p. 12.

154   *"The mosquitoes in the Everglades are fearful"*: Herman Walker letter to Thomas Will, 1/12/1912, TWP.

155   *one quipped that he couldn't visit his land*: "Everglade Fruit in Court," *Kansas City Star*, 11/22/1914.

155   *The Everglades became so synonymous*: Phillip Weidling and August Burghard, *Checkered Sunshine: The Story of Fort Lauderdale, 1793–1955* (Gainesville: University of Florida, 1966), p. 51.

155   *"some poor, deluded victim"*: Margaret C. Topham letter to U.S. Department of Agriculture, "The Everglades of Florida Hearings," p. 1263.

155   *"one of the biggest land swindles in history"*: *Washington Times*, 12/8/1911, quoted in "Everglades of Florida Hearings," p. 953.

155   *Even the fund's trustees*: "The Everglades of Florida," 1911 report, p. 119.

155   *Still, a parade of current and former Florida officials*: Governor Park Trammell, former Governor Gilchrist, Everglades Drainage District chief engineer Fred Elliot, and State Chemist Rufus Rose all testified.

155   *"agitations and misrepresentations"*: "State Backs Up Land Men," *Kansas City Star*, 11/21/1914.

155   *"I read something not so long ago"*: Hanna and Hanna, *Lake Okeechobee*, p. 147.

156   *"the action of Mr. Bolles"*: Ibid., p. 148.

156   *"If the people want to be humbugged"*: *New York Times*, 2/9/1912, quoted in Blake, *Land into Water—Water into Land*, p. 117.

156   *It came out that Secretary Wilson*: "The Everglades of Florida Hearings," pp. 19, 775–81, 833–34, 1318–19; "The Everglades of Florida Hearings, Majority Report," U.S. House Committee on Expenditures in the Department of Agriculture, 62nd Cong., 2nd sess., 1912, pp. 2–3. Wilson reportedly told underlings that he was not running his department for the protection of fools who bought land without seeing it. Purcell, "Plumb Lines, Politics and Projections," p. 179.

156   *Wright confessed that on at least four occasions*: "The Everglades of Florida Hearings," pp. 1348–51, 1465–1509, 1755–59; "Majority Report," pp. 4–8.

156   *The press also chronicled the entertaining feud*: Dovell, *The History of the Everglades of Florida*, p. 315. Gilchrist shot back in a long diatribe published in Florida's papers: "From the top of Mr. Clark's Mt. McKinley bump of egotism, it is natural to suppose that everyone else is a little pinhead." Kabat, *Albert W. Gilchrist*, p. 150.

157   *The hearings revealed*: "The Everglades of Florida Hearings," pp. 338–91, 1039–43; Morgan, "The Florida Everglades Incident," pp. 22–23; Arthur E. Morgan, *Dams and Other Disasters: A Century of the Army Corps of Engineers in Civil Works* (Boston: Porter Sargent Publisher, 1971), pp. 372–75.

157   *"With only a cursory examination in the field"*: "Majority Report," p. 2.

157   *"completely incompetent as an engineer"*: "Everglades of Florida Hearings," p. 353. Elliott finally skimmed a copy of Wright's report while working out west and toned down its enthusiasm before sending it back to Washington for publication. But after giving it a closer read, Mor-

gan refused to release even the revised report and telegraphed his chief to return to Washington. Elliott soon concluded that Morgan was right, and that his disobedience had saved the bureau from embarrassment. Morgan later became one of America's top drainage engineers, and the first chairman of the Tennessee Valley Authority.

157   *"I don't want you to say anything more about the Everglades":* "Everglades of Florida Hearings," p. 1026.

157   *"The Everglade interest is all-powerful":* Charles Elliott letter to Arthur Morgan, 3/11/1912, AMP.

157   *Wright's most obvious mistake:* "Everglades of Florida Hearings," pp. 358–61, 410–18; Daniel W. Mead, et al., "Report of the Everglades Engineering Board of Review," Report for Everglades Land Sales Co., Everglades Land Co., and Everglades Sugar and Land Co., 11/12/1912, UFA, pp. 16–18; Meindl, "The Role of James O. Wright," p. 693.

158   *Wright's second error:* "Everglades of Florida Hearings," pp. 361–65; Mead, et al., "Report of the Everglades Engineering Board," pp. 37–39; McCally, *The Everglades,* pp. 101–02; Isham Randolph, et al., "Report of the Florida Everglades Engineering Commission to the Board of Commissioners of the Everglades Drainage District and the Trustees of the Internal Improvement Fund," Senate Document 379, 63rd Cong., 2nd sess., 1913, p. 56.

158   *Wright also dramatically underestimated:* "Everglades of Florida Hearings," pp. 343–49; Mead, et al., "Report of the Everglades Engineering Board," p. 40; "The Engineering Plans for Draining the Florida Everglades," *Engineering News* 67, no. 13, 3/28/1912.

158   *ludicrously low dredging costs:* "The Engineering Plans for Draining the Florida Everglades," *Engineering News,* 67, no. 13, 3/28/1912; "Report of Everglades Engineering Board of Review," p. 40.

159   *Elliott never recovered his health:* Morgan, "The Florida Everglades Incident," p. 27.

159   *Wright blamed his newfound notoriety: Minutes,* vol. 9, pp. 504–05.

159   *"more firmly convinced than ever":* James Wright, "Why Was Wright's Report on the Everglades Suppressed?" 1912 pamphlet, FSA, pp. 13–14.

159   *independent review by three hydraulic engineers:* They were Daniel Mead, later the president of the American Society of Civil Engineers, Leonard Metcalf, founder of the renowned Boston engineering firm Metcalf & Eddy, and Allan Hazen.

159   *"totally inadequate to accomplish the drainage":* "Report of the Everglades Engineering Board of Review," pp. 18, 25, 31–36.

159   *Echoing Captain Rose's suggestions:* Ibid., pp. 8–9. Charles Elliott also recommended a "progressive" or gradual approach to drainage, saying it made sense to focus on the most valuable farmland first.

160   *They urged the state to rely exclusively on direct east–west canals:* Ibid., pp. 9, 31–33.

160   *Everglade Magazine warned:* "Special Announcement to All Purchasers," *Everglades Magazine* 3, no. 11 (March 1913).

160   *even Bolles revised his brochures:* "The Garden of the Glades," 1914 Okeechobee Fruit lands Company brochure, FSA.

160   *"The Plunderer: A Story of the Florida Everglades":* Country Gentleman, quoted in "Unfair Propaganda Regarded the Everglades Being Disseminated by Country Gentleman," *Palm Beach Post,* 8/22/1919.

160   *"in many localities in the North":* Hanna and Hanna, *Lake Okeechobee,* p. 150.

160   *Isham Randolph:* "The name of Isham Randolph attached to any enterprise was a guarantee of honesty, integrity and technical efficiency," one admirer wrote. *The National Encyclopedia of American Biography,* vol. 19, 1926, p. 359.

161   *Its strongest recommendation was a familiar one:* Randolph, et al., "Report of the Florida Everglades Engineering Commission," p. 5.

161   *"Without that canal":* Isham Randolph to William S. Jennings, 3/2/14, WJP.

161   *it could power a hydroelectric plant:* Randolph, et al., "Report of the Florida Everglades Engineering Commission, pp. 5–6.

161   *an extensive latticework of additional canals:* Ibid., pp. 11–14.

161   *"revert to the swamp conditions which now prevail":* Ibid., p. 63.

161   *"entirely practicable":* Ibid., p. 5.

161   *It predicted that muck soils:* Ibid., pp. 62–63.

161   *"in the Everglades violent floods are inconceivable":* Ibid., p. 54.

162   *"worth . . . every dollar":* Ibid.

162   *"we are actually losing thousands of settlers":* "Back to Broward," 1915, Back to Broward League pamphlet, RTE, p. 16.

162   *"They Came to Boost":* Ibid., p. 16.

162   *"Not during the present administration"*: Ibid., p. 6.
162   *Wright's replacement was a thirty-four-year-old Tallahassee native:* FEF, Elliot Family Papers, M86-
       038; Catherine Parramore, "Fred C. Elliot, Man of Vision," *Journal of the Florida Engineering
       Society,* 13, no. 5 (April 1960); "Temporary State Job Lasts 38 Years," *Miami Herald,*
       6/12/1947.
163   *he was sure that the reclamation of the Everglades:* "True Condition of Affairs Regarding the Ever-
       glades," *Miami Evening Metropolis,* 9/25/1912; "Chief Engineer Elliot's Reclamation Address,"
       Palm Beach Weekly News, 7/11/1913.
163   *"The wonderful lands which you are now rescuing":* "Chief Engineer Elliot's Reclamation
       Address," p. 1.
163   *"notwithstanding the catastrophes which are liable to occur":* Ibid., p. 8.
164   *"a life of ease, plenty and independence":* Newhouse, *History of Okeelanta,* p. 5.
164   *His son Lawrence:* Will, *A Cracker History of Okeechobee,* pp. 185–87.
164   *"They came in swarms":* Will, *A Cracker History of Okeechobee,* p. 187.
164   *"farming here is not the Cock-sure thing":* Thomas Will to C. A. Huff, 9/7/15, TWP, Box 4.
164   *Realtors handed out snapshots of a quaint sign:* Little, *Pioneering in the Everglades,* p. 54.
165   *"Jim says don't kill any":* Ruth Robbins Beardsley, *Pioneering in the Everglades* (Fort Myers Beach,
       FL: Island Press, 1973), pp. 35–36.
165   *"If 'n a man was to put his mind to it":* Will, *A Cracker History of Okeechobee,* p. 96.
165   *the southerners who hunted and fished:* Ibid., p. 3.
165   *Three weeks after Newhouse arrived:* Newhouse, *A History of Okeelanta,* p. 15.
166   *settlers made sure to burn every available copy:* Dovell, *A History of the Everglades of Florida,* p. 363.
166   *The usually upbeat Thomas Will began to worry:* Thomas Will letter to Harold Bryant, 7/4/1916,
       Will Papers, Box 6.
166   *For several relatively dry years:* McCally, *The Everglades,* p. 130.
166   *Lake Okeechobee retreated:* Dovell, *A History of the Everglades of Florida,* p. 375.
166   *celebratory headlines in . . . the* Palm Beach Post: 5/10/19, 4/11/19, 3/31/19, 10/2/18,
       2/15/17.
167   *Within a year, Okeechobee:* Hanna and Hanna, *Lake Okeechobee,* pp. 190, 232.
167   *Moore Haven became the largest town:* Will, *A Cracker History of Okeechobee,* pp. 189–94. A Miami
       judge who opened a sugar mill in Moore Haven announced a plan to supply 100 candy fac-
       tories and 1,000 candy stores, but it soon went bust.
167   *"I have watched the development of the Everglades":* "Nation's Wealthiest Developers Heavy Investors
       in Everglades," *Palm Beach Post,* 1/26/1919.
167   *No investor was more enthusiastic:* Will Irwin, "The Rise of Fingy Conners," *Colliers Magazine,* July
       1908. Republished online at www.buffalonian.com; Hanna and Hanna, *Lake Okeechobee,* pp.
       222–29; Will, *A Cracker History of Okeechobee,* pp. 229–34.
168   *"There are no rules in his fighting":* Irwin, "The Rise of Fingy Conners."
168   *"Balmy sunshine, wonderful climate":* Hanna and Hanna, *Lake Okeechobee,* p. 223.
168   *Before the settlers arrived:* Lawrence E. Will, *Swamp to Sugar Bowl: Pioneer Days in Belle Glade* (Belle
       Glade: The Glades Historical Society, 1984), pp. 11–12; Will, *A Cracker History of Okeechobee,*
       pp. 34–37; John Kunkel Small, "Narrative of a Cruise to Lake Okeechobee," *American Museum
       Journal* 18 (December 1918), pp. 685–700.
168   *"picturesque beyond description":* Small, "Narrative of a Cruise to Lake Okeechobee," pp. 698–99.
168   *"and when farmers found this out":* Will, *A Cracker History of Okeechobee,* p. 187.
169   *"The natural features of that region":* Small, "Narrative of a Cruise to Lake Okeechobee," p. 691.
169   *One member of Small's party:* Ibid.
169   *"All the glamour and mystery":* Rothra, *Florida's Pioneer Naturalist,* p. 139.
170   *"I was grieved at the loss":* Little, *Pioneering in the Everglades,* p. 80.
170   *his favorite place was a hammock island:* Rothra, *Florida's Pioneer Naturalist,* pp. 84–87.
170   *"My eyes," he once wrote:* Simpson, "Paradise Key," p. 5.
170   *"Their great smooth white stems":* Rothra, *Florida's Pioneer Naturalist,* p. 86.
171   *the conservationists who ultimately saved it were women:* Linda D. Vance, *May Mann Jennings: Florida's
       Genteel Activist* (Gainesville: University Press of Florida, 1985), pp. 80–85.
171   *"If the park tract is so dense and useless":* Vance, *May Mann Jennings,* p. 86. Marjory Stoneman
       Douglas somewhat cattily suggested that Mrs. Jennings got involved in the Paradise Key fight
       because her husband wanted a state road built to boost the value of their landholdings in the
       area, but she doesn't provide any evidence of that. Douglas, *Voice of the River,* p. 136.
172   *Miami had the world's highest per capita consumption:* C. H. Ward, "The Lure of the Southland,"
       1915 Miami tourism pamphlet, FSA.

172 *Simpson observed how massive quarries:* Rothra, *Florida's Pioneer Naturalist*, p. 156.
172 *And a dynamic midwestern entrepreneur named Carl Fisher:* Mark S. Foster, *Castles in the Sand: The Life and Times of Carl Graham Fisher* (Gainesville: University Press of Florida, 2000); Abraham D. Lavender, *Miami Beach in 1920: The Making of a Winter Resort* (Charlestown, SC: Arcadia Publishing, 2002).
172 *By 1920, "Crazy Carl":* Lavender, *Miami Beach in 1920*, pp. 13–14, 109. In 1920, Fisher opened the beach's first luxury hotel, the Flamingo, and a pedestrian mall modeled on the Rue de la Paix in Paris, Lincoln Road. He also hosted America's first international polo match. The *Miami Metropolis* reported that year that "South Beach Night Life Is Gay and Interesting," which, in a manner of speaking, is still true today.
173 *"The jungle itself seemed to protest":* Foster, *Castles in the Sand*, p. 157.
173 *"as beautiful a stream as ever flowed":* Charles Richard Dodge, "Subtropical Florida," 1894 article, reprinted in Oppel and Meisel, *Tales of Old Florida*, p. 25.
173 *Meanwhile, 34,000 acres of the Everglades:* Fred C. Elliot, "Biennial Report to the Board of Commissioners of the Everglades Drainage District, 1927–1928," FEP, p. 76.
173 *"The drying up of the Glades":* John King quoted in Fred Sklar, Chris McVoy et al. "Hydrologic Needs: The Effects of Altered Hydrology on the Everglades," South Florida Water Management District, Everglades Interim Report, 1998 p. 2-1.
173 *some of the Everglades had already lost:* Fred Elliot speech for Governor John Martin, 10/1/1926, FEF.
173 *This was not only the result of subterranean fires:* McCally, *The Everglades*, pp. 143–44.
174 *"Drainage and burning have become such a fad":* John Kunkel Small, *From Eden to Sahara: Florida's Tragedy* (Lancaster, PA: The Science Press Printing Co., 1929), p. 85.
174 *John Gifford issued the first call:* John Gifford, "Looking Ahead: Views on Everglade Topics," *Tropic Magazine* 1, no. 4 (July 1914).
174 *May Mann Jennings defended the Everglades drainage project:* May Mann Jennings letter to Minnie-Moore Wilson 5/12/1915, p. 2 RTE.
174 *"Only Florida's climate is safe":* Charles Torrey Simpson, *Out of Doors in Florida: The Adventures of a Naturalist Together with Essays on the Wild Life and the Geology of the State* (Miami: E. B. Douglas, 1923), pp. 136–37.
174 *"There is something very distressing":* Simpson, *In Lower Florida Wilds*, pp. 140–41.

11   Nature's Revenge

176 *"What's the matter with the Everglades?":* "Overflow in 1925 Threatens if Main District Canals Are Not Opened to Dispose of Lutevuls Discharge," *Everglades News*, 11/21/1924.
176 *South Florida enjoyed one of history's wildest land booms:* Kenneth Ballinger, *Miami Millions: The Dance of the Dollars in the Great Florida Land Boom of 1925* (Miami: Franklin Press, 1936): Charles Donald Fox, *The Truth About Florida* (New York: Charles Renard Corp., 1925); William Frazer and John J. Guthrie Jr., *The Florida Land Boom: Speculation, Money and the Banks* (Westport, CT: Quorum Books, 1995); Derr, *Some Kind of Paradise*, pp. 175–97; Parks, *Miami*, pp. 105–12; Paul S. George, "Brokers, Binders and Builders: Greater Miami's Boom of the Mid-1920s," *Florida Historical Quarterly* 65, no. 1 (July 1986); Vic Knight, "The Florida Land Boom: A Promoter's Dream," *South Florida History* (summer–fall 1994); Homer K. Vanderblue, "The Florida Land Boom," *Journal of Land and Public Utility Economics,* May and August 1927; Frederick Essary, "Have Faith in Florida!" *The New Republic,* 10/14/1925.
176 *"Was there ever anything like this migration to Florida?":* Tallahassee Democrat, 10/16/1925.
177 *boosters joked that it would soon be possible:* Stockbridge and Perry, *Florida in the Making*, pp. 211–12.
177 *Madcap drivers routinely flouted traffic laws:* Parks, *Miami*, p. 107. Police officers began to deal with parking scofflaws by removing the front seats of their cars.
177 *Crime became so rampant in Miami:* Ballinger, *Miami Millions*, p. 58.
177 *A veteran who had swapped an overcoat:* Maury Klein, *Rainbow's End: The Crash of 1929* (Oxford: Oxford University Press, 2001), p. 93.
177 *A Miami entrepreneur bought and resold a lot:* The entrepreneur was Mitchell Wolfson, the founder of the Wolfsonian Museum. George, "Brokers, Binders and Builders," p. 57.
177 *A screaming mob snapped up 400 acres:* Nixon Smiley, *Knights of the Fourth Estate: The Story of the Miami Herald* (Miami: E. A. Seemann Publishing, 1974), p. 67.
177 *"Hardly anybody talks of anything but real estate":* Parks, *Miami*, p. 116.

178   *"The majority of these depicted an entirely mythical city":* T. H. Weigall, *Boom in Florida* (London: John Lane the Bodley Head Limited, 1971), p. 112.

178   *"thousands of newly arrived Florida land owners":* Fox, *The Truth About Florida,* p. 23.

178   *"Florida? Wonderful!"* Gertrude Mathews Shelby, "Florida Frenzy," *Harper's Monthly* 152, no. 26, p. 177.

179   the Miami Herald *shattered the world's newspaper advertising record:* Smiley, *Knights of the Fourth Estate,* p. 54.

179   *"Are you aware of the fact that Real Estate":* "Buy Them for the Kiddies," *Miami Herald,* 3/13/1924.

179   *by 1925, it had thirty high-rises:* Ballinger, *Miami Millions,* pp. 107; Hanna and Hanna, *Florida's Golden Sands,* p. 341.

179   *One hotel leased its dining room:* Frazer and Guthrie, *The Florida Land Boom,* p. 98.

179   *"rehearsed the mosquitoes":* Polly Redford, *Billion-Dollar Sandbar* (New York: E. P. Dutton & Co., 1970), p. 45.

179   *an equally energetic builder named George Merrick:* Parks, *Miami,* p. 108; Derr, *Some Kind of Paradise,* pp. 188–90; Klein, *Rainbow's End,* p. 91.

179   *He paid $100 a week:* Douglas, *Voice of the River,* p. 108; Weigall, *Boom in Florida,* p. 90.

180   *"the Most Richly Blessed Community":* Parks, p. 120.

180   *"The wealth of south Florida":* Jack E. Davis, "Conservation Is Now a Dead Word: Marjory Stoneman Douglas and the Transformation of American Environmentalism," *Environmental History* 8 (January 2003), p. 59.

181   *its name inspired by the Seminole word for hammock:* That word was Opatishawockalocka. Catherine Lynn, "Dream and Substance: Araby and the Planning of Opa-locka," *Journal of Propaganda and Decorative Arts,* Florida Theme Issue, 1998, p. 163.

181   *Ernest "Cap" Graham:* EGP; Interviews with Bob Graham and William Graham.

181   *he later entered politics to take on the gangsters who controlled Hialeah:* Cap Graham was elected to the state Senate in 1936 and immediately introduced two bills taking on the notorious Hyde-Slayton Gang—one abolishing the city of Hialeah, the other reestablishing it and giving Graham the power to appoint a new mayor and city council. In 1940, when Graham ran for reelection, the gang leader Red Slayton sent him a postcard from the state penitentiary. Bob Graham says the postcard read: "Mr. Graham, I wish I could vote for you for reelection because you're the only honest politician I've ever known. You said if you'd elected you'd run us out of Hialeah, and you did it."

181   *she was once assigned to cover an enlistment ceremony:* Douglas, *Voice of the River,* pp. 112–13.

181   *two poems celebrating the "greatness" of the highway:* Davis, "Conservation Is Now a Dead Word," p. 60.

182   *After unleashing a tirade of profanity:* Lawrence Will recorded his response as: "Dag-nabbit, I can't even set foot on the blasphemous property to see what in the hooraw I've bought!" But Will acknowledged that Conners actually used words a bit stronger than "dag-nabbit," "blasphemous," and "hooraw." Will, *A Cracker History of Okeechobee,* p. 231. Alfred Jackson Hanna and Kathryn Abbey Hanna noted that Conners had a particular talent for foul language: "His profanity ranged through all the gradations known to but few notorious masters of that ungentle art." *Lake Okeechobee,* p. 222.

182   *When he opened the Conners Highway:* Hanna and Hanna, *Lake Okeechobee,* pp. 226–28.

182   *"The barriers of America's last frontier":* Palm Beach Post, 7/5/1924, quoted in ibid., p. 227.

182   *Ads for a planned Everglades subdivision:* Smiley, *Knights of the Fourth Estate,* p. 64.

182   *Ads for Caterpillar bulldozers:* "Conquering the Everglades," Holt Manufacturing Company advertisement, *Saturday Evening Post,* 6/30/1923.

182   The Herald *staged a $100 contest for its readers:* Margarita Fichtner, "The Hidden Jewel," *Miami Herald,* 9/15/2002.

182   *"The Everglades is calling":* "Come South, Young Man," *Miami Herald,* n.d.

182   *"The Everglades has lost population":* "Overflow in 1925 Threatens," *The Everglades News,* 11/21/1924.

183   *"I only hope the old rule":* Hanna and Hanna, *Lake Okeechobee,* p. 275.

183   *In 1922, the region was almost entirely underwater:* Little, *Pioneering in the Everglades,* pp. 90–114.

183   *"We began to realize":* Ibid., p. 113.

183   *"The fact is, gentlemen":* Ibid., p. 114.

183   *"when finished," Fred Elliot wrote:* Fred C. Elliot, "Draining the Everglades," *The Florida Magazine,* June–August 1924, Everglades Drainage District reprint, FEP, p. 4.

183 *"absolute insurance against any future overflow"*: "The Reclaimed Everglades," *Palm Beach Post,* 1/15/24.

184 *The Army Corps had blossomed:* Office of History, *The History of the US Army Corps of Engineers;* Reuss and Walker, "Financing Water Resources," Grunwald, "An Agency of Unchecked Clout"; Marc Reisner, *Cadillac Desert: The American West and Its Disappearing Water* (New York: Penguin Books, 1987); George Buker, *Sun, Sand and Water: A History of the Jacksonville District, U.S. Army Corps of Engineers, 1821–1975,* Washington, pp. 11–15, 91–97.

184 *the bombastic Seminole War veteran Andrew Humphreys:* Barry, *Rising Tide,* pp. 32–7, 42–5, 47–9. Here's Barry's best anecdote of the general's egomania: At Fredericksburg, Humphreys led a disastrous charge, losing more than one thousand men in fifteen minutes. His response: "Oh, it was sublime!" He told a friend: "I felt more like a god than a man." Ibid., pp. 48–9.

184 *The Corps evaded two congressional directives:* Hearings before the House Committee on Flood Control, 70th Cong., 2nd sess, 1929, pp. 247–248. The first appropriation was only $1,000, but the second was $40,000, more than enough for a decent survey.

185 *"I wish to say that gloom seems to be on every hand"*: *Minutes,* vol. 15, 9/8/23 letter from W. A. McRae, p. 120.

185 *"neglect of duty, inability, incompetence"*: "Address of John W. Martin, Governor of Florida, on the Everglades Drainage Problem," 10/28/1926, UFA, p. 11.

185 *"To hell with them"*: J. B. Johnson, "Outline of Situation and Conditions in Everglades Drainage District," 1926 pamphlet, FEP, p. 6.

185 *"then I consider the entire Glades proposition hopeless"*: Ernest Graham letter to Fred Elliot, 7/16/1923, EGP.

185 *"We might make a similar demand"*: J. M. Griffin, C. A. Walsh et al., letter to Elliot, 10/24/1924, FEP.

185 *The people of the Everglades had no more faith:* Hanna and Hanna, *Lake Okeechobee,* p. 276.

186 *Elliot exuded authority:* For example: "Brief of Work Performed Under Various Governors," 10/1/1926.

186 *This was the heyday of the American engineer:* Barry, *Rising Tide,* p. 264.

186 *"The most charitable conclusion"*: "Clean Out the Canals," *Everglades News,* 12/5/1924, p. 2.

186 *"The lake is truly at a level so high"*: Quoted in Ted Steinberg, *Acts of God: The Unnatural History of Natural Disasters in America* (Oxford: Oxford University Press, 2000), p. 60.

186 *In fact, he believed water shortages:* Fred Elliot, "Improvement of Our Rivers Against Flood and Waste," 1/9/1924 memorandum, p. 2.

186 *"safer from flood or overflow"*: "State Drainage Engineer Is Delighted with Results: Declares All Danger of Future Floods Has been Completely Eliminated by Dyking." *Hialeah Herald,* 12/5/1923. HMSF, Flood Control newsclips. In fairness to Elliot, the boosterish *Herald* may have exaggerated his hubris; the reporter's credibility was not enhanced by identifying the chief engineer as "E. B. Elliott" instead of "F. C. Elliot."

187 *"Throughout the country the delusion"*: "Even Florida Is Not Fool-Proof," *Forbes,* 10/1/1925.

187 *"You are going to Florida to do what?"*: George, "Brokers, Binders and Builders."

187 *Even a mild hurricane that grazed south Florida in July:* Steinberg, *Acts of God,* p. 51.

187 *"There is more risk to life"*: Ibid., p. 51.

187 *"nobody seemed to be alarmed"*: Lawrence Will, *Okeechobee Hurricane and the Hoover Dike: Killer Storms in the Everglades* (Belle Glade: The Glades Historical Society, 1990), pp. 13–14.

187 *That night, Miami was pummeled:* Jay Barnes, *Florida's Hurricane History* (Chapel Hill: University of North Carolina Press, 1998), pp. 11–126; Eliot Kleinberg, *Black Cloud: The Great Florida Hurricane of 1928* (New York: Carroll & Graf, 2003), pp. 26–30; Robert Mykle, *Killer 'Cane: The Deadly Hurricane of 1928* (New York: Cooper Square Press, 2002), pp. 84–88; Smiley, *Knights of the Fourth Estate,* pp. 70–82; Will, *Okeechobee Hurricane and the Hoover Dike,* pp. 25–35.

188 *"The intensity of the storm"*: Kleinberg, *Black Cloud,* p. 154.

188 *Gray threw open his door and screamed:* Smiley, *Knights of the Fourth Estate,* pp. 77–78.

188 *"Scores of men, women and children were drowned"*: Barnes, *Florida's Hurricane History,* p. 120.

188 *One carpenter grabbed his family:* Will, *Okeechobee Hurricane and the Hoover Dike,* p. 21.

188 *A railroad agent drowned:* Ibid., p. 28.

189 *When the Herald's new city editor:* Smiley, *Knights of the Fourth Estate,* pp. 80–81.

189 *"the poor people who suffered"*: Steinberg, *Acts of God,* p. 58.

189 *but official spin continued to portray the storm as a minor inconvenience:* Steinberg, *Acts of God,* pp. 54–57.

189 *One booster took out full-page ads:* Muir, *Miami, U.S.A.,* p. 153.

189 *It was a negligent homicide:* "The Dead Accuse," *Everglades News,* 9/24/1926, quoted in Dovell, *A History of the Everglades of Florida,* p. 426.

189 *"The first thing to do":* C. C. Morgan, "Everglades Drainage: How Are We Going to Get It?," Punta Gorda Publishing Co., pamphlet 1927.

190 *"reckless and foolish":* "Martin Sees Solution Up to Citizens," *Palm Beach Times,* 1/13/1927, PBCHS.

190 *"Of course, the Drainage Commissioners are easier":* Johnson, OK "Outline of Situation and Conditions in Everglades Drainage District," p. 1.

190 *an Everglades reclamation conference:* The reclamation conference was a vivid illustration of the power of Florida's businessmen, who discussed the future of the Everglades as if they were the official decision-makers. The conference was not even held in Florida; it was in Baltimore, the home of S. Davies Warfield, Florida's leading railroad baron at the time.

190 *"There is no better drainage engineer than Elliot here":* "Conference on Florida Everglades Reclamation at the Continental Building in Baltimore," May 1927 transcript, HMSF, p. 36. During the conference, Elliot proclaimed that "with the exception of a storm such as we had in 1926, the lake is already in satisfactory condition." Well, yes, with the exception of that.

190 *Elliot soon proposed a new $20 million plan of attack:* Elliot's plan was generally endorsed by a friendly board of engineers appointed by the drainage district—and preapproved by Elliot. "Report of the Everglades Engineering Board of Review to the Board of Commissioners of the Everglades Drainage District." Fred Elliot "Memorandum Re Studies for Flood Control, Irrigation, Etc." 10/28/1927, p. 2, FEF.

190 *a taller, wider, and sturdier dike:* Elliot also suggested that settlers could build homes on top of the dike, a recommendation that fortunately was never followed.

190 *Finally, Elliot called for the federal government:* Fred Elliot, "A Waterway Across Florida," presentation to Florida Engineering Society, FEF, 4/21/1928; Fred Elliot, "Memorandum of War Department Hearing at Moore Haven," FEF, 10/19/1927; Fred Elliot, "Memorandum for Mr. Ahern," FEF, 11/4/1927; Fred Elliot, memorandum to accompany letter from Governor John W. Martin to Hon. Frank R. Reid, 10/18/1927.

190 *"until the resources of local interests":* House Document 215, 70th Cong., 1st sess, 1928, p. 50; Blake, *Land into Water—Water into Land,* p. 143. Jadwin did recommend about $640,000 for dredging on the Caloosahatchee, but only for navigation purposes.

190 *"floods such as occurred there in 1926":* "Huge Glades Reclamation Project Is Explained by Elliot at Session Here," *Palm Beach Times,* 4/23/1928, PBCHS.

190 *throw[ing] the brick at Santa Claus:* "Everglades Plan Upheld by Martin in Labor Speech," *Palm Beach Post,* 9/6/1927.

191 *"clear thinking, straight shooting and careful administration":* Fred C. Elliot letter to Everglades Drainage District Board of Commissioners, 7/6/1928, FEF, p. 6.

191 *"There have been hardships.":* Elliot Report on the Everglades, 5/2/1928, p. 22, Fred Elliot Papers.

191 *"This is a serious time for the Everglades":* Elliot letter to Board of Commissioners, 7/6/1928, Fred Elliot Papers.

191 *"the most dishonest plan of bond-selling.":* "Again—and Still," *Everglades News,* 1/21/1927, p. 2.

191 *The hostility in the upper Glades became so intense:* Will, *Okeechobee Hurricane and the Hoover Dike,* p. 177.

191 *"We are in far more danger":* "Again—and Still," *Everglades News,* 1/21/1927, p. 2.

191 *The upper Glades sold $11 million:* Howard Sharp letter to Glenn Skipper, 2/14/1929, HHP: Campaign and Transition, Trips, Lake Okeechobee. The most prominent crops were peppers, tomatoes, and beans. Nathan Mayo, "Possibilities of the Everglades," *Quarterly Bulletin of the Florida Department of Agriculture* 37 (October 1926), p. 22; "Civilization Is Quickly Taking Backwoods Lands," *Palm Beach Post,* 6/10/1928.

191 *"Folks don't do nothin' down dere":* Zora Neale Hurston, *Their Eyes Were Watching God* (originally published in 1937; Urbana, IL: University of Illinois Press: 1978), p. 192.

192 *Howard Sharp, now a county commissioner:* "Voices Cried for Drainage Before the Storm," *Belle Glade Herald,* 9/14/1978, p. 20.

192 *"Fred C. Elliot of Tallahassee":* Howard Sharp, "Elliot Doesn't Expect Flood: He Never Does," *Everglades News,* 7/27/1928; "Voices Cried for Drainage Before the Storm," *Belle Glade Herald,* 9/14/1978, p. 19.

192 *"Advocates of a high lake level":* "Voices Cried for Drainage Before the Storm," *Belle Glade Herald,* 9/14/1978, p. 20.

192   *The storm of 1928:* Kleinberg, *Black Cloud;* Mykle, *Killer 'Cane;* Barnes, *Hurricane History,* pp. 127–40; The *Belle Glade Herald* ran a special section on the fiftieth anniversary of the storm, 9/14/1978; Jeff Klinkenberg, "A Storm of Memories," *St. Petersburg Times,* 7/12/1992; Zora Neale Hurston's novel *Their Eyes Were Watching God* is still the most vivid account of the hurricane, and possibly of any hurricane.

192   *"The suffering throughout is beyond words":* Kleinberg, *Black Cloud,* p. 113.

193   *"I had thought our storm experiences very trying":* Little, *Pioneering in the Everglades,* p. 140.

193   *One family rode out the storm in a treetop:* Kleinberg, *Black Cloud,* p. 110.

193   *"Louder and higher and lower and wider":* Hurston, *Their Eyes Were Watching God,* pp. 234, 238–37.

194   *"The complete devastation was simply unbelievable":* Chester Young, "The Cleaning Up of Bodies Recalled," *Belle Glade Herald,* 9/14/1978.

194   *Governor Martin refused to activate the National Guard:* Kleinberg, *Black Cloud,* p. 130.

194   *But a grisly tour through the Everglades changed his mind:* Ibid., pp. 154–55; Mykle, *Killer 'Cane,* p. 205.

194   *"Without exaggeration," he wrote in a telegram:* Kleinberg, *Black Cloud,* p. 154.

194   *"Two Thousand Lives Pay the Price of Politics":* Ibid., p. 191.

195   *One man reportedly thrust the bones of a drowned friend:* Hearings of the Senate Commerce Committee, 71st Cong., 2nd sess., 5/9/1930, p. 22.

195   *The legislature finally agreed:* In a private memorandum, Elliot later warned Governor David Sholtz that it would be pure folly for the state to try to take back control of "the wreck of the Everglades." He wrote that "failure would result and discredit come to the Trustees thereby." "Memorandum for Governor Sholtz," FEF, circa 1933.

195   *Elliot had the gall to claim a measure of vindication:* Elliot memo on "Effect of September Hurricane on Lake Okeechobee," 10/5/1928, FEF. It is worth noting that while Elliot had proposed a dike twenty-seven feet above sea level, the wind tide from the 1928 storm was estimated at 29.6 feet above sea level.

195   *Elliot and the commissioners also lashed out:* Elliot draft resolution, November 1928, FEF. It was true that navigation guidelines had required higher lake levels, but Elliot had generally supported those higher levels, and had made it clear in the past that drainage interests would trump navigation interests. He also made it abundantly clear in his memoranda that he retained final decision power over water levels in the Everglades. In any case, slightly lower levels would not have prevented the tragedy. "Memorandum," 10/25/1927, FEF.

195   *"That tent disgorged":* Will, *Okeechobee Hurricane and the Hoover Dike,* pp. 166–68.

195   *"I've heard it advocated":* Hearings before the House Committee on Flood Control, 70th Cong., 2nd sess. 1/10/29–2/1/29, p. 247–48.

195   *After spending $18 million:* Elliot, "Biennial Report, 1927–1928," pp. 5–10.

196   *The hopelessly impolitic Attorney General Davis:* U.S. House Committee on Flood Control, Hearings on Flood Control in Florida and Elsewhere," 70th Cong., 2nd sess., 1929, pp. 145–46.

196   *In 1848, when Senator Westcott first proposed to drain the Everglades: Congressional Globe,* 12/20/1848, p. 69.

12   "Everglades Permanence Now Assured"

PAGE

197   *"There is nothing like it in the world":* Hearings Before the House Committee on Public Lands on Everglades National Park, 71st Cong., 3rd sess., December 1930.

197   *Five months after the 1928 hurricane:* "Hoover Visit to Clewiston a Big Success," *Clewiston News,* 2/22/1929; "Hoover Ends His Trip to Glades," *South Florida Developer,* 2/22/1929. Hoover also spent a half hour at the Southern Sugar Company's new mill, which had been imported from Pennsuco, and was later taken over by the U.S. Sugar Corporation. Thomas Edison and Henry Ford visited the same day.

197   *The lobbying continued that night:* L. C. Speers, "Florida Flood Need Impresses Hoover," *New York Times,* 2/17/1929, p. 1.

198   *The Okeechobee hurricane had claimed:* The Red Cross reported 246 deaths in Mississippi, but John Barry suggested in *Rising Tide* that the flood killed at least 500. The official death toll in the Everglades was 1,836, but the National Hurricane Center recently upgraded that figure to at least 2,500.

198   *tears welled in Hoover's eyes:* "Hoover Came to Lake, Residents Grateful but Wait and Wonder," *Everglades News,* 2/22/1929; Kleinberg, *Black Cloud,* p. 198. President Hoover's daily calen-

dar, Herbert Hoover Presidential Library and Museum: www.ecommcode2.com/hoover/calendar/home.cfm. Hoover conducted his presidential transition out of J. C. Penney's Miami Beach mansion; he spent most of the time fishing.

198 *he was an indefatigable man of action:* Richard Norton Smith, *An Uncommon Man: The Triumph of Herbert Hoover.* (New York: Simon & Schuster, 1984), pp. 21, 24, 39.

198 *As commerce secretary he had urged:* In his speech at the Hoover Dike dedication in January 1961, ex-president Hoover said he visited the lake after the 1926 storm as well. "In those days we gave these wicked manifestations no endearing names of gentle women," he recalled. (Speech in Clewiston Museum archives.)

198 *For decades, the agency had insisted:* John McPhee noted archly that the Corps had made pronouncements that the river was finally under control "before the great floods of 1884, 1890, 1891, 1897, 1898, and 1903, and . . . again before 1912, 1913, 1922, and 1927." *The Control of Nature,* New York: Farrar, Straus and Giroux, 1989.

198 *"to prevent the destructive effect":* Barry, *Rising Tide,* p. 175.

198 *Grudgingly, Jadwin submitted a Corps plan:* Barry argues that the response to the Mississippi flood "set a precedent of direct, comprehensive and vastly expanded federal involvement in local affairs," essentially setting the stage for the New Deal. Ibid., p. 407.

199 *"protection must be designed":* Hearings before the House Committee on Flood Control, on Flood Control in Florida and Elsewhere, 70th Cong., 2nd sess., 1929, p. 239. The bill authorizing the dike actually classified it as a navigation project, to avoid setting an exorbitant precedent for future flood control projects. The bill directed the Corps to dredge a navigation channel along the southern perimeter of the lake—and if the dirt excavated to create the channel happened to end up in a dike alongside it, well, so much the better. Buker, *Sun, Sand and Water,* p. 104.

199 *Hoover thwarted his efforts:* Jadwin originally proposed that the federal government should only pay 37.5 percent of the project's cost, which probably would have scuttled the project. But after he reached retirement age in 1929, Hoover's handpicked replacement raised the federal share to 60 percent. By the time Hoover signed the bill, the share was 80 percent. The actual share turned out to be even higher.

199 *The combined population of Belle Glade and Pahokee:* Dovell, *A History of the Everglades of Florida,* pp. 526–29; "Our Farmers Win Recognition," *Everglades News,* 11/29/43, Special Army "A" Edition.

199 *in Clewiston:* Clarence R. Bitting letter to Governor Holland, 2/26/1943, GSH, Everglades Drainage District folder; Josiah Ferris Jr., "The Everglades—Agro-Industrial Empire of the South," speech to the Lakeland Kiwanis Club, UFA.

199 *a lucrative crop in the Everglades:* Wright told the House committee investigating his engineering miscues that "the settlement and salvation of the Everglades is sugar cane." "Everglades of Florida Hearings," p. 157.

200 *"It has been demonstrated beyond the peradventure of any doubt":* Clarence Bitting, "The Everglades: Agro-Industrial Empire of the South," U.S. Sugar Corporation pamphlet, 1944.

200 *"Everglades Permanence Now Assured":* Florida Grower, 44, no. 4, April 1936.

201 *"Everglades Drainage Found":* "Everglades Drainage Found Too Well Done; Fires in the Dried Soil Have Ruined $40,000,000 Land," *New York Times,* 10/1/1939, p. 53.

201 *"The saw grass country lies prostrate.":* Thomas Will letter to Old Everglades Buyers, 11/18/1931, TWP, quoted in Meindl, "Past Perceptions of the Great American Wetland," p. 393. Hanna and Hanna, *Lake Okeechobee,* p. 277.

201 *"This has cost me a professional career":* Thomas Will letter to W. L. Alexander, 9/24/1936, TWP, quoted in Dovell, "Thomas Elmer Will, Twentieth Century Pioneer," p. 47.

201 *"Citizens of Florida":* John O'Reilly, "The Everglades, Where Drainage Threatens Wildlife with Extinction," *New York Herald Tribune* 4/9/1939.

201 *generating so much acrid smoke:* Alden H. Hadley, "Reminiscences of the Florida Everglades," *The Florida Naturalist* 4 no. 2 (January 1941), p. 29.

202 *Loggers had cut down 90 percent:* John H. Davis, Jr., *The Natural Features of Southern Florida, Especially the Vegetation and the Everglades* (Tallahassee: Florida Geological Survey, 1943), quoted in Derr, *Some Kind of Paradise,* p. 116.

202 *Fishermen hauled in so many mullet:* Brown, *Totch,* p. 89.

202 *their nets left Biscayne Bay:* Beard, *Everglades National Park Project,* p. 54.

202 *gigging 200 tons of frogs:* "A Preliminary Evaluation Report on the Effects on Fish and Wildlife Resources of the Everglades Drainage and Flood Control Project," U.S. Fish and Wildlife Service Region 4, October 1947.

202 *"They are going deeper and deeper":* Minutes of Everglades National Park Commission meeting, 10/21/1946, ECP, 19422, p. 4.

202 *"Our beautiful streams could not be left alone":* Charles Torrey Simpson, *Florida Wildlife: Observations on the Flora and Fauna of the State and the Influence of the Climate and Environment on Their Development.* (New York: The MacMillan Co., 1932), p. 114.

202 *"which contained more birds":* Beard, *Everglades National Park Project,* p. 50.

203 A National Geographic *writer flew over the burning Everglades:* John O'Reilly, "South Florida's Amazing Everglades," *National Geographic* 77, no. 1 (January 1940), p. 139.

203 *"The Breathmaker made the Everglades":* Author interview with Buffalo Tiger.

203 *they had used a kind of schoolboy logic:* Douglas, *The Everglades,* p. 286.

203 *A brilliant U.S. Geological Survey hydrologist:* Garald G. Parker, et al., *Water Resources of Southeastern Florida, with Special Reference to the Geology and Groundwater of the Miami Area* (Washington: U.S. Government Printing Office, 1955). This is considered the definitive guide to south Florida's hydrogeology at: http://sofia.er.usgs.gov/publications/papers/wsp1255/.

204 *Meanwhile, U.S. Department of Agriculture scientists:* "Soils, Geology and Water Control in the Everglades Region," U.S. Soil Conservation Service, Division of Drainage and Water Control, March 1948 report, p. 97.

204 *The editor of the popular* Rivers of America *series:* Douglas, *Voice of the River,* pp. 190–91.

204 *she liked to say that she channeled the energy and emotion:* Ibid., p. 128. "I've done very well without it, thank you," Douglas wrote.

204 *"I was hooked with the idea":* Ibid., p. 190.

205 *"There are no other Everglades in the world":* Douglas, *The Everglades,* p. 5.

205 *"The endless acres of sawgrass":* Ibid., 349.

205 *"where all forms of life cease to fear man":* Ernest Coe, "The Land of the Fountain of Youth," reprinted from *American Forests and Forest Life,* ECP, 718.

206 *first proposed a national park:* Ernest F. Coe, *Story of the Everglades National Park Project.* Unpublished manuscript commissioned by National Park Service, 1950, ECP, 22888, p. 2.

206 *Fairchild warned that the children of the twenty-first century:* Fairchild letter to Spessard Holland, 3/6/1941, GSHP, Everglades National Park folder; Rothra, *Florida's Pioneer Naturalist,* p. 182, May Mann Jennings speech to the Florida Chamber of Commerce, 2/23/1939, GSHP, Everglades National Park folder.

206 *Simpson wrote a searing article:* Rothra, *Florida's Pioneer Naturalist,* p. 182.

206 *"saved for all time":* May Mann Jennings speech to the Florida Chamber of Commerce, 2/23/1939, GSHP, Everglades National Park folder.

206 *a single-minded, Yale-educated landscape architect:* ECP; Coe, *Story of the Everglades National Park Project;* Theodore Pratt, "Papa of the Everglades National Park," *Saturday Evening Post,* 8/9/1947; Marjory Stoneman Douglas, "The Forgotten Father," *Audubon Magazine,* 1974. Republished online at: www.evergladesonline.com/50year/forgot.htm.

206 *he began sloshing around the Everglades:* Pratt, "Papa of the Everglades National Park," p. 48.

206 *Coe fell madly in love with this "great empire of solitude":* Ernest Coe letter to President-elect Hoover, 2/13/1929, HHP, Campaign and Transition.

206 *"It is the spirit of the thing":* Hearings before the House Public Lands Committee, 71st Cong., 3rd sess., December 1930, p. 69.

207 *Coe's obsession with an Everglades park:* The park was originally proposed as Tropic Everglades National Park; Coe's advocacy group was called the Tropic Everglades Park Association; the commission was the Tropic Everglades Park Commission. The park service wisely slashed the "Tropic," which was unnecessary and technically inaccurate.

207 *"The blaze that had been lighted in him":* Douglas, "The Forgotten Father."

207 *Douglas recounted how her father:* Douglas, *Voice of the River,* p. 135.

207 *Coe fired off thousands of letters:* ECP. See also: Coe letters to Holland, 7/25/1940, 9/21/1941, GSHP.

207 *His commission employed more stenographers:* Coe had four stenographers; the attorney general's office had three. May Mann Jennings, lobbying for funds for the commission, warned Coe that legislators and the governor wanted the commission "cut to the bone." May Mann Jennings letter to Coe, 7/3/1937, ECP, 19959.

207 *Critics groused that he must be on the payroll:* Thomas Pancoast letter to Governor Holland, 6/14/1941, GSHP, Everglades National Park Commission folder.

207 *"When a fellow like that gets up before a meeting":* J. H. Meyer letter to Spessard Holland, 9/8/1941, GSHP, Everglades National Park folder. Meyer was the commission's abstractor, dealing with title searches and other real estate matters. He was also a Democratic political

operative, and warned Governor Spessard Holland that Coe was "a Republican and a Yankee and a Christian Scientist, and other things I can't write."

207 *"It is a safe assertion that had this park been in existence":* Coe letter to Hoover, 2/13/1929.

207 *And without bothering to consult any Seminoles:* Coe letter to John Collier, Bureau of Indian Affairs, 3/27/1935, ECP, 784.

207 *When a governor finally did ask:* Billy Cypress, "Miccosukee: 'Pohaan Checkish.' (Leave Us Alone)," *Miami Herald,* 2/19/1996.

208 *he arranged for the federal committee evaluating the park:* Coe, *Story of the Everglades National Park Project,* pp. 5–6; Douglas, "The Forgotten Father." Douglas had to join Coe in the observer's coop. She didn't get airsick, but she was glad when the flight was over.

208 *Coe then led the committee on a three-day boat tour:* Coe, *Story of the Everglades National Park Project,* pp. 7–24.

208 *And the committee swiftly recommended his plan:* "The Proposed Everglades National Park," Senate Document 54, 72nd Cong., 1st sess., 1/22/1932.

208 *Coe's expansive boundaries:* Coe was so committed to his proposed boundaries that he emblazoned them on the Everglades National Park Association's stationery.

208 *its entertainment value would be only part of its appeal:* Hearings before the House Public Lands Committee, December 1930, p. 23.

208 *a well-intentioned scientist pulled a king snake out of his bag:* The familiar version of this story is that park opponents brought a sackful of snakes into the chambers, but the transcript shows that Dr. Howard Kelly brought one, explaining to the committee that "I brought this to show you what a nice, big, kindly creature a king snake is." Hearings before the House Public Lands Committee, December 1930, p. 56. Marjory Stoneman Douglas claimed that Congresswoman Ruth Bryan Owen, the Great Commoner's daughter and a staunch supporter of the park, threw the snake around her neck to show it was harmless, but that wasn't in the transcript.

209 *"The Everglades section is almost impassable":* William S. Kenney, "Park Plan Disputed," *New York Times,* 7/8/1934.

209 *The Izaak Walton League:* Report of Land Committee on and Boundaries meeting, 6/30/1936, ECP, 19476a. Commercial fishermen and spongers also raised concerns about their industries, but federal officials promised they would not be affected. Coe's minutes note that the Izaak Walton League's representative "was requested by chairman Copeland to refrain from personal remarks which he directed against Ernest F. Coe." Transcripts of Everglades National Park Commission meetings, 12/2/1936 and 1/1/1937, ECP 19387a and 19391b; "Opinions Conflict on Park Boundaries," *Miami Herald,* 6/28/1936.

209 *The chairman of the commission's boundaries committee:* Report to Everglades National Park Commission by Committee on Lands and Boundaries, 10/19/1936, Ernest Coe memo on report, 12/1/1936, ECP 19421.

209 *"the child of Mr. Coe's brain":* Transcript of Everglades National Park Commission meeting, 4/3/1937. "I want to make it clear that I am an employee of the Collier interests, but the mere fact that I am such does not prejudice me in the slightest degree," Copeland said. "You may not believe it, but it is true: If it had not been for me that area in Collier County would never have been included in the Park."

209 *He insisted that any reduction:* The Interior Department had final say over the boundaries, but it always encouraged the state to reach a consensus first. Coe secretly urged Secretary of the Interior Harold Ickes to set the maximum boundaries and tell Florida to take it or leave it. "Any considerable curtailment," he wrote, will "seriously jeopardize the purposes for which this Park can stand." Coe letter to Ickes, 6/8/1937, ECP, 19870.

209 *"to eliminate the Key Largo and marine gardens area":* Coe letter to Governor Spessard Holland, 12/11/1942, GHP, Everglades National Park folder.

209 *May Mann Jennings found Coe's intransigence "absurd":* Jennings letter to Governor Spessard Holland, 1/5/1941, GSHP, Everglades Park Commission folder; Jennings letter to Thomas Pancoast, 6/6/1937, ECP, 19938; Jennings letter to park commission, 6/6/1937, ECP, 19939; Jennings letter to Coe, 6/13/1937, ECP 19949. Coe offered to come to Tallahassee to help lobby for money, but Jennings warned that nothing would be worse for their cause; Copeland came instead. Governor Fred Cone was particularly antagonistic to Coe, criticizing his expensive salary and his fancy title of "executive secretary."

209 *"He antagonized the [Izaak] Walton League":* Meyer letter to Governor Holland, 9/8/1941, GSHP, Everglades National Park folder. In 1946, Coe fell and broke his hip, an accident that

the historian Nixon Smiley suggests was "a happy event for the future of the park" by keeping him out of the political mix. Smiley, *Knights of the Fourth Estate,* p. 222.

209   *Coe kept firing off letters:* Pratty, "Papa of the Everglades National Park," p. 49.

210   *"I am about to die waiting":* Transcript of Everglades National Park Commission meeting, 1/11/1937, ECP, 19391, p. 12.

210   *Spessard Lindsey Holland was born in 1892:* Memorial Addresses and Other Tributes on the Life and Contributions of Spessard L. Holland," Senate Document 56, 92nd Cong., 2nd sess.; Spessard Holland, "Outline for Biography of S. L. Holland" and "Unpublished Autobiography," Miscellaneous Manuscripts, Box 70, UFA; Virginia Holland Gallemore, oral history interview, Spessard Holland biographical file, UFA; Charles Stafford, "Sen. Spessard L. Holland: Statesman and Southerner," *Floridian,* 10/11/1970, reprinted in "Spessard Holland: Now and Always His Own Man," Holland and Knight pamphlet, 1971. The late Chesterfield Smith graciously spoke to me about Holland before his death.

211   *He was his state's most popular:* Bill McBride, "Remembering the Legend," in "Spessard Holland: Now and Always His Own Man." Holland did lose one friend to politics. His closest ally in the Senate was Cap Graham, who helped run his gubernatorial campaign in 1940. But when Graham ran to succeed Holland in 1944, Holland stayed neutral, and Graham held a grudge for years. Interview with Bob Graham.

211   *Zora Neale Hurston once wrote an essay:* Carla Kaplan, *Zora Neale Hurston: A Life in Letters* (New York: Doubleday, 2002), pp. 760–64.

211   *He could be transported by the beauty of a red-shouldered hawk in flight:* Ernest Coe letter to Governor Holland, 6/13/1943, GSHP, Everglades National Park folder.

212   *"I do not believe any plan for conservation will get very far":* "Conservation Is Objective of Governor," *Polk County Record,* 1/3/1941, special Holland inaugural edition, Miscellaneous Manuscripts, Box 70, UFA.

212   *Everglades National Park, in Holland's view:* Minutes of Everglades National Park Commission meeting, 10/21/1946, ECP, 19433, pp. 42–43.

212   *In fact, after oil was discovered beneath the Big Cypress:* Minutes of the Everglades National Park Commission meeting, 10/21/1946, pp. 30–34. "Nobody in Washington, Tallahassee or anywhere else can tell you that there is going to be a park until this exploration and production of oil is behind us," Holland said. "None of us would want to preclude the state or private owners from the possibility of producing any oil actually there."

213   *The work Holland did to cut the deals:* One of Holland's most important achievements as governor was the refinancing of the Everglades Drainage District's debt, which involved long and delicate negotiations with the lead bondholders. Holland eventually persuaded them to accept 33 cents on the dollar, financed by the federal Reconstruction Finance Corporation, which freed up the district's land to be donated to the park. Holland also negotiated an agreement on mineral rights below the park, reserving the state's right to collect royalties in case the feds every decided to drill for oil there.

213   *"I want to see the project advanced":* Ickes letter to Coe, 7/21/1942, ECP.

213   *He wrangled a get-to-know-you meeting with five leading Pork Choppers:* Nixon Smiley, "Poker Game Helped Found Everglades Park," *Miami Herald,* 12/3/1967, p. 1; John Pennekamp, "Talk Before the Miami Rotary Club," 8/11/1955, JPP; Smiley, *Knights of the Fourth Estate,* pp. 223–25.

214   *On December 6, 1947:* "Dedication Ceremonies for Everglades National Park," 12/6/1947, pp. 5–9, SHP.

214   *"Here are no lofty peaks":* Ibid., pp. 10–13.

13   Taming the Everglades

216   *"There never was a country more fabulous":* David Halberstam, *The Fifties* (New York: Fawcett Columbine, 1993), p. 116.

216   *That year, Americans broke the sound barrier:* David McCullough, *Truman* (New York: Simon & Schuster, 1992), p. 691.

216   *it had yet to pass a fence law:* During the settlement of Ernest Graham's estate, when his sons were asked to calculate the cost basis of his cattle herd, they explained that Cap had simply paid some cowboys to round up wild cattle in the area. Bob Graham says he'll never forget the

expression of disbelief on the face of the IRS agent, who could not fathom that wild cows still roamed Dade County in the late 1930s.

218 *"Everglades Is Unconquered Despite Man's Great Fight":* *Miami Herald,* 11/2/1947.

218 *There were no mass casualties:* George Buker calculates that if the Corps had not been able to lower the lake through the St. Lucie, it could have reached a level higher than the 1928 storm. Buker, *Sun, Sand and Water,* p. 105.

218 *Spreading across five million acres:* Stuart McIver, "The Great South Florida Flood," *Sunshine,* 9/9/1990, p. 22.

218 *"We've never had a water situation":* Barnes, Florida's *Hurricane History,* p. 176.

219 *"Our St. Lucie River":* Edwin A. Menninger letter to Governor Holland, 3/5/1948, SHP, Box 287, Folder 61.

219 *"I was maligned, threatened, waylaid":* Lamar Johnson, *Beyond the Fourth Generation* (Gainesville: University Press of Florida, 1974), p. 140.

219 *the "Crying Cow" report:* "Tentative Report of Flood Damage," Everglades Drainage District, 12/12/1947, SFWMD.

220 *"I want control of the Missouri River!":* Reisner, *Cadillac Desert,* p. 183.

220 *The Corps had its critics:* Harold Ickes, foreword, in Arthur Maass, *Muddy Waters: The Army Engineers and the Nation's Rivers* (Cambridge, MA: Harvard University Press, 1951), pp. ix–xiv.

220 *"This nation has a large and powerful adversary":* McPhee, *The Control of Nature,* p. 7.

221 *"Hideous":* Waters of Destiny, U.S. Army Corps of Engineers film. Nanciann Regalado of the Corps gave me a copy of this unforgettable film.

221 *"Florida's economists view this soil and water surgery":* "New Glades Lake to Dwarf Big Okeechobee," *Palm Beach Post,* c. 1950, at PBCHS County.

222 *"The easy solution, of course":* Jeanne Bellamy, "Taming the Everglades: A Report on Water Control." *Miami Herald* pamphlet, 1948.

222 *The C&SF project incorporated elements:* "Comprehensive Report on Central and Southern Florida Project for Flood Control and Other Purposes," House Document 643, 80th Cong., 2d sess., 5/6/1948.

224 *The Army Corps claimed that it "would produce":* "Comprehensive Report," pp. 2, 57.

224 *The U.S. Fish and Wildlife Service said it would:* "Preliminary Evaluation Report on the Effects of Fish and Wildlife Resources of the Everglades Drainage and Flood Control Project." Fish and Wildlife Service, Region 4, October 1947.

224 *"the first scientific, well-thought-out plan":* Marjory Stoneman Douglas, "What are they doing to the Everglades?" unpublished essay c. 1948, RTE, p. 1.

224 *"With near-perfect timing":* Jesse Mock, "Florida Flood Control and Waterway Resume," 6/24/1948, Ervin News Service, SHP, Box 287, Folder 61, Flood Control; "Senator Holland Gets Credit for Flood Program," *Melbourne Times,* 7/2/1948.

224 *one eighteen-mile canal:* Blake, *Land into Water,* p. 183.

224 *his friends at U.S. Sugar:* R. Y. Patterson letter to Governor Holland, 1/21/1948, Holland letter to Patterson, 1/28/1948, SHP, Box 287, Folder 61, Flood Control Program. Patterson, a U.S. Sugar executive, also served as the head of the Everglades Drainage District's water control committee.

224 *in one 1948 speech to the state's cattlemen:* "Unity Important to Get Flood Control Program, Says Holland," *Florida Cattleman,* September 1948, p. 15.

225 *"Everybody . . . will benefit from this dramatic control":* William Roy Shelton, *Land of the Everglades: Tropical Southern Florida* (Tallahassee: Florida Department of Agriculture, 1957), p. 36.

225 *Holland meticulously choreographed a Senate hearing:* Senate Subcommittee on Flood Control and Improvement of Rivers and Harbors, Committee on Public Works, Hearing on Flood Control—Central and Southern Florida," 80th Cong., 2nd sess., 1948, pp. 138–279.

225 *"I have not heard of any opposition, Senator":* Ibid., pp. 195–196.

225 *"I do not recall that I have ever attended":* L. Boyd Finch, "The Florida Swamp That Swallows Your Money," *Harper's,* February 1959, HMSF, p. 80.

225 *"CONSERVATION IN ACTION":* Central and Southern Florida District pamphlet, ACEHQ.

225 *"keep the water of the Everglades in balance":* Marjory Stoneman Douglas, "What Are They Doing to the Everglades?," p. 9.

226 *one Audubon Society official:* Oliver Griswold, "Have We Saved the Everglades?," *The Living Wilderness* 13, no. 27 (winter 1948–49), p. 10.

226 *"This is the wish of the majority of the people":* A Report on Water Resources of Everglades National Park, Florida," Central and Southern Flood Control District, 5/22/1950, SFWMD, p. 13.

226 *Florida's education department,* Henry F. Becker, ed., *Florida: Wealth or Waste?* (Tallahassee: Florida State Department of Education, c. 1953).

226 *Aldo Leopold, a founder of the Wilderness Society:* Aldo Leopold, *A Sand County Almanac* (New York: Oxford University Press, 1949).

227 *Lyons warned that the costly "Hollandizing" of south Florida:* Ernest Lyons, " 'Flood Control' Destroys Last Natural Frontier," in "Florida's Problem: How Much 'Water Control' Is Boondoggle?" *Stuart News* publication, 1949, SHP, p. 7.

227 *South Florida started out with a marvelous flood control plan:* Lyons, "'Flood Control' Destroys Last Natural Frontier," pp. 6–8.

228 *"What is a species more or less among engineers?"* Leopold, *A Sand Country Almanac,* p. 100.

228 *"The engineers think only in terms of ditches":* Blake, *Land into Water,* p. 179.

229 *"It appears to me that the federal government":* John M. DeGrove, *The Central and Southern Florida Flood Control Project: A Study in Intergovernmental Cooperation and Public Administration* (University of North Carolina dissertation, 1958; University Microfilms, Ann Arbor, 1985), pp. 302–303. The congressman was Edward Boland of Massachusetts.

229 *by 1965, five years after its scheduled completion date:* "A Few Facts and Figures About the Flood Control District," publication of C&SF Flood Control District publication, January 1969, SFWMD.

229 *"There is no point quoting statistics":* Jack Kofoed, *The Florida Story* (Garden City, NY: Doubleday & Co., 1960), pp. 110, 267.

229 *Actually, statistics do give a sense:* Charlton W. Tebeau and Ruby Leach Carson. *Florida: From Indian Trail to Space Age, Vol. 1* (Delray Beach, FL: The Southern Publishing Co., 1965), pp. 94–95, 149–50; Luther J. Carter, *The Florida Experience: Land and Water Policy in a Growth State* (Baltimore: Published for Resources for the Future by Johns Hopkins University Press, 1974), pp. 5–6, 28–29; Rothchild, *Up for Grabs,* p. 92; Redford, *Billion-Dollar Sandbar,* p. 235.

230 *"The River of Grass . . . is retreating":* "Everglades Pay Dirt!" Atlantic Coast Line Railroad Company pamphlet, 1948, FSA.

230 *Over the next five years:* Frank J. Coale, "Sugar Production in the EAA," in A. B. Bottcher and F. T. Izuno, eds., *Everglades Agricultural Area: Water, Soil, Crop and Environmental Management* (Gainesville: University Press of Florida, 1994), p. 225.

231 *Aerojet General Corporation moved to the Homestead area:* Juanita Greene, "Learn from Aerojet," *Miami Herald,* 7/28/2004; Gene Marine, *America the Raped: The Engineering Mentality and the Devastation of a Continent* (New York: Avon Books, 1969), pp. 42–43.

231 *"remarkable strides in water control":* Shelton, *Land of the Everglades,* p. 34.

232 *"the real estate boys read the bill":* Halberstam, *The Fifties,* p. 134.

232 *"like rows of pole beans":* Parks, *Miami,* p. 147.

232 *"Live in the Path of Progress!":* This was an advertisement for the new Palm Lakes development in Hialeah by the Sengra Development Corporation. "Sengra" was shorthand for Senator Graham; it was Cap Graham's family company. "The Graham Companies: 70 Years, Celebrating a Family Tradition of Service, 1932–2002," Graham Companies booklet, 1/25/2002, p. 22. Courtesy Bob Graham.

232 *One dredging firm: Turning Swamps into Dollars,* Ellicott Machine Corp., Baltimore, 1958 pamphlet.

232 *In the mid-1950s:* L. Alan Eyre, "Land Reclamation and Settlement of the Florida Everglades," in *This Changing World,* pp. 41–42, UFA.

233 *"We've been blessed":* "The Graham Companies," p. 27.

233 *"Nonsense," you protest:* "Conservation in Action."

233 *The assessed value of land within the flood control district":* Tom Huser, "Into the Fifth Decade: The First Forty Years of the South Florida Water Management District, 1949–1989," SFWMD, p. 59.

234 *two Baltimore brothers named Leonard and Julius Rosen:* The story of the Rosen brothers and Gulf American is recounted in Carter, *The Florida Experience,* pp. 232–40; Rothchild, *Up for Grabs,* pp. 82–101; Susan Orlean, *The Orchid Thief* (New York: Ballantine, 1998), pp. 116–26.

234 *"Have you ever seen a bald sheep?":* Rothchild, *Up for Grabs,* p. 84.

234 *"a rich man's paradise":* Orlean, *The Orchid Thief,* p. 117.

234 *"Lot number 72 is sold!":* Carter, *The Florida Experience,* p. 235; Rothchild, *Up for Grabs,* p. 87.

235 *"The wilderness has been pushed aside":* Niki Butcher, "History of Big Cypress National Preserve," reprinted at www.friendsofbigcypress.org.

235 *"As long as the sun shines":* Kofoed, *The Florida Story,* p. 282.

14   Making Peace with Nature

PAGE

239   *"We must build a peace in south Florida":* Remarks of Governor Reubin O'D. Askew, 9/22/1971, GRAP, Series 949, Carton 16, Water Management Conference.

239   *Salt invaded the wells:* Charles D. Schilling, "Florida Flood Control—Fact or Fiction," *Salt Water Sportsman* 19, no. 10 (October 1958); "Remarks of Reubin O'D. Askew," 9/22/1971; Arthur R. Marshall, "Remarks for Presentation to the Governor," 4/13/1971, ARMP.

240   *the National Park Service's most endangered property:* Undersecretary of the Interior Russell Train testified in 1969 that "Everglades National Park has the dubious distinction of having the most serious preservation problems facing the National Park Service today." Gary A. Soucie, "The Everglades Jetport—One Hell of an Uproar." *Sierra Club Bulletin* 54, no. 7 (July 1969), p. 4.77

240   *a phenomenon chronicled in articles:* Peter Farb, "Disaster Threatens the Everglades," *Audubon* (September/October 1965), pp. 302–309; Fred Ward, "The Imperiled Everglades," *National Geographic* 141, no. 1 (January 1972); Richard Rhodes, "The Killing of the Everglades," *Playboy* (January 1972); William Ross McCluney, ed. *The Environmental Destruction of South Florida* (Coral Gables: University of Miami Press, 1971).

240   *The veteran drainage engineer Lamar Johnson:* Lamar Johnson, "A Survey of the Water Resources of Everglades National Park," July 1958 report, p. 2.

240   *"I found no Eden":* Farb, "Disaster Threatens the Everglades," p. 303.

240   *"This beautiful part of the world":* Ward, "The Imperiled Everglades," p. 3.

240   *"Time is running out for the Everglades":* Patricia Caulfield, *Everglades,* with an essay by John G. Mitchell (New York: A Sierra Club/Ballantine Book, 1971), p. 34.

241   *"What a liar I turned out to be!":* Pat Cullen, "Saving the Everglades: 'We Need a Careful Balance," *Palm Beach Post,* 4/5/1971.

241   *America experienced an extraordinary awakening:* Philip Shabecoff, *Earth Rising: American Environmentalism in the 20th Century* (Washington, D.C.: Island Press, 2000), pp. 4–7; Robert Gottlieb, *Forcing the Spring: The Transformation of the American Environmental Movement* (Washington: Island Press, 1993), pp. 81–114.

241   *In 1969, a secret poll:* Elizabeth B. Drew, "Dam Outrage: The Story of the Army Engineers," *Atlantic Monthly,* 1970, p. 52.

241   *Pundits joked that every congressman now claimed to be an ecologist:* Philip Wylie, "Against All Odds, the Birds Have Won," *New York Times,* 2/1/1970.

241   *In his State of the Union address:* Gottlieb, *Forcing the Spring,* pp. 108–109.

242   *"We were worried about losing the garden clubs":* Interview with John Whitaker. He said that even though President Nixon's eyes used to glaze over when hearing about pollution standards, he was genuinely passionate about national parks. Whitaker said that as a boy growing up in California, it had infuriated Nixon that wealthier families could afford to drive to Yellowstone and the Grand Canyon.

242   *The resulting backlash included some overwrought alarmism:* In *The Population Bomb,* Paul Ehrlich warned that at least 100 million people would die of starvation in the 1970s, "a mere handful compared to the numbers that will be starving by the end of the century." (New York: Ballantine Books, 1971, p. 3.) In fact, only a few million people have starved—and not because of any worldwide food shortages. C. Richard Tillis predicted the nine-degree temperature spike in "The Spaceship Earth," reprinted in McCluney, ed., *The Environmental Destruction of South Florida,* p. 2.

242   *The eloquent prophet of this petrochemical age:* Carson, Rachel, *Silent Spring* (originally published 1962; Boston: Houghton Mifflin, 1987). Carson was guilty of some hyperbole as well; DDT would come in handy today to combat malaria in the developing world. But in general, her warnings about pesticides were farsighted and desperately needed at the time.

242   *"We still talk in terms of conquest":* "Rachel Carson Dies of Cancer," *New York Times,* 4/15/64.

243   *The backlash against man's assault on nature:* Mark Reisner's *Cadillac Desert* is my favorite account of Corps follies. In 1971, ninety-year-old Arthur Morgan, the engineer who had exposed James Wright's frauds as a young man in Florida, published a vicious attack on the Corps called *Dams and Other Disasters.*

243   *"The rather sudden general awareness of the science of ecology":* Drew, "Dam Outrage," p. 52.

243   *Supreme Court Justice William O. Douglas wrote:* William O. Douglas, "The Public Be Dammed," *Playboy,* July 1969, p. 143.

243 *Senator Gaylord Nelson of Wisconsin, the founder of Earth Day:* Drew, "A Dam Outrage," p. 51. Nelson noted that at least beavers have a purpose for their dams, and don't ask taxpayers to foot the bill.

243 *It was pushing a gigantic dam:* Reisner, *Cadillac Desert,* pp. 208–9; Blake, *Land into Water,* pp. 199–215.

243 *It was digging the Mississippi River Gulf Outlet:* Michael Grunwald, "Canal May Have Worsened City's Flooding," *Washington Post,* 9/14/2005.

243 *justifying each pork barrel project by predicting miraculous increases:* For example, the Corps justified taming the Missouri River by predicting 12 million tons of annual barge freight; the actual peak was 3.3 million tons. Michael Grunwald, "A River in the Red."

243 *One cartoonist routinely depicted:* George Fisher, *God Would Have Done It If He'd Had the Money* (Little Rock: Arkansas Wildlife Federation Conservation Foundation, 1983), cover, p. 13.

243 *"Their mommies obviously never let them":* Al Burt, "The Elocutioner," *Miami Herald,* 3/18/1984.

244 *"To anyone who has ever so much as heard the word 'ecology'":* Marine, *America the Raped,* p. 36.

244 *"a leading contender for first place":* Dasmann, *No Further Retreat,* pp. 2–3.

244 *"our beautiful state of Florida":* June Cleo and Hank Mesouf, *Florida: Polluted Paradise* (Philadelphia: Chilton Books, 1964), p. ix.

244 *Politically, Florida was still the land of laissez-faire:* Tom Ankersen, *Coping With Growth: The Emergence of Environmental Policy in Florida* (Unpublished master's thesis, University of South Florida, 1982), p. 1; Blake, *Land into Water,* pp. 195–96; Carter, *The Florida Experience,* pp. 4–9.

244 *Florida's environmental movement grew stronger:* Author and University of Florida oral history interviews with Reubin Askew, Bob Graham, John Jones, and Nathaniel Reed. Author interviews with Joe Browder, Juanita Greene, Jay Landers, Charles Lee, and Joe Podgor. Ankersen, *Coping with Growth,* p. 34; Carter, *The Florida Experience,* pp. 155–67, 265–312; Blake, *Land into Water,* pp. 198, 203–09.

245 *Audubon's abrasive but effective southeastern representative:* Author interviews with Joe Browder; Charles Stafford, "Joe Browder: A Political Animal Who's at Home in the Swamp," *Floridian,* 8/22/1971; Judith Bauer Stamper, *Save the Everglades* (New York: Steck-Vaughn Company, 1993).

245 *He eventually landed a job:* Browder's best scoop was his exclusive footage of preparations for the Bay of Pigs invasion; sometimes it paid to have a dad in the CIA. Browder also produced a documentary called *Come Hell or High Water* about the impending collapse of south Florida's water resources. Incidentally, while he was moonlighting for Audubon, he was also serving on a national journalistic ethics committee.

245 *Florida had its own version:* Al Burt, *Becalmed in the Mullet Latitudes: Al Burt's Florida* (Miami: Miami Herald Publishing Co., 1983), pp. 201–203; Jeffery Kahn, "Light Now Shines on Biologist's Lonely Quest," *Palm Beach Post,* 9/10/1984; Author interviews with Joe Browder, John Jones, Timothy Keyser, Charles Lee, Joe Podgor, and Nathaniel Reed; Thomas Ankersen oral history interview with Arthur R. Marshall Jr., University of Florida College of Law, Center for Governmental Responsibility.

246 *"It is time—well past time":* Arthur R. Marshall, "The Future of South Florida's Salt and Freshwater Resources," reprinted in McCluney, ed., *The Environmental Destruction of South Florida,* p. 18.

246 *"I once offered all I could":* Arthur Marshall letter to John Jones, C. 1977, courtesy John Jones.

246 *local officials tried to get him fired:* W. H. Carmine letter to Representative Paul Rogers, 10/25/1961, ARMP, Box 1, Folder 33: Fish and Wildlife Service.

246 *"Certainly some find my views disputatious":* Arthur Marshall letter to Governor Reubin Askew, 8/19/1975, ARMP, Box 8, Folder 1: St. Johns Water Management District.

246 *"In Florida, it has always been said":* Burt, *Becalmed in the Mullet Latitudes,* p. 202.

247 *"If you don't synthesize knowledge":* Robert H. Boyle and Rose Mary Mechem, "Anatomy of a Manmade Drought," *Sports Illustrated,* 3/15/1982.

247 *"Ignoring the principles of the environment":* Marshall was talking about Audubon's Charles Lee. Arthur Marshall oral history interview.

247 *Douglas called him "the leading man":* Douglas, *Voice of the River,* pp. 226–27.

247 *Marshall testified as a private citizen:* Thomas T. Ankersen, "Law, Science and Little Old Ladies," Florida Humanities Council Forum, Summer 1995, pp. 18–23. John Jones says that while Marshall was officially on his own, Jones and the Florida Wildlife Federation had recruited him to speak.

247 *He lambasted south Florida's car-dependent culture:* Arthur Marshall oral history interview.

248   *"Do not treat Art Marshall lightly":* "Flood Body Shake-up Urged," *Tampa Tribune,* 9/24/1971.

248   *"a snowballing degeneration of major resources":* Arthur Marshall, "Statement for Presentation to the Governor and Cabinet of Florida," 4/13/1971, ARMP, Box 1, Folder 18.

248   *And he was not the kind of scientist:* Arthur Marshall letter to Nathaniel Reed, July 1976, ARMP, Box 1, Folder 43, Nathaniel P. Reed Correspondence.

248   *One was a U.S. Supreme Court decision:* David R. Colburn and Lance deHaven-Smith, *Government in the Sunshine State* (Gainesville: University Press of Florida, 1999), pp. 38–42; Carter, *The Florida Experience,* pp. 43–47. One example of Pork Chop power: Florida's racetrack receipts were split evenly among the state's sixty-seven counties, even though Dade had more than a million residents and several northern counties had just a few thousand.

249   *The other tectonic shift in Florida politics:* Edmund F. Kallina, *Claude Kirk and the Politics of Confrontation* (Gainesville, University Press of Florida, 1993); Ralph De Toldano and Philip V. Brennan, Jr., *Claude Kirk: Man and Myth* (Moonachie, NJ: An Anthem Book, 1970). Nathaniel Reed, Kirk's environmental brain, regaled me with hours of stories about Claudius Maximus. Kirk's University of Florida oral history interview is also great fun, from his descriptions of his first wife's breasts to his declaration that all of his successors as governor were "fraudulent" to his speculation that if he had been named Albert instead of Claude, he would have been an A student instead of a C student (pp. 2, 9, 33).

249   *The aide was Nathaniel Pryor Reed:* "The Dedicated Years," a scrapbook compiled by Alita Reed, courtesy Nathaniel Reed. Carter, *The Florida Experience;* Kallina, *Claude Kirk and the Politics of Confrontation,* p. 151–67.

250   *"I would take up whatever crusade":* Kirk oral history interview.

250   *"You mean when I take a crap in Palm Beach":* Author interview with Reed.

250   *"Anybody who says you can achieve environmental quality":* Jim Long, "Pollution Fighter Nat Reed: Florida's Mr. Clean Shifts His Attack to Community Sewage Disposal," *Florida County Government,* May/June 1970, in "The Dedicated Years."

250   *"The chamber of commerce types would raise holy hell":* Author interview with Reed.

251   *Dredge-and-fill permits plummeted 90 percent:* Kallina, *Claude Kirk and the Politics of Confrontation,* p. 155.

251   *During a drought in 1967:* Blake, *Land into Water,* pp. 189–90.

251   *One of the enduring myths:* House Document 638. Modern Corps leaders routinely whitewash their agency's environmental history, claiming they only followed orders and did what the nation wanted. For example, General Robert Griffin, the agency's former director of civil works, gave PowerPoint presentations featuring a slide titled "National Priorities: Then and Now." One side featured the "Crying Cow" report, with the caption "1947: Move the Water, Prevent Floods." The other side features the Corps plan to restore the Everglades, decorated by a gator, a heron, and a panther, with the caption "1999: Protect Wetlands, Restore Ecosystems." The truth is more nuanced than that.

251   *One Corps hydrologist:* Carter, *The Florida Experience,* p. 101.

251   *"You didn't join the Corps":* Author interview with Richard Bonner.

252   *"Instead of a lush wetland wilderness":* Schilling, "Florida Flood Control: Fact or Fiction?"

252   *A boat company purchased full-page ads:* The company was Boston Whaler. "Into the Fifth Decade," p. 47.

252   *a response one critic compared:* Farb, "Disaster in the Everglades."

252   *"If we don't get water":* John O'Reilly, "Water Wanted for a Parched Park," *Sports Illustrated,* 6/7/1965.

252   *Even Lamar Johnson:* Johnson, *Beyond the Fourth Generation,* p. 211.

253   *"In short, we can have our cake and eat it, too":* Joe J. Koperski, 1/4/1968 memo, "Water Supply to ENP Based on Present Level of Demand for Other Users," ACEHQ. In a blame-the-victim letter to the Corps, Hodges wrote that if the park did not support the plan with no water guarantee, it would be responsible for its own demise: "Their failure to do so may jeopardize the future of Florida and the life of Everglades National Park." 4/11/1968 letter, ACEHQ.

253   *several conservation-minded congressmen:* Democratic Congressman Dante Fascell of Miami led the charge in the House. Democratic Senators Edmund Muskie of Maine and Gaylord Nelson of Wisconsin pushed in the Senate.

253   *Publicly, Holland still claimed to oppose the guarantee:* The otherwise excellent accounts of this era by Luther Carter and Nelson Blake both contend that the water guarantee park passed over Holland's objections. But as Reed points out, Congress didn't do anything regarding Florida in those days over Holland's objections.

253   *a one-time exception for the Everglades:* Reed came to Washington to testify on behalf of the guarantee, and Holland told him that he would be asked one question: This is only for the Everglades, correct? And he was to answer: Yes, sir. Reed did so, and then the hearing was gaveled to a close. Afterward, Senator Henry Jackson of Washington told Reed: "You just got away with murder." Author interview with Reed.

253   *The Senate tried to make it clear:* The Senate report on HR 15166 reiterates "the national interest in the preservation of the Everglades National Park." Florida retained its sovereignty to distribute water among its own users however it chose, but it had to let the federal project supply the national park.

253   *"the pressures for making":* This quotation is from Senator Muskie, a loyal defender of Everglades National Park. Carter, *The Florida Experience,* p. 124.

254   *"In my opinion, many of the expensive structures":* Frank Craighead, "The Water Situation in Everglades National Park," 1/13/1969 memorandum, courtesy Joe Browder.

254   *"The only 'out' I see":* Garald Parker, "The Truth About the Everglades," 1973 Friends of the Everglades pamphlet, reprint of Parker letter to Arthur Marshall, MSDP.

254   *The Dade County Port Authority bragged:* Soucie, "The Everglades Jetport"; Carter, *The Florida Experience,* pp. 187–227; Blake, *Land into Water,* pp. 216–22; Jim DeFede, "Destiny's Child," *Miami New Times,* 12/21/2000. The backers of the project actually called it the "Everglades jetport," an illustration of their ecological cluelessness.

255   *"I don't need you next week":* Author interview with Reed.

255   *"The answer to that question is under study":* Author interviews with Reed, Browder.

256   *"I guess they want war":* Author interview with Reed.

256   *So they didn't bother to disguise their hostility:* John MacDonald, "Last Chance to Save the Everglades," *Life,* 9/5/1969; Soucie, "The Everglades Jetport"; Rhodes, "The Killing of the Everglades."

256   *"we will do our best to meet our responsibilities":* Rhodes, "The Killing of the Everglades."

256   *Hickel camped out in the Everglades:* William M. Blair, "Hickel and Kirk Reach Accord on a Plan to Save Everglades," *New York Times,* 3/16/1969. Interview with Reed.

257   *"Development of the proposed jetport":* "Environmental Impact of the Big Cypress Jetport," 1969 Interior Department Study, p. 1. Dade County also commissioned a study on the jetport by former Interior Secretary Stuart Udall, who had been working with developers to mold environmentally acceptable compromises. Udall hired the Everglades ecologist Frank Craighead to lead the study, and according to Joe Browder, Craighead nearly quit because of threats to his independence. Eventually, Craighead concluded that there was no possible compromise that would allow an acceptable jetport in Big Cypress, and Udall said so to Dade County.

257   *an influential* Life *article:* MacDonald, "Last Chance to Save the Everglades."

257   *"It happens to Indians year after year":* Homer Bigart, "Naturalists Shudder as Officials Hail Everglades Jetport," *New York Times,* 8/11/1969.

257   *Finally, Browder visited the most famous Everglades advocate:* Douglas, *Voice of the River,* pp. 224–26; Stamper, *Save the Everglades,* p. 31.

258   *"made her look like Scarlett O'Hara":* John Rothchild wrote this brilliant description in his introduction to *Voice of the River,* the autobiography he ghostwrote for Douglas.

258   *"Nobody can be rude to me":* Margarita Fichtner, "Marjory Stoneman Douglas," *Miami Herald,* 9/22/1985.

258   *"Natural assets and wildlife preserves":* Wylie, "Against All Odds, the Birds Have Won."

258   *But Joe Browder, who was now a Washington lobbyist:* Congressman Dante Fascell and Senator Lawton Chiles shepherded the bill through Capitol Hill. Browder and Chiles also went to visit newly retired Senator Holland, who agreed to lobby his old colleagues, but first crossed out the Okaloacoochee Slough. The slough was recently preserved as well.

258   *Browder secretly tipped off the White House:* Ehrlichman, a former land-use attorney in Seattle, had proposed the land-use restrictions. But Browder placed his call to John Whitaker, who worked for Ehrlichman but was skeptical of the land-use plan. "It was pure politics, but we did the right thing," Whitaker recalled. Clark Hoyt, "Big Cypress Issue Heads for Showdown," *Miami Herald,* 5/6/1971. The Ehrlichman strategy change was the turning point, but it did not produce an immediate bill before Nixon's second term, when the administration became preoccupied with Watergate. The Florida Wildlife Federation's John Jones recalls that Senator Alan Bible refused to let the Big Cypress bill out of his committee unless the state came up with $40 million. Governor Askew then worked with

State Senator Bob Graham and House Speaker Richard Pettigrew to come up with the cash out of the state's environmental land program. But Bible still refused to let the bill go, so Senator Jackson, the chairman of the full committee, yanked the bill out of Bible's subcommittee and brought it to the floor for a vote.

259 *Three inches of rain fell:* "Into the Fifth Decade," pp. 71–72; "Fiery Ordeal of the Everglades," *Life*, 5/7/1971; Philip D. Carter, "Drought-Ravaged South Florida Faces an Environmental Disaster," *Washington Post*, 5/14/1971.

259 *In many ways, he was Kirk's polar opposite:* Colburn and deHaven-Smith, *Government in the Sunshine State*, p. 68.

260 *his 1984 campaign for the presidency:* Askew ran as "A Different Democrat," opposing the nuclear freeze movement and defending right-to-work laws anathema to Democratic labor unions. Primary voters preferred a more traditional Democrat, Walter Mondale; Askew dropped out after finishing last in the New Hampshire primary.

260 *"the world's first and only desert":* Remarks of Governor Reubin Askew, 9/22/1971, GRAP, Series 949, Box 16, Water Management Conference.

260 *Their fourteen-page report dripped with Marshall's influence:* "Final Draft, Governor's Conference on Water Management in South Florida," GRAP, Series 949, Box 16, Water Management Conference.

261 *"I'm 81. I won't live to see this through":* "Flood Body Shake-up Urged," *Tampa Tribune*, 9/24/1971.

261 *"It is not offbeat or alarmist":* Ankersen, Coping with Growth, p. 79.

261 *State Senator Bob Graham of Miami Lakes and House Speaker Richard Pettigrew:* The most controversial of the four major growth management bills was the Environmental Land and Water Management Act, which included the provisions for state regulation of Areas of Critical State Concern and Developments of Regional Impact. Graham and Askew finally got it through the Senate by agreeing that it would only take effect if the voters approved the bond bill as well. Graham says that Senate president Jerry Thomas planned to bury the bond bill, but he was off the floor when it came over from the House. Graham immediately interrupted and asked for unanimous consent to take up the bill, and managed to get a vote before Thomas could stop him without public embarrassment. "I've always thought that if Thomas had been in that chair a few minutes earlier, we never would have gotten the most important parts of the program enacted," Graham says.

261 *kept sprouting in the Everglades floodplain:* Askew had tried to enact strict wetlands regulations, and his environmental aide, Jay Landers, had lined up five votes for the bill on the nine-member Senate Natural Resources Committee. But shortly before the vote, a tenth member was mysteriously added to the committee, and the bill deadlocked. "That was my welcome to Tallahassee," Landers says.

262 *"Managing growth in Florida":* Author interview with Reubin Askew.

262 *One poll found:* The Democratic pollster Patrick Caddell conducted the poll. Blake, *Land into Water*, p. 192.

262 *one Big Cypress property rights group:* "Regarding Death for Those Leading the Fight to Steal Our Land," East Collier County Landowners Committee flyer, 7/25/1975, courtesy Joe Browder. An alligator poacher named Gator Bill, who befriended Browder during the jetport fight, warned the recalcitrant landowners that if anything happened to Browder, they would be fed to the gators. Nothing happened.

263 *"If man cannot live with a living Everglades":* Joe Browder, "The Everglades, The Jetport and the Future," reprinted in McCluney, ed., *The Environmental Destruction of South Florida*, p. 40.

15   Repairing the Everglades

PAGE

264 *"The Everglades is trying to tell us something":* Carter, "Drought-Ravaged South Florida Faces an Environmental Disaster."

264 *"The Everglades is not just stressed":* Arthur Marshall, Statement for Presentation to the Governor," 4/13/1971, ARMP.

265 *In the Everglades, Marshall observed:* Arthur Marshall, "A Critique of Water Management inFlorida," South Florida Water Management District, 11/20/1980, ARMP.

265 *The head, heart, and body of the ecosystem:* Von Drehle, "Bury My Heart at Southwest 392nd Terrace."

266   *"The Everglades ecosystem as we know it"*: "Report of the Flood Control District Special Study Team on the Florida Everglades," August 1970, ARMP, p. 35. Marshall was only one of the authors, but his voice is unmistakable in this section.

266   *One nineteenth-century visitor described the natural Kissimmee*: Villers Stuart, *The Equatorial Forests and Rivers of South America; also in the West Indies and the Wilds of Florida* (London: John Murray, 1891, courtesy Robert Mooney).

266   *Its basin also included a web of lush marshes*: Henri Dauge, "Mr. Wegg's Party on the Kissimmee," *Harper's New Monthly Magazine,* 1886, reprinted in Oppel and Meisel, eds., *Tales of Old Florida,* Carr, *The Everglades,* pp. 24–25.

266   *it was still distinguished: Kissimmee Valley Gazette,* 12/29/1899, quoted in Mike Thomas and Joe Kilsheimer, "State Charts the Current of Fate for River," *Orlando Sentinel,* 5/29/1983.

266   *As late as 1958:* "A Detailed Report of the Fish and Wildlife Resources in Relation to the Corps of Engineers Plan of Development, Kissimmee River Basin," U.S. Fish and Wildlife Service, December 1958, p. 20.

267   *Fish and Wildlife protested:* "A Detailed Report of the Fish and Wildlife Resources," p. 20.

267   *So the Corps manipulated its analysis:* Governor Askew's Division of Planning discovered these manipulations, but its report was suppressed to avoid controversy. Mike Thomas, "Files Point to Canal Built by Mistake," *Orlando Sentinel,* 5/30/1983.

267   *"I've just returned from the deathbed of an old friend":* Steven Trumbull, "Progress to Doom Pretty River?," *Miami Herald,* 10/3/1965. Juanita Greene recalls that when Trumbull returned to the *Herald* newsroom, he was absolutely livid.

267   *The Corps spent ten years and $35 million:* Carter, *The Florida Experience,* pp. 104–7; interviews with Lou Toth.

267   *almost everyone agreed that it never should have been started:* Douglas, *Voice of the River,* p. 229; Natalie Angier, "Now You See It, Now You Don't," *Time,* 8/6/1984.

267   *"The Kissimmee Valley was fantastic country":* Bubba Mills, "The Great Valley," reprinted in "For the Future of Florida—Repair the Everglades," Friends of the Everglades pamphlet (spring 1981), p. 4.

268   *Waterfowl declined 92 percent:* Toth interview. "Kissimmee River Restoration," South Florida Water Management District brochure.

268   *"One year they found eight ducks":* Interview with John Jones.

268   *"The river is still there":* Arthur Marshall letter to Marjory Stoneman Douglas, 10/3/1976, MSDP, Box 23, Folder 8.

269   *She hated hunting, but she liked Jones:* Douglas, *Voice of the River,* p. 229; Jones interview.

269   *"It was a sad mistake to tamper":* "Let It Be," *Florida Sportsman,* August 1978.

269   *"I believe we are on the road":* Arthur Marshall letter to Marjory Stoneman Douglas, 8/29/1976, MSDP, Box 27, Folder 61. Marshall had lost his funding for his center at the University of Miami, and he was hired as a consultant on the Kissimmee study. "I am too cantankerous and uncompromising to be the project director," he told Douglas. "Dear God! I don't know how to pussyfoot my way into reality, and besides I need the money."

269   *"We shall see the Kissimmee River flowing sweet":* Arthur Marshall letter to Marjory Stoneman Douglas, 3/21/1977, MSDP, Box 27, Folder 61.

270   *Marshall had a holistic vision for repairing the entire ecosystem:* "For the Future of Florida," Arthur Marshall, Statement to South Florida Water Management District, 11/20/1980, ARMP; Arthur Marshall, Statement to SFWMD, 6/11/1981, ARMP.

270   *He thought of summer rains as paychecks:* Arthur Marshall, "Repairing the Florida Everglades," 6/11/1971, University of Miami, ARMP.

270   *"the BIG issue":* Arthur Marshall, memo to Coalition to Repair the Everglades, 1982, ARMP, Box 1, Folder 5.

271   *"No one really knew if Einstein was correct":* Al Burt, "The Marshall Plan," reprinted in RTE, http://everglades.fiu.edu/marshall/marsh008/marsh008.html.

271   *"The good news is on the cover, Governor":* Author and University of Florida oral history interviews with Estus Whitfield. After serving Graham for eight years, Whitfield stayed to serve Republican Bob Martinez and Democrat Lawton Chiles.

271   *"The sad fact is that Florida":* Robert H. Boyle and Rose Mary Mechem, "There's Trouble in Paradise," *Sports Illustrated,* 2/9/1981, pp. 82–96.

272   *Daniel Robert Graham:* My description of Graham borrows heavily from a profile I wrote for *The Washington Post Magazine,* although the profile focused on his worries about terrorism. Michael Grunwald, "Running Scared," 5/4/2003. Other sources on Graham are his lengthy University of Florida oral history interview; S.V. Date, *Quiet Passion: A Biography of Senator Bob Graham*

(New York: Penguin, 2004); D. Robert Graham, *Workdays* (Miami: Banyan Books, 1978); David Von Drehle, *Among the Lowest of the Dead* (New York: Times Books, 1995).

272 *Cap Graham, was a proud exploiter of the Everglades:* Cap Graham had an uncanny knack for advocating positions that later turned out to be environmentally friendly, even though his positions were never motivated by environmental concern. He opposed the Cross Florida Barge Canal because he thought it would hurt south Florida's military bases. He opposed Fred Elliot's expansive drainage plans and the Central and Southern Florida flood-control project because he thought they would unfairly burden Dade County's taxpayers. He wanted additional culverts to restore flows through the Tamiami Trail because water was stacking up behind the highway and flooding his fields. Bob Graham calls his father "an accidental environmentalist." Howard Kohn and Vicki Monks noted the contrast between father and son in their insightful (but often inaccurate) "Greetings from the Everglades," *Mother Jones,* December 1987.

272 *"Professional pols viewed Graham's candidacy":* Von Drehle, *Among the Lowest of the Dead,* pp. 9–10.

273 *Today, he is best known:* Graham says he first picked up the notebook habit from his father, who used to jot down notes about his cattle. He started scribbling full-time during his first gubernatorial campaign, when he was working on a book about his workdays and wanted to remember details. He says his notebooks are like PalmPilots, and he does use them to jot down ideas and names. That said, it is hard to see any reason to keep track of chocolate Slim-Fasts except for an obsessive-compulsive personality.

273 *unveiling a "hit list" of nineteen water projects:* The "hit list" debacle was a turning point for Carter and the Corps. Carter proposed it over the objections of his domestic policy adviser, Stuart Eisenstadt. He dispatched his lobbyists, Jim Free and Frank Moore, to twist arms on Capitol Hill. They finally found enough Democrats willing to risk their political careers by sustaining Carter's veto—and then the president suddenly caved to House Speaker Thomas (Tip) O'Neill, once again without telling Eisenstadt. Not only had Carter alienated Congress by attacking its pork, he had shown that he could be rolled. Author's interviews with Eisenstadt, Free, and Moore.

274 *Jones soon received an invitation:* Graham, Jones, Tim Keyser, and Whitfield all had different recollections of this meeting, but they agreed on the basic outline.

274 *landowners booed and yelled at her:* Davis, "Conservation Is Now a Dead Word," p. 53.

274 *In 1983, he announced his "Save Our Everglades" program:* Ken Klein, "Florida Governor Announces Plan to Protect Everglades," Associated Press, 8/9/1983; Kerry Gruson, "Plan Urged to Save Everglades Ecology," *New York Times,* 8/10/1983; Minutes from Meeting on Everglades Revitalization, 6/14/1983, MSDP; Jeffery Kahn, "Restoring the Everglades," *Sierra,* September/October 1986.

275 *Most of Save Our Everglades came straight out of the Marshall Plan:* "Save Our Everglades Report Card," 11/4/1983, HMSF, Everglades folder.

275 *Former senator Gaylord Nelson:* Philip Shabecoff, "Program Aims to Rescue Everglades From 100 Years of the Hand of Man," *New York Times,* 1/20/1986.

275 *"The Everglades is a national park down there by Miami":* Barry Bearak, "Reflooding a Riverbed," *Los Angeles Times,* 12/27/1983.

276 *"Now we're less confident in technology":* Minutes from Meeting on Everglades Revitalization, 6/14/1983, MSDP, Box 25, Folder 26.

276 *And the Corps refused to dismantle its C-38 Canal:* "Central and Southern Florida Project, Kissimmee River, Final Feasibility Report and EIS," U.S. Army Corps of Engineers, September 1985. In this odd report, the Corps argued that the state's plan to restore the Kissimmee, "while generally beneficial for environmental concerns, do not contribute to the nation's economic development," and were therefore ineligible for federal assistance. The report never explains why it took the Corps seven years to realize this.

276 *"The national environmental groups":* Interview with Charles Lee.

276 *"The Corps didn't have a lot of credibility":* Author interview with Hatch, who is still considered the greenest Corps commander ever.

276 *Senator Graham made sure the Corps:* In its 1990 Water Resources Development Act, Congress directed the Corps to have a report on the Kissimmee ready by April 1992, an unusually fast timetable. But the Corps report claimed that the state's recommended plan, which would have filled more than half the canal, would force the removal of 350 homes and businesses along the river. Lou Toth, the Kissimmee River expert at the South Florida Water Management District, says this forced the state to scale back its plan. "Central & Southern Florida, Environmental Restoration of the Kissimmee River, Final Integrated Feasibility Report and EIS," December 1991.

277 *"We are making fair headway":* Arthur Marshall, "A Report Card," August 1983. ARMP, Box 1, Folder 5.

277 *"We have fashioned balanced bipartisan legislation":* *Congressional Record,* 11/7/1989, p. 27735.

278 *Webb was a veteran of western water wars:* Webb became friends with Joe Browder at Interior, where Browder was working as a political aide. Browder then encouraged his move into Everglades activism after Webb's wife, Mary Doyle, became the dean of the University of Miami Law School. After Webb died of brain cancer in 1997, Doyle worked at Interior under President Clinton, focusing on Everglades restoration. Interviews with Joe Browder, Mary Doyle, George Frampton, Charles Lee, Terrence Salt. Cecile Betancourt, "Glades Champion James D. Webb Dies at Age 60," *Miami Herald,* 1/3/1997.

278 *Communities, farms, and the Everglades:* "Everglades in the 21st Century," Everglades Coalition work plan, June 1992, courtesy Joe Browder.

278 *"Jim knew it was going to be tough":* Author interview with George Frampton.

279 *"Look, we're engineers":* Author interview with Terrence Salt.

16   Something in the Water

PAGE

280 *"Tragically, the ecological integrity":* Memorandum in Support of the Motion of the United States for Partial Summary Judgment on Liability, *U.S. v. South Florida Water Management District,* 11/10/1990, ELC.

280 *One day in the early 1980s:* Interview with Reed. Cattails are native to the Everglades, but they were usually only found around nutrient-rich gator holes in the natural ecosystem.

281 *Big Sugar joined Big Tobacco and Big Oil:* "Big, Bad Sugar," *St. Petersburg Times,* 10/1/1989; "Big Sugar," St. Petersburg Times, 10/1/1989; "Florida's Sugar Daddies," *Petersburg Times,* 7/26/1990; "Sugar Growers Reap Bonanza in the Glades," *Orlando Sentinel,* 9/18/1990; Alec Wilkinson, *Big Sugar* (New York: Alfred A. Knopf, 1989); McCally, *The Everglades.*

281 *Thanks to lavish campaign donations:* The sugar industry donated about $11 million to congressional candidates from 1979 to 1994, including about $3 million from sugarcane growers. The rest came from the sugar beet and corn syrup industries. "The Politics of Sugar," The Center for Responsive Politics, www.opensecrets.org.

281 *where the industry owed its existence to government support:* The Everglades has excellent nitrogen-rich soil, but it is more susceptible to freezes than tropical nations, and history showed that farming there was too risky without government flood control. The University of Florida's Institute of Food and Agricultural Services, a longtime supporter of the industry, rated south Florida only "fair" as a place to grow sugar. T. J. Schueneman, "An Overview of Florida Sugarcane," University of Florida Institute of Food and Agricultural Sciences, July 2002, http://edis.ifus.ufl.edu/sc032.

281 *Big Sugar received no direct subsidies:* The price supports are enforced through "nonrecourse loans" to processors that obligate the government to buy sugar whenever prices drop below an established "market stabilization price." The State Department's nation-by-nation quotas for foreign sugar imports—and massive U.S. tariffs on any sugar imported in addition to those quotas—ensure that U.S. sugar prices are well above the world market price. José Alvarez and Leo C. Polopolus, "The History of U.S. Sugar Protection," University of Florida Institute of Food and Agricultural Sciences, June 2002, www.edisxifus.ufl.edu/sco19; Aaron Schwabach, "How Protectionism Is Destroying the Everglades," *Environmental Law Reporter,* December 2001.

282 *"The closest most of them got to the actual crop":* Carl Hiaasen, *Strip Tease* (New York: Warner Books, 1993), p. 13.

282 *foreign-service trainees were sent there:* Alec Wilkinson, *Big Sugar,* p. 173.

282 *Marjory Stoneman Douglas wrote Governor Graham:* Marjory Stoneman Douglas letter to Governor Bob Graham, 9/27/1984, MSDP, Box 14, Folder 21. Douglas also believed that tomato farming should be banned in southern Dade County to protect the Everglades, and suggested the ban would reduce illegal immigration. "Burger King is already buying all its tomatoes from Mexico. . . . It is stupid in our country to try to be self-sufficient. Keep Mexicans in Mexico by giving them something to do. Sugar should be raised in Puerto Rico, Haiti and other West Indian islands, which would keep the refugees at home." 6/6/1985 letter, MSDP.

282 *"Those of you who may not know the power":* Florida Audubon Society fund-raising letter, 11/15/1990, courtesy Malcolm Wade.

283 *"along with all the pesticides, fertilizers, dead cats":* Douglas, *Voice of the River,* p. 228.

283 *the company's official history:* Joseph J. McGovern, "The First Fifty Years," U.S. Sugar Corporation publication, 1981. Author interview with Wade.

283 *The water management district's sugar-friendly board:* Memorandum from J. W. Dineen to D. Morgan, 6/16/1971, Memorandum in Support of the Motion of the United States for Partial Summary Judgment on Liability, 11/19/1990, *United States v. SFWMD,* ELC.

284 *The scientist who best documented this:* Affidavit of Ron Jones, *U.S. v. SFWMD,* ELC; John Dorschner, "Swamp Warrior," *Miami Herald Tropic Magazine,* 4/28/1996. Jones accepts most Darwinian science regarding the adaptation and evolution of species, but he believes that God initially created life on earth.

284 *"Cattails were the grave markers on the Everglades":* Author interview with Ron Jones.

285 *Lehtinen was another child of the swamp:* Rebecca Wakefield, "Lehtinen for Mayor," *Miami New Times,* 5/22/2003; Michael Grunwald, "Water Quality Is Longstanding Issue for the Tribe," *Washington Post,* 6/24/2002; "In the Trenches," *Florida Trend,* 7/1/2003; Von Drehle, "Bury My Heart at Southwest 392nd Terrace."

285 *"Sometimes, Dad would reject a ship":* Author interview with Dexter Lehtinen.

285 *More than three decades later, Lehtinen would spend his own money:* Dexter Lehtinen, "The Wounds That Never Heal," *Army Times* advertisement, 9/6/2004.

286 *"I never understood that":* Author interview with Dexter Lehtinen.

286 *He began meeting secretly:* The late Peter Rosendahl, a scientist at Everglades National Park who later worked for the sugar industry, secretly sent some of his phosphorus data to a group of sportsmen, who gave it to Lehtinen. A group of environmentalists, including Jim Webb, also met with Lehtinen. But the greatest influence on Lehtinen was park superintendent Michael Finley, who had tried to warn the water management district about phosphorus, and had been ignored. Author interview with Jack Moller.

286 *Lehtinen knew his new bosses:* Lisa Gibbs, "Federal Suit to Protect Everglades Bogs Down," *Legal Times,* 7/8/1991; James Hagy, "Watergate," *Florida Trend,* March 1993.

287 *Even Army Corps officials:* George E. Buker, *The Third E: A History of the Jacksonville District of the U.S. Army Corps of Engineers, 1975–1998,* U.S. Army Corps of Engineers, Office of History, p. 102. Author interviews with Dexter Lehtinen, Terrence Salt.

287 *he was hounded by internal Justice Department investigations:* Lehtinen was essentially forced out of office in 1992 after an investigation by the Justice Department's Office of Professional Responsibility into his alleged politicization of the office and retaliation against internal critics. The investigation was undoubtedly ginned up by his enemies in his own office and Main Justice, but Lehtinen did make some poor decisions that left himself vulnerable. When he finally quit, several dozen career prosecutors in his office held a party in a Miami bar, playing James Brown's "I Feel Good" over and over on the jukebox.

287 *So the state hired a New York law firm:* Gibbs, "Federal Suit to Protect Everglades Bogs Down."

288 *"The litigation thus far has failed dismally":* Ibid.

288 *"Millions of dollars have been spent":* Ibid.

288 *a lobbyist whose penchant for compromise:* Craig Pittman, "Everglades Deal Could Be Unmaking of Activist," *St. Petersburg Times,* 4/7/2002.

288 *led by one of south Florida's best-connected lawyers:* This was Jim Garner, a fixer from Fort Myers.

289 *"Because the health of the Everglades":* "Florida's Sugar Daddies," *St. Petersburg Times,* 7/26/1990.

289 *they launched a public relations blitz:* Barnaby J. Feder, "Sugar Growers Seek Cleaner Image," *New York Times,* 12/3/1991; Rosalind Resnick, "Nothing Sweet About Sugar," *Florida Trend,* March 1991; Jeff Klinkenberg, "Showdown in the Everglades," *St. Petersburg Times,* 9/27/1992.

289 *They trumpeted the findings of their own scientists:* Curtis Richardson of Duke University has been the lead researcher for the sugar industry. He has revised his initial findings, and now argues that the standard should be 15.6 parts per billion, but most of the scientific community has settled on 10. In the five years after 1988, the sugar industry spent more than $17.5 million on outside scientists. De'Ann Weimer, "Hired Science," *Palm Beach Daily Business Review,* 5/27/1994.

289 *"I don't believe it":* Von Drehle, "Bury My Heart at Southwest 392nd Terrace."

289 *"If it please the court":* Hearing Transcript, 5/21/1991, U.S. v. SFWMD, ETC; author interviews with Carol Browner, Dexter Lehtinen, Tim Searchinger.

290 *Chiles suddenly abandoned his strategy:* Carol Browner and other Chiles aides hastened to explain that the governor was not admitting guilt, and state officials continued to deny responsibility for the pollution during settlement talks. But it was hard for the army to keep fighting after the general publicly surrendered.

291   *"We are no longer going to spend millions":* Jeff Hardy, "Tentative Settlement Reached in Everglades Clean up," United Press International, 7/11/1991.

292   *"we were ready to string them up":* Author interview with Malcolm Wade.

292   *"The environmentalists wanted":* Author interview with George Wedgworth.

292   *One executive warned an Audubon leader:* Author interview with John Flicker.

292   *"Bottom line: Gridlock reigns":* James R. Hagy, "Watergate," *Florida Trend,* March 1993.

293   *And now the long-dormant southwest coast:* Michael Grunwald, "Growing Pains in Southwest Fla.," *Washington Post,* 6/25/2002.

293   *"There was no place left":* Author interview with Andy Eller.

294   *"Mother Nature is having a nightmare":* Joy Williams, "The Imaginary Everglades," *Outside,* January 1994, p. 95.

294   *In his best-seller:* Al Gore, *Earth in the Balance: Ecology and the Human Spirit* (New York: Penguin Books, 1993), p. 340. As Gore acknowledged, he had voted for the supports as a Tennessee senator, because farm-state senators tended to stick together.

294   *where her pet dog was killed:* Paul Anderson, *Janet Reno: Doing the Right Thing* (New York: John Wiley & Sons, 1994), p. 23.

294   *"Christmas has come early":* Jeff Klinkenberg, "A Great Day for the Everglades," *St. Petersburg Times,* 2/28/1993. According to the Everglades Coalition's official history, "optimism was the mood, in reaction to a new administration in Washington." Everglades Coalition Past Conferences, www.evergladescoalition.org.

295   *"the ultimate test case":* Robert McClure, "Interior Chief Vows to Help Everglades," *Sun-Sentinel,* 2/23/1993.

295   *"That's where I'm at my best":* Author interview with Bruce Babbitt.

296   *The interior secretary is not usually a link:* Author interviews with Bruce Babbitt, Ed Dickey, George Frampton.

296   *"I knew we could fight sugar":* Author interview with Bruce Babbitt.

297   *A week after the conference:* DeAnn Weimer, "How Big Sugar Got Its Everglades Deal," *Miami Daily Business Review,* 8/13/93.

297   *"The enviros were obsessed with phosphorus":* Author interview with George Frampton.

297   *At a news conference in the Interior Department auditorium:* "Statement of Principles," ELC; Everglades Litigation Collection; Tom Kenworthy, "Everglades Revival Plan Unveiled," *Washington Post,* 7/14/1993; Rita Beamish, "Sugar Growers Agree to Everglades Cleanup Plan," Associated Press, 7/13/1993; Karl Vick, "Agreement Would Clean Up Everglades," *St. Petersburg Times,* 7/14/1993; Paul Anderson and Lori Rozsa, "Glades Cleanup Pact Hailed, Reviled," *Miami Herald,* 7/14/1993.

298   *"Somebody listened to sugar":* Weimer, "How Big Sugar Got Its Everglades Deal."

298   *"It has the potential":* Vick, "Agreement Would Clean Up Everglades."

299   *Babbit was surprised and annoyed:* Babbit was especially irritated by the complaints that sugar should pay the entire bill. That just wasn't going to happen. At a news conference with Governor Chiles on March 17, George Frampton said that before starting the negotiations, he had asked a host of experts what the federal government could hope to recover from the sugar industry if it won at trial. He said the answers ranged from a low of $120 million to a high of $280 million. "That made us feel like dopes," Wade recalled in an interview. "The Campaign to Preserve the Florida Everglades and Preserve Thousands of Farm Jobs in South Florida," U.S. Sugar report, 1994.

299   *"I just thought he was a big grower":* Author interview with Bruce Babbitt.

299   *"the single most important test":* David K. Rogers, "Babbitt: Saving Glades Crucial," *St. Petersburg Times,* 1/16/1994.

299   *He did make a point of chatting with migrant workers:* A few workers held signs that read "Bobbitt Babbitt," a reference to Lorena Bobbitt, a Virginia woman who had recently gained national notoriety by slicing off her husband's penis.

299   *"We are dealing with people who have shown":* Clean Water Action memo to Joe Browder and Tom Martin, 1/27/1994, courtesy Joe Browder.

300   *But the sugar growers walked out:* There were two immediate causes of the walkout. One was a last-minute declaration by Carol Browner's EPA that it would require permits for the filter marshes. The other was a science report by the federal task force that suggested one option for restoration would be a flowway through the Everglades Agricultural Area—and included a map wiping out the town of South Bay. But many of the negotiators suspect that the industry was looking for an excuse to go to the state legislature. Author interviews with Bob Johnson, Richard Ring, Malcolm Wade.

300   *the sugar industry began spreading even more campaign cash:* "The Battle of the Everglades," Center-for Responsive Politics, www.opensecrets.org. Karl Vick, "Big Sugar: A Sweet Deal Under Fire," *St. Petersburg Times,* 5/15/1994.

300   *Barley approached Florida's best-connected law firms:* George Barley memo to Joe Browder and Tom Martin, 2/17/1994. Courtesy Joe Browder.

300   *"National Democratic Party operatives":* Joe Browder memo to Everglades Coalition leadership, 4/7/1994, courtesy Joe Browder.

301   *"Six bullets kills you":* Brian Nelson, "The History of the Taming of the Everglades," *CNN Future-Watch,* 11/5/1994.

301   *"I disapprove of it wholeheartedly":* Marjory Stoneman Douglas letter to Governor Chiles, 2/26/1994, courtesy Joe Browder.

301   *"the Clinton administration has joined with the Florida Legislature":* Everglades Coalition letter, 5/3/1994, quoted in "A Decade of Progress," U.S. Sugar Corporation, www.ussugar.com/environment.

301   *Ron Jones called it:* William Booth, "The Everglades Forever?" *Washington Post,* 5/3/1994.

301   *Joe Browder, a lifelong Democrat:* Joe Browder memo to Everglades Coalition leadership, 6/15/1994, courtesy Joe Browder.

301   *"I'm determined to make this":* George Barley letter to Joe Browder, 2/25/1994, courtesy Joe Browder.

301   *Everglades Forever was so unpopular:* John H. Cushman Jr., "Florida Adopts Bill on Everglades Pollution," *New York Times,* 5/4/1994; Kirk Brown, "Chiles Signs Everglades Act, Trades Barbs with Protesters," *Palm Beach Post,* 5/4/1994.

302   *So far, the marshes have kept more than 2,000 tons:* South Florida Water Management District reports.

302   *"I could never understand":* Author interview with Sam Poole.

302   *"We didn't take them too seriously":* Author interview with George Frampton.

303   *"Everglades Forever was a defining moment":* Author interview with Bruce Babbitt.

17   Something for Everyone

PAGE

304   *"The model is consensus":* Governor Jeb Bush speech to Everglades Coalition, January 1999, courtesy Allison DeFoor II. Some of the reporting for this chapter first appeared in my June 2002 series of four articles about Everglades restoration in the *Washington Post.*

304   *But the region was also on the verge:* "The Initial Report of the Governor's Commission on a Sustainable South Florida," October 1995, www.state.fl.us/everglades/gcssf/gcssf.html; "Imaging the Region: South Florida Via Indicators And Public Opinions," Florida Atlantic University/Florida International University Joint Center for Urban and Environmental Problems, 2001; Michael Grunwald, "Hotenfreude," *New Republic,* 3/8/2004; Grunwald, "Growing Pains in Southwest Florida."

305   *students in Broward County were lining up:* Steve Harrison, "School Crunch Turns Lunch into Brunch," *Miami Herald,* 8/25/2001.

305   *One newspaper published a cautionary vision:* Robert McClure, "When the Everglades Was Paved Over, a Vital South Florida Resource Was Destroyed," *Fort Lauderdale Sun-Sentinel,* 12/10/1995.

305   *"It is easy to see that our present course":* "The Initial Report of the Governor's Commission on a Sustainable South Florida." Author and University of Florida oral history interviews with Stuart Appelbaum, Richard Pettigrew, Terry Rice, Stuart Strahl, and Malcolm Wade. Author interviews with Maggy Hurchalla and Roy Rodgers.

305   *The region's highway mileage:* "Imaging the Region."

305   *Hurricane researchers calculated:* Author interview with Christopher Landsea; Elliot Kleinberg, "'26 Storm Would Have Caused $80 Billion in Damage Today," *Palm Beach Post,* 5/30/1999; Michael Grunwald, "Water World," *New Republic,* 3/1/2004.

307   *"This couldn't be your father's Corps of Engineers":* Author interview with Appelbaum.

307   *The atmosphere in Florida was even more toxic:* Author and University of Florida oral history interviews with Bob Graham, Nathaniel Reed, and Malcolm Wade. Author interviews with Mary Barley, Joe Browder, Robert Coker, John Flicker, Paul Tudor Jones (by e-mail), and Fowler West.

308   *"Rest Easy":* George Barley's funeral book, courtesy Mary Barley.

308   *Jones was a high-energy, ultracompetitive alpha male:* Cyril T. Zaneski, "Soured on Big Sugar, Bro-
       ker Boosts Glades," *Miami Herald,* 9/22/1996; Neil Santaniello, "Glades Crusader Has Heart,
       Cash," *South Florida Sun-Sentinel,* 9/22/1996; Lisa Schuchman, "Millionaire Bets on Everglades
       Future," *Palm Beach Post,* 9/30/1996; David Olinger, "The Savior in Question," *St. Petersburg
       Times,* 10/19/1996.

308   *He thought most environmentalists were nice people:* Paul Tudor Jones e-mail to author.

308   *"What's next: Special taxes on golf courses?":* Robert McClure, "Sugar Industry and Environmen-
       talists Lash Out in Campaign," *Sun-Sentinel,* 1/14/1996.

309   *Alfonso Fanjul was so angry:* Kenneth Starr, *The Starr Report: The Official Report of the Independent
       Counsel's Investigation of the President.* Lewinsky told the grand jury that during their February
       19 breakup, the president took a call from a sugar grower whose name was "something like
       Fanuli." White House phone records confirmed that Clinton spoke to Alfonso Fanjul from
       12:42 P.M. to 1:04 P.M.

309   *"I think it's fair to say":* Author interview with Bob Graham.

309   *Meanwhile, a new eco-war was erupting:* Author interviews with Browder, Shannon Estenoz, Alan
       Farago, Flicker, Al Gore, Graham, Tom Jensen, Barbara Lange, Charles Lee, Kathleen McGinty,
       Reed, Tim Searchinger, Brad Sewell, Stewart Strahl, and Ron Tipton. Jim Defede did a superb
       job covering the airport controversy for *Miami New Times.*

309   *Dade County awarded the group a no-bid lease:* Jim Defede, "Flying Blind: Would Your County
       Commissioners Approve a Deal Worth $500 Million Without Having Basic Information?,"
       Miami New Times, 2/1/1996.

310   *"No, I'm on the* other *side":* Author interview with Alan Farago.

310   *"when this chapter of Everglades history":* Joe Browder memorandum to Nathaniel Reed,
       2/22/1997, courtesy Joe Browder. Browder says a lobbyist for Paul Tudor Jones once warned
       him not to mention the airport before a meeting at the White House, saying it would jeopar-
       dize Jones's relationship with the Clinton administration.

311   *After listening to Audubon's ever-pragmatic Charles Lee:* The intern was Kyle Lonergan; he was whis-
       pering to Barbara Lange of Friends of the Everglades.

311   *both parties jumping back on the bandwagon:* Author interviews with Babbitt, Ed Barron, Mitchell
       Berger, Peter Deutsch, Mark Foley, George Frampton, Gore, Jensen, C.K. Lee, McGinty, Clay
       Shaw, and West.

311   *his staff was already pulling:* Tom Jensen coordinated much of this work for Gore during the
       1995 government shutdown, working with George Frampton at Interior and Michael Davis
       at the Pentagon.

311   *He also pledged to purchase at least 100,000 acres:* The coalition had demanded 100,000 acres for
       water storage, in addition to the 40,000 acres that were already reserved for water quality fil-
       ter marshes. The administration later claimed that Gore had only meant 100,000 acres total.
       But its press release for the Gore plan refutes that, calling for "at least 100,000 acres of land
       in Everglades Agricultural Area for water storage." The Clinton/Gore Administration's Ever-
       glades Restoration Plan, Principles and Elements, 2/19/1996.

311   *Speaker Gingrich soon embraced the Everglades:* One politician who brought the Everglades to Gin-
       grich's attention was a smart but troubled Miami-Dade County commissioner named Arthur
       Teele Jr., a black Republican with a strong environmental record. George Barley and Joe Brow-
       der first piqued Teele's interest in the Everglades, and Vice President Gore once asked Teele
       to show Gingrich some polling data about the popularity of the Everglades. In 2005, Teele
       was indicted for corruption, and committed suicide in the lobby of the *Miami Herald* building.

311   *"Newt told me: 'This is great politics!'":* Author interview with Foley.

312   *his commission approved a conceptual plan:* "The Conceptual Plan of the Governor's Commission
       for a Sustainable South Florida," August 1996.

312   *"We had to come up with something for everyone":* Michael Grunwald, "A Rescue Plan, Bold and
       Uncertain," *Washington Post,* 6/23/2002.

312   *"When we . . . preserve places like the Everglades":* Transcript of President Clinton's Weekly Radio
       Address, 10/12/1996, White House Press Office.

312   *"This is a great issue!":* Author interviews with Stuart Strahl and Ron Tipton.

312   *But Congress did approve Dole's $200 million gift:* The money was included in the Freedom to
       Farm Bill, which also extended the federal sugar program for another seven years. The bill
       was also supposed to wean major commodities farmers off federal assistance, but it turned into
       a spectacular agricultural bonanza.

313   *"The Everglades is one of the greatest environmental treasures":* Author interview with Al Gore.

313   *The Corps was a behemoth:* Missions of the U.S. Army Corps of Engineers, www.usace.army.mil; Michael Grunwald, "Generals Push Huge Growth for Engineers," *Washington Post,* 2/24/2000; Grunwald, "As Corps Widens Reach, a Cleanup Turns Messy," *Washington Post,* 5/22/2000; Grunwald, "An Agency of Unchecked Clout," *Washington Post,* 9/10/2000; Grunwald, "Working to Please Hill Commanders," *Washington Post,* 9/11/2000.

313   *"The Corps has nothing going on":* Buker, *The Third E,* p. 113.

313   *subsequent investigations would reveal:* The Pentagon inspector general, the General Accounting Office, and the National Academy of Sciences all produced reports blasting the Corps. And I was destroying the nation's forests for an overlong series on the Corps in the *Post.* Grunwald, "How Corps Turned Doubt Into a Lock," *Washington Post,* 2/13/2000; Grunwald, "A Race to the Bottom," *Washington Post,* 9/12/2000; Grunwald, "Snake River Dams: A Battle Over Values," *Washington Post,* 9/12/2000; Grunwald, "Pentagon Rebukes Army Corps," *Washington Post,* 12/7/2000; Grunwald, "Army Corps Suspends Del. River Dredging Project," *Washington Post,* 4/24/2002.

314   *Exhibit A for the skeptics:* Author interviews with Billy Cypress, Bob Johnson, Dexter Lehtinen, Joette Lorion, Stuart Pimm, Terry Rice, and Brad Sewell.

314   *"Glades Plan Turning into River of Morass":* Cyral T. Zaneski, *Miami Herald,* 5/31/1999.

314   *"They think they are fighting a holy war":* Michael Grunwald, "An Environmental Reversal of Fortune," *Washington Post,* 6/26/2002.

314   *The bureaucratic sniping soon devolved:* To participants on just about every side of the Everglades wars, the Cape Sable seaside sparrow is a symbol of everything wrong with Everglades water management. They just disagree about why. Someday, someone will write a book about the sparrow; for this book, it's enough to say that the sparrow lives in the park, that it's in big trouble, and that it's a very complex problem.

315   *One congressman complained:* The congressman was James Hansen of Utah, a property rights ideologue who stacked his hearing with critics of the park. The only environmentalist invited to testify was Joette Lorion, who now represented the Miccosukee Indians, and had publicly resigned from the environmental community because she disagreed with its efforts to buy out the Eight-and-a-Half-Square-Mile Area. William E. Gibson, "Glades Buyout Plan Dealt a Blow," *Fort Lauderdale Sun-Sentinel.* 4/28/1999.

315   *and invited thirty other agencies:* Most environmentalists acknowledge that the Everglades Coalition simply dropped the ball on the Restudy. It was distracted by the penny-a-pound and airport fights, and by a number of battles in the state legislature, and the restoration project began taking shape before most environmentalists knew what was going on. There were a few environmentalists on the Governor's Commission, but they had an uphill battle to fight to put teeth into the plan, and some of them seemed a bit intoxicated by the notion of consensus.

315   *In October, the team submitted:* U.S. Army Corps of Engineers, Central and Southern Florida Project Comprehensive Review Study, "Draft Integrated Feasibility Report and Programmatic Environmental Impact Statement," Jacksonville, FL, October 1998; U.S. Army Corps of Engineers, Central and Southern Florida Project Comprehensive Review Study, "Final Integrated Feasibility Report and Programmatic Environmental Impact Statement," April 1999. Author and University of Florida oral history interviews with Stuart Appelbaum, Don Carson, Michael Collins, Michael Davis, Tom MacVicar, John Ogden, Terry Rice, Richard Ring, Terrence Salt, Stuart Strahl, Malcolm Wade, and George Wedgworth. Author interviews with Henry Dean, Bob Johnson, Greg May, Christopher McVoy, Michael Ornella, Richard Punnett, Nanciann Regalado, Carol Sanders, and Tom Van Lent.

316   *Vice President Gore unveiled the plan:* Craig Pittman, "Gore Unrolls New Blueprint for the Everglades," *St. Petersburg Times,* 10/14/1998.

316   *the Comprehensive Everglades Restoration Plan:* U.S. Army Corps of Engineers: "Rescuing an Endangered Ecosystem: The Plan to Restore America's Everglades," 7/1/1999; "Final Integrated Feasibility Report and Programmatic Environmental Impact Statement." Michael Grunwald, "In Everglades, a Chance for Redemption," *Washington Post,* 9/14/2000; Grunwald, "A Rescue Plan, Bold and Uncertain," *Washington Post,* 6/24/2002.

317   *"a Disney Everglades":* Author interview with Punnett.

317   *"The entire south Florida ecosystem":* "Rescuing an Endangered Ecosystem," p. 14.

319   *Even the inventor of the models:* Author interview with Tom MacVicar.

319   *"There are unique and significant uncertainties":* "Final Integrated Feasibility Report and Programmatic Environmental Impact Statement," Section O, p. 6.

319   *"The ways in which this ecosystem will respond":* "Final Integrated Feasibility Report," p. xiii.

319 *The cochair of the science team:* Author interview with Ronnie Best.
319 *its plan to convert mined-out limestone pits:* Michael Grunwald, "Between a Rock and a Hard Place," *Washington Post,* 6/24/2002.
320 *"Maybe this plan is premature":* Grunwald, "A Rescue Plan, Bold and Uncertain."
320 *On December 31, 1998:* Park superintendent Richard Ring was out of town for the holidays, so Robert Johnson, the head of the park's science staff, brought the comments to deputy superintendent Lawrence Belli to sign. "Is this controversial?" Belli asked. "Everything we do is controversial," Johnson replied. Belli signed. Ring was furious when he returned to work, but he stood by the comments.
320 *"does not represent a restoration scenario":* Everglades National Park, "Comments to U.S. Army Corps of Engineers," 12/31/1998, p. 20. Author interviews with Johnson, Stuart Pimm, Ring, Tim Searchinger, and Van Lent.
320 *"There is insufficient evidence":* Ibid., p. 20.
321 *"it is difficult to identify":* Ibid., p. 22.
321 *"The Corps gave the cities and the ag guys":* Author interview with Bob Johnson.
323 *"reestablish the natural sheet flow through the Everglades":* "Rescuing an Endangered Ecosystem," p. 11.
323 *"largely retains the fragmented management":* Everglades National Park, "Comments to U.S. Army Corps of Engineers," p. 20.
324 *"There are serious failings in the plans":* Stuart Pimm letter to Bruce Babbitt, 1/28/1999. Pimm's co-authors were Edward O. Wilson of Harvard University, Paul Ehrlich of Stanford University, Gary Meffe of the University of Florida, Gordon Orions of the University of Washington, and Peter Raven of the Missouri Botanical Garden. Author interview with Pimm.
324 *"It's not that there are gaping holes in this plan":* Cyril T. Zaneski, "Big Ecological Guns Fault Plan for Everglades," *Miami Herald,* 1/30/1999.
324 *The Everglades Coalition was again divided:* The Nature Conservancy, National Wildlife Federation, and National Parks Conservation Association were generally inclined to support Audubon and the administration. Author interviews with Joe Browder, Shannon Estenoz, Alan Farago, John Flicker, Richard Grosso, David Guggenheim, Barbara Lange, Charles Lee, Tim Searchinger, Brad Sewell, Stuart Strahl, Ron Tipton, and Mark Van Putten.
324 *"I urge everyone to be very careful":* Tom Adams e-mail, 3/8/1999.
324 *"cries from the fringe":* Zaneski, "Big Ecological Guns Fault Plans for the Everglades."
324 *Audubon had more scientists and staff :* The Washington leadership of the National Wildlife Federation, National Parks Conservation Association, the Nature Conservancy, and World Wildlife Fund tended to support the administration.
324 *he believed the coalition would only marginalize itself:* Author interview with Strahl.
325 *"Regardless of my concerns":* Paul Tudor Jones, e-mail to author.
326 *"We didn't think it was so radical":* Author interview with Tim Searchinger.
326 *"It matters not at all who gets credit for this":* Pimm briefing paper for meeting with Babbitt, 2/22/1999.
327 *The Restudy team was sent back to work:* The team also revamped its implementation schedule to try to provide much greater environmental benefits within the first decade, accelerating the entire project so that almost everything would be done within twenty years. Several Audubon officials proclaimed that the plan was now ready. But skeptics such as Tim Searchinger argued that there was still little environmental progress in the first $4 billion worth of projects, and that there was no reason to expect Congress to follow the newly accelerated timetable.
327 *"a series of improvements":* "Draft Summary of the 2010 and 2015 Case Studies," 6/4/1999, p. 5. Reprinted at www.everglades.org.
327 *"There is one more change":* Colonel Alfred Foxx e-mail to Gary Hardesty, 6/17/1999.
327 *"Even though I understand":* Michael Ornella e-mail to Michael Magley, 6/11/1999.
327 *it promised that 80 percent:* "Rescuing an Endangered Ecosystem."
328 *"Let's get it done!"* Larry Lipman, "Gore Urges $7.8 Billion for Everglades," *Palm Beach Post,* 7/2/1999.
328 *Shuster's aides began calling the bill:* Author interview with Michael Strachn.
328 *and that consensus began to unravel:* "Stakeholder Concerns with the Chief of Engineers Report," presentation to Governor's Commission for the Everglades, 2/3/2000.
329 *"back-room, closed-door, secret deals":* "Statement of Dexter Lehtinen Regarding Back-Room Secret Deals on the Everglades," 9/23/1999, courtesy Joette Lorion.

329 *Sugar growers secretly financed an anti-CERP campaign:* Dan Morgan, "Think Tanks: Corporations' Quiet Weapon," *Washington Post,* 1/29/2000; Glenn Spencer, "The Final Integrated Feasibility Report on the Everglades Restudy: Awash in Uncertainty," Citizens for a Sound Economy policy paper, 5/5/1999.

329 *after a stormy meeting within the Republican caucus:* Senator Frank Murkowski of Alaska challenged Majority Leader Trent Lott, asking whether Smith had been promised the chairmanship to return to the party. Lott said no. Smith's and Lott's aides confirm that Smith had already been talking to Lott about returning before Chafee's death, although the chance for an open job obviously hastened his return. Smith won a 5–4 vote of the Republicans on the committee, and Inhofe decided not to challenge him within the full caucus. Incidentally, the swing vote was Christopher Bond of Missouri, who turned against Smith and endorsed his GOP primary opponent in 2002. Author's interviews with John Czwartacki, James Inhofe, Bob Smith, George Voinovich.

330 *"Nobody was looking out for the taxpayers":* Author interview with George Voinovich.

330 *Jeb Bush was a conservative policy wonk:* Ellen Debenport, "Jeb Bush: Gladiator for Change," *St. Petersburg Times,* 5/27/1994; Ellen Debenport, "The Bush Brothers Aren't Twins," *St. Petersburg Times,* 10/2/1994; Peter Wallsten, "Two Paths, One Prize," *St. Petersburg Times,* 10/18/1998; Mark Leibovich, "The Patience of Jeb," *Washington Post,* 2/23/2003. Author interviews with Michael Collins, Allison DeFoor, and David Struhs.

330 *Bush told the Everglades coalition:* Bush speech to Everglades Coalition, January 1999.

330 *He also infuriated environmentalists:* Lori Rozsa, "Buyout of Land in Glades Cancelled," *Miami Herald,* 6/24/1999.

331 *On September 21, 1999:* Michael Grunwald, "How Enron Sought to Tap the Everglades, *Washington Post,* 2/8/2002; David Fleshler and Neil Santaniello, "Glades Offered Financial Help, But Some Fear Firm May Want Rights to Water," *Sun-Sentinel,* 11/11/1999; "Money-for-Water Plan Good for Fla., Company Says," *Palm Beach Post,* 11/12/1999; "Liquid Assets: Enron's Dip into Water Business Highlights Pitfalls of Privatization," Public Citizen policy report, March 2002.

331 *Accompanied by one of Florida's top Republican fixers:* That was Jim Garner, the Fort Myers attorney and former chairman of the water management district's board. Cathy Vogel, a former spokeswoman for the district, had written the privatization paper. Azurix's CEO was Rebecca Mark, later one of the key players in the Enron debacle. She was also joined by John "Woody" Wodraska, a former water management district executive director who was now a top Azurix executive.

331 *"It shows some outside-the-box thinking":* Brian Yablonski memo to Allison DeFoor, 11/12/1999. This memo was scribbled on a copy of a news article: Robert King, "Money-for-Water Plan Good for Fla., Company Says," *Palm Beach Post,* 11/12/1999.

331 *"I want to be perfectly clear on this":* DeFoor replied to Yablonski on the same article.

18   Endgame

I am particularly indebted to my sources for this chapter, because many of them patiently granted me hours of highly detailed interviews and thousands of documents at a time when I thought much more of this book would be about contemporary Everglades battles. Many of their stories and insights did not make it into the text, and I feel guilty about that. I can only say that I learned more about how Congress works from my interviews with staffers such as John Czwartacki, Stephanie Daigle, Jo-Ellen Darcy, Tom Gibson, Ben Grumbles, C. K. Lee, Chelsea Maxwell, Catherine Ransom, Michael Strachn, and Richard Worthington than I learned during my short and ignominious stint covering Congress for *The Washington Post.* Senators Bob Graham, Slade Gorton, James Inhofe, Connie Mack, Robert Smith and George Voinovich also took time to talk to me about the Everglades, as did Congressmen Peter Deutsch, Norm Dicks, Mark Foley, Porter Goss, Ralph Regula, and Clay Shaw. Former Vice President Al Gore granted me an hourlong interview and did not seem to mind my questions about events in 2000 that must have brought back awful memories; from the Clinton administration, chief of staff John Podesta, Interior Secretary Bruce Babbitt, EPA administrator Carol Browner, Council on Environmental Quality chairman George Frampton and his predecessor Kathleen McGinty were also helpful, as were Patricia Beneke, Michael Davis, Mary Doyle, General Robert Flowers, Don Jodrey, Bill Leary, Richard Ring, Terrence Salt, Peter Umhofer, and Joseph Westphal. From the Everglades Coalition, I am especially indebted to Tom Adams,

Mary Barley, Joe Browder, Shannon Estenoz, Alan Farago, John Flicker, Richard Grosso, David Guggenheim, Paul Tudor Jones, Larry Kast, Barbara Lange, Charles Lee, Nathaniel Reed, Brad Sewell, Tim Searchinger, Stuart Strahl, and Fowler West. From the State of Florida, I am grateful to Ernie Barnett, Michael Collins, Kathy Copeland, Henry Dean, Allison DeFoor II, and David Struhs. I was also assisted by Mitchell Berger, Chairman Billy Cypress, Robert Dawson, Ed Dickey, Dexter Lehtinen, Joette Lorion, Tom MacVicar, Fred Rapach, Terry Rice, Malcolm Wade, and George Wedgworth.

PAGE

333   *"We view this as the most important year":* "Everglades 2000: A Time to Act," Everglades Coalition conference agenda, 1/7/2000.

333   *"Action taken to restore the Everglades":* Ibid.

334   *the Sierra Club attacked him:* Sean Scully, "GOP Grants Chastened Smith Gavel of Environmental Panel," *Washington Times,* 11/3/1999; Colleen Luccioli, "Smith Named Chair of Senate Environment Committee," *Environment and Energy Daily,* 11/8/1999.

334   *"John Chafee was strongly committed":* Statements of Robert Smith, Dexter Lehtinen, Malcolm Wade, Nathaniel Reed, David Struhs, and Geroge Voinovich; Senate Committee on Environment and Public Works, Everglades field hearing, 1/7/2000.

335   *"Both parties are sticking":* Robert P. King, "Everglades a Prize, Pawn in Presidential Race," *Palm Beach Post,* 1/29/2000.

335   *The good feelings only went so far:* The orchid thief was Robert Coker. When I told Jones a sugar executive had taken one of his plants, he said he wasn't surprised: "They have been stealing from the people of Florida, the migrant workers, and the natural resources of our state since Castro's rise made them sugar daddies."

335   *Before the Naples conference ended:* Cyril T. Zaneski, "Babbitt Opposes Airport in S. Dade," *Miami Herald,* 1/8/2000. Babbitt helped to gin up the primary alternative to the airport when he secretly encouraged the Collier family to submit a mixed-use proposal. He had worked with the Colliers when he was governor of Arizona, facilitating a land swap in which the Colliers gave the federal government land for the Big Cypress expansion in exchange for a valuable parcel in downtown Phoenix. Now he hoped to kill two birds with one stone: The Air Force could give the air base to Interior, which would trade it to the Colliers in exchange for the family's oil rights in Big Cypress.

335   *Dade County's backroom deals:* Farago quit the board of the Tropical Audubon Society after its leaders insisted on printing a pro-airport editorial in its newsletter alongside Farago's anti-airport editorial.

336   *Vice President Gore's advisers:* Tony Coelho was the Gore adviser most committed to the dream of winning Cuban-American support.

336   *in fact, rumors were flying:* Interviews with Alan Farago, Gus Garcia. In an October 30, 2000, e-mail, Farago referred to rumors that the Cuban American National Foundation "tried to make a deal with Clinton that they would deliver the kid if the Administration would guarantee Homestead would be transferred as an airport."

336   *The developers were paying:* Shawn Zeller, "Where Lobbyists Outnumber the Gators," *The National Journal,* 10/28/2000.

336   *But Farago noticed that Dade County's flight plans:* In 1997, Farago watched Dade County officials assure the club's leadership that the flight plans for the new airport would avoid Ocean Reef. Farago then handed out copies of the flight plans the county had filed with the Federal Aviation Administration—directly over Ocean Reef.

336   *Ocean Reef's residents ultimately decided:* The turning point was a county hearing on the airport, when one hundred residents arrived early in chartered buses and dutifully entered their names on the speakers' list. Then they had to wait for hours while dozens of politicians made windy speeches about the desperate need for the airport and the selfishness of its opponents. When a few residents were finally allowed to speak, they were heckled and booed. The hearing was clearly a sham, and the buses that headed back to Ocean Reef that night carried some angry snowbirds. The community soon passed a tax assessment to fight the airport. Jim DeFede, "Who Owns HABDI?" *Miami New Times,* 11/23/2000.

337   *The usually mild-mannered Senator Mack yelled at him:* Author interviews with Connie Mack and George Voinovich.

337   *In February, they threatened:* Amy Driscoll, "EPA chief cancels S. Florida rally," *Miami Herald,* 2/25/2000; Don Chinquina e-mail to Joe Browder, 2/24/2000.

337   *"Al Gore spilled blood for these people":* Author interview with Mitchell Berger.

338   *He wasn't convinced that the airport:* In a 2005 interview, Gore sounded like he still thought the airport plan made sense. When asked about Senator Graham's proposal of a "green airport," roundly denounced by environmentalists as a Trojan Horse, Gore replied: "Yes! Exactly!"

338   *But it soon became clear:* "Every stakeholder in Florida other than the environmentalists has concerns about referencing the Chief's Report," Senator Smith's staff noted. Document courtesy Stephanie Daigle.

338   *"They wrote us a letter":* The Mack aide was C. K. Lee.

339   *Senators Mack and Graham wrote a letter:* Bob Graham and Connie Mack letter to Lt. General Joe N. Ballard, 7/30/1999; Ballard letter to Graham and Mack, 9/27/1999; Graham and Mack letter to assistant Army secretary Joseph Westphal, 11/9/1999; Westphal letter to Graham and Mack, 1/24/2000. The administration tried to claim that it had never intended to guarantee the 79 billion gallons—actually 245,000 acre-feet—but that just wasn't true.

339   *They had such a common vision:* The staff of Dawson & Associates included five former leaders of the Corps, three former chairmen of committees that fund the Corps and a slew of former regulators. Michael Grunwald, "Growing Pains in Southwest Fla.," *Washington Post,* 6/25/2002.

340   *Dexter Lehtinen warned:* Cyril T. Zaneski, "Glades Funding Faces Deadline," *Miami Herald,* 5/30/2000. The Miccosukee Tribe has been a stalwart defender of Everglades water quality, but it has often tangled with the Everglades Coalition and Everglades National Park over water flows. Some environmentalists blame the tribe's white advisers—Lehtinen, biologist Ron Jones, and former Corps colonel Terry Rice, who share a history of antagonism with Park officials—but Miccosukee leaders say that's patronizing. It's a sad situation, because despite the bad blood, the tribe and the park would seem to share a common interest in restoring natural flows through the Everglades.

340   *anything less, he told Congress:* Senate Environment and Public Works Hearing, 5/11/2000.

340   *He just wanted to pass a bill:* During his years out of government, Frampton had served as Gore's private attorney, and he returned to the White House to be Gore's environmental wingman. But once he got there, Gore disappeared on the campaign trail, and Frampton found himself working for Clinton instead. Much to Frampton's surprise, Clinton developed a keen interest in the environment during his last years in office.

340   *The only administration official:* Babbitt initially told his aides: "Guarantee the Water. The rest is just plumbing." But it soon became clear the water would not be guaranteed.

340   *Babbitt faxed a heated letter:* Bruce Babbitt letter to George Frampton, 3/14/2000. Babbitt and Frampton had a competitive relationship, and Interior sources believe Babbitt informed Frampton on a Friday that the letter was coming on Monday so that his former aide could stew about it all weekend. The White House asked Interior officials to destroy all copies of the letter, but one copy is in the author's possession.

341   *Audubon issued one statement:* National Audubon Society press release, Statement of Daniel P. Beard on the Smith-Graham-Mack Everglades Restoration Bill, 6/27/2000.

341   *our feeling was:* Author interview with John Flicker.

341   *"I was focused like a laser beam":* Author interview with Larry Kass.

342   *But as Baucus listened to testimony:* Senate Environment and Public Works hearing, 5/11/2000.

342   *An aide to the senator kept passing him notes:* Jo-Ellen Darcy, the longtime committee aide to Senator Baucus, says his off-the-reservation anti-Everglades speech was the worst moment of her Hill career. She chased her boss down a corridor after the hearing, asking: "What the hell was that?"

342   *But the state had only used that power once:* The one time was a reservation by the St. Johns Water Management District for Payne's Prairie. Henry Dean, the director of the St. Johns district at the time, later took over the South Florida Water Management District. But Interior officials and environmentalists such as Shannon Estenoz persuaded the Senate staff that the reservations were a "rusty tool."

343   *After months of roller-coaster negotiations:* Bob Dawson wrote a letter to the Senate declaring that the draft version of CERP did not reflect the Governor's Commission consensus, in part because it provided guarantees for existing water users but not future water users. But Richard Pettigrew, the former chairman of the commission, wrote a response—drafted by Audubon's Tom Adams—that the opposite was true. Dawson apologized, and Florida newspapers wrote editorials trashing him, but the brouhaha helped break the logjam, and the sugar industry and other economic interests got most of what Dawson had asked for. Jennifer Sergent, "Alliance Opposes Plan for Glades," *Stuart News,* 8/17/2000. "Hold Off Late Ambush of Everglades Project," *Palm Beach Post,* 9/4/2000.

343   *In early September:* C. K. Lee, Senator Mack's staff, thought the deal was done, but he didn't
       like the tone of Bush aide Leslie Palmer's voice, so he called back to make sure. She said the
       governor still wanted a better deal in the House, because the Senate deal didn't serve his con-
       stituents. "What constituents?" Lee asked. "The senator has the same constituents!" But Lee
       and Mack say the call to Bush cleared things up immediately. "I can't disagree with anything
       you said," Bush told Mack.

343   *Graham and Mack fought off substantive measures:* The senators actually tried to substitute an
       amendment calling on the administration to hurry up and approve an airport, but that didn't
       fly, either.

343   *Robert Johnson, the head of Everglades National Park's science staff:* Michael Grunwald, "In Ever-
       glades, a Chance for Redemption." *Washington Post,* 9/14/2000.

344   *"Of course, we should all live long":* Stuart Pimm letter to Mark Van Putten, 9/15/2000.

344   *"This is an historic agreement":* National Audubon Society press release, statement of Stuart Strahl
       on Compromise Agreement on S2797, 9/6/2000.

345   *"If we can agree to support the Everglades":* Author interviews with Tom Adams and Bob Dawson.

345   *When Senator Inhofe tried to persuade:* Author interview with James Inhofe.

345   *After a Corps economist blew the whistle:* His name was Don Sweeney, and he had led a $60 mil-
       lion study of the project until he concluded it was a boondoggle. His bosses then removed
       him from the study but continued to copy him on e-mails directing his successor to cook the
       books. Grunwald, "How Corps Turned Doubt Into a Lock."

345   *"In 1983, restoring the natural health":* *Congressional Record,* 9/25/2000, p. S9144.

345   *"All of a sudden, we come along":* *Congressional Record,* 9/21/2000, p. S8916.

346   *"This arrangement may not be perfect":* Ibid., p. S8925.

346   *"If you have any doubts":* *Congressional Record,* 9/25/2000, p. 9147.

346   *But in the House, only one issue mattered:* Several aides to Speaker Hastert spoke on condition of
       anonymity. They all agreed that "it was all about Clay Shaw."

347   *Kathleen McGinty, Gore's top environmental adviser:* Jim DeFede, "Who Owns HABDI?" *Miami
       New Times,* 11/23/2000.

347   *"Tell them to go fuck themselves":* Author interviews with Joe Browder, Donna Brazile, Norris
       McDonald. Brazile did not recall using those words words; she said her job was "ass-kissing,
       not ass-kicking." But McDonald said that's the message he received from her office.

347   *Joe Browder, a lifelong Democrat:* Joe Browder e-mail to Blanca Mesa et al., 6/8/2000, courtesy
       Joe Browder.

347   *he had commissioned a Democratic pollster:* The pollster was Mark Mellman. Carl Pope, the exec-
       utive director of the Sierra Club, gave the results to the Gore campaign.

348   *"Until the Administration and in particular":* Nathaniel Reed e-mail, 8/8/2000.

348   *"Is there any trust in this room?"* Author interviews with Mitchell Berger, Kathleen McGinty, Sam
       Poole, Karsten Rist.

348   *"Tell him that only a true friend":* The activist Don Chinquina of Tropical Audubon. DeFede,
       "Who Owns HABDI?"

349   *"This is bullshit!":* Author interviews with Robert Smith and four Republican congressional
       aides who requested anonymity.

349   *"Control of the House is in Bob Smith's hands":* Tom Gibson e-mail to Steve Ellis, 10/2000.

349   *Congress finally passed Altoonaglades:* At the last minute, the bill was almost set aside before the
       Senate could vote on it. Democratic Senator Blanche Lincoln wanted the bill to authorize the
       Delta Regional Council, a economic development program for the Mississippi Delta region,
       and she thought she had written Senator Smith to request the provision. When she never
       heard back, she put a hold on the entire bill. "Nice try, Bob," Senator Lott told Smith on the
       Senate floor. "We'll do the Everglades next year." But Smith refused to back down. The Sen-
       ate recessed for an hour, and Smith's staff figured out thaat Lincoln had sent her request to the
       wrong Senator. The bill was back on track.

350   *"waffling as usual":* Mildrade Cherfiles, "Nader Makes Campaign Stop in Miami," Associated
       Press, 11/4/2000; Sean Cavanaugh, "1,000 Attend Nader's Rally in S. Florida," *South Florida
       Sun-Sentinel,* 11/5/2000.

350   *"There are no airports":* DeFede, "Who Owns HABDI?"

350   *"Ninety-nine and forty-four-hundredths percent pure":* Author interview with Gore.

350   *"Oh, I don't thnk the airport was a major factor":* Ibid.

350   *Senator Graham woke up in Miami Lakes:* Senator Graham graciously provided me with his note-
       book from December 11, even though he knows that nothing good happens when he gives
       up his notebooks.

352 *"You've got to watch those guys":* Author interview with Michael Davis and Robert Smith.

352 *President Clinton, shooting the breeze:* The president was talking to C. K. Lee and Catherine Ransom.

352 *"I'd be happy to speculate about the Supreme Court!":* Federal News Service transcript of White House Driveway Stakeout, 12/11/2000.

352 *"We worked together to save a national treasure":* Ibid.

352 *"Marvin? He's doing well":* Ibid.

353 *South Florida was suffering through in one of its worst droughts:* Michael Grunwald, "A Rescue Plan, Bold and Uncertain," *Washington Post,* 6/22/2002; E-mails courtesy Paul Gray.

354 *In fact, Graham's notes reveal:* He watched MSNBC from 2:10 until 2:25. From 5:55 until 6:10, he bought the milk and picked up three shirts. His notes from 4:25 read: "Jorge Mas re: HAFB." He later dressed for bed, read Roll Call, and then fell asleep at 2:25 a.m. Notes courtesy Bob Graham.

354 *"This is a victory for common sense":* The quotation is from Brad Sewell of NRDC. "President Clinton, Air Force Reject Everglades Airport," 1/16/2001 press release. Courtesy Alan Farago.

Epilogue   The Future of the Everglades

PAGE

357 *The evidence is just above Lake Okeechobee:* Lou Toth showed me the restored Kissimmee River. While wading in the broadleaf marshes that had recolonized the basin's cattle pastures just a year after the restoration started, I stupidly asked him what the floodplain would look like in ten years. "Like this!" he replied. Michael Grunwald, "An Environmental Reversal of Fortune," *Washington Post,* 6/26/2002; "Kissimmee River Restoration," South Florida Water Management District pamphlet.

358 *"They just don't get it":* Grunwald, "An Environmental Reversal of Fortune."

358 *The Shark Slough restoration:* The warring agencies and politicians have finally reached a compromise on the Eight-and-a-Half Square Mile Area, agreeing to buy out one-third of the community and build a levee to protect the rest, but at press time they were still fighting over money and design.

358 *By 2003, the Everglades Forever Act:* Michael Grunwald, "Sugar Plum," *New Republic,* 5/12/2003. Robert P. King, "Cattails Spur Everglades Debate," *Palm Beach Post,* 8/26/2003.

359 *but with obvious enforcement loopholes:* The bill established new criteria for measuring phosphorus runoff that will not necessarily require a 10 ppb standard at the point of release. It also relaxed the enforcement standards, essentially excusing the state from penalties as long as it uses the best available cleanup technology. But Judge Moreno has suggested that he will insist on performance, as opposed to good-faith efforts.

359 *Officials at Justice, Interior, and EPA:* In not-for-attribution interviews in 2003, federal officials described Everglades Whenever to me as "shameful," "sinful," and worse. But they could not give their names. President Bush's instructions were: Don't embarrass my brother.

359 *"Score one for Big Sugar":* Lesley Clark, "Glades Cleanup Setback Predicted," *Miami Herald,* 9/24/2003.

359 *But in June 2005:* Judge Federico Moreno, "Order Requiring Special Master to Hold a Hearing on the Issue of Remedies," 6/1/2005.

359 *"It's different from what we told Congress":* Gary Hardesty, "5-Yr. Report to Congress, HQUSACE Guidance," internal Army Corps of Engineers memorandum, 3/7/2005.

360 *"we haven't built a single project":* Ibid.

360 *One former Corps leader has predicted:* The leader was Michael Parker, the assistant Army secretary overseeing the Corps until President Bush fired him for complaining publicly about his budget. Grunwald, "A Rescue Plan, Bold and Uncertain."

360 *But he has also practically seized control:* Michael Grunwald, "Fla. Steps in to Speed up State-Federal Everglades Cleanup," *Washington Post,* 10/13/2004.

361 *a water management district scientist named Christopher McVoy:* Robert P. King, *Palm Beach Post,* "Is Everglades' Flowing Past the Key to Its Future?" 4/30/2002.

361 *even Nathaniel Reed:* Daniel Cusick, "Stakeholders Worry Restoration Is In Jeopardy," Greenwire, 3/25/2005.

361 *"I get so depressed":* Author interview with Reed.

362 *Environmentalists sued:* "R.I.P. Lake Okeechobee?" *Palm Beach Post,* 6/12/2005.

362 *And the lake's coffee hue:* Suzanne Wentley, "Lake Okeechobee a Mess After Wilma," *Stuart News,* 11/3/2005.

362 *So are the red tides that have increased 1,500 percent:* The scientist Larry Brand determined this figure.

362 *The Collier family has retained the right:* In 2002, President Bush announced a $120 million plan to buy out the Collier oil rights in Big Cypress. But a 2005 study by the Department of Interior's inspector general found that the price was wildly inflated, and the deal was put on hold.

363 *Meanwhile, new attention is being paid:* The National Academy of Sciences warned in 2002 that CERP might send more nitrogen into Florida Bay, and that restoration could make the bay cloudier. "Florida Bay Research Programs and Their Relationship to the Comprehensive Everglades Restoration Plan," Committee on Restoration of the Greater Everglades Ecosystem, August 2002.

363 *Postwar Everglades suburbs:* In 2003, Weston and Miramar were the sixth- and seventh-fastest-growing cities in America. Noah Bierman and Tim Henderson, "South Florida's Sprawl Quickly Nearing Limit; Western Cities Near Build-Out," *Miami Herald,* 7/10/2003.

363 *Southeast Florida's office sprawl:* Robert E. Lang, "Beyond Edge City: Office Sprawl in South Florida," The Brookings Institution Survey Series, March 2003.

364 *"We are permitting in SW Florida":* E-mail from Richard Harvey to John Hall; Grunwald, "Growing Pains in SW Fla."

364 *recently retired WCI Communities CEO Al Hoffman:* I interviewed Hoffman in 2002; on his office wall was a note from President Bush: "You are the man!" Hoffman described the Florida panther as a "bastardized species," complained about regulators "who think the world will end if they can't protect that little tree," and predicted that development would continue to plow into the Everglades "as sure as the sun is coming up tomorrow."

365 *Even Governor Bush's developer-friendly Growth Management Commission:* "A Liveable Florida for Today and Tomorrow," Florida Growth Management Study Commission, February 2001.

365 *"The developers are very, very powerful":* Author interview with Mel Martinez.

365 *the reader reaction was furious:* Daniel de Vise, "Broward has had enough, readers say," *Miami Herald,* 12/30/2001. The letters quoted were from Ray McLeery, Heather Hack, Coleen Werner, and John Hoover.

367 *"It is parched today":* Author interview with Azzam Alwash.

367 *The World Bank has cited CERP:* The World Bank considered making ecosystem restoration its top environmental focus in 2003, but officials decided that they did not want to give developing countries incentives to destroy nature and fix it later. Author interview with Robert Watson.

368 *"This will be modeled":* Nelson Hernandez, "Governors' Bay Strategy Counting on Federal Funds," *Washington Post,* 12/10/2003.

368 *The Corps helped ravage those marshes:* Michael Grunwald and Susan Glasser, "The Slow Drowning of New Orleans," *Washington Post,* 10/9/2005.

368 *For four decades, scientists and local critics:* Michael Grunwald, "Canal May Have Worsened City's Flooding," *Washington Post,* 9/14/2005; Joby Warrick and Michael Grunwald, "Investigators Link Levee Failures to Design Flaws," *Washington Post,* 10/24/2005.

368 *America's skewed water resources priorities:* Michael Grunwald, "Money Flowed to Questionable Projects," *Washington Post,* 9/8/2005.

369 *Before the war in Iraq:* Bob Woodward, *Plan of Attack* (New York: Simon & Schuster, 2004), p. 150.

Acknowledgments

I FIRST SLOGGED INTO THE EVERGLADES in August 2000, as a reporter for *The Washington Post.* I was writing a long series of articles about the U.S. Army Corps of Engineers, and the Corps was embarking on an $8 billion effort to restore the Everglades. It struck me as a muggy, unpleasant place to spend an afternoon, which I later learned was a common reaction for summer visitors. But I was fascinated by the idea of the Corps trying to repair its abusive relationship with nature, and I returned to the Everglades in 2002 to write another long series of articles about the restoration project.

I mention all this partly to explain how I stumbled into this topic, but mostly to acknowledge my extraordinary debt to my indulgent bosses at *The Post,* including Phil Bennett, Steve Coll, Jackson Diehl, Len Downie, Tom Frail, and Liz Spayd. They encouraged me to convert my strange obsessions into acres of newsprint. And when I decided that the saga of the Everglades was bigger than the Corps, and that journalism alone could not do it justice, they gave me two years off to become a historian. I'm lucky to work at *The Post,* and all of us at *The Post* are lucky to work for Don Graham, who is the ideal CEO. Maybe it's because he's the grandson of an Everglades pioneer.

There's one more reason I mentioned my first swamp slog: It's my way of acknowledging that I'm not originally an Everglades guy, or a Florida guy, or even a nature guy. I grew up on Long Island, and my idea of the outdoors was a tennis court. So I'm indebted to hundreds of people who educated me about the Everglades, got their feet wet helping me understand it, and steered me toward the documents I needed to tell its story. My sources are listed in the notes, but I want to thank Joe Browder, Allison DeFoor, Shannon Estenoz, Alan Farago, Joe Knetsch, and Tim Searchinger for their advice as well as their information. I am especially grateful to the wise-beyond-his-years Ben Mathis-Lilley, who provided excellent research assistance, and will someday write a book much smarter than this one. At Simon & Schuster,

I was honored to work with Serena Jones, Roger Labrie, Emily Takoudes, and my brilliant editor, Alice Mayhew. Clyde Butcher generously provided photographs; his work is available at www.clydebutcher.com. Thanks also to Andrew Wylie, my agent, and the Brookings Institution, where I was a guest scholar in 2004, as well as Jonathan Abel, Gail Clement, Kelly Crandall, Bob Mooney, Shane Runyon, and Gene Thorp.

I could not have written this book without more than a little help from my friends, especially Gary Bass, Peter Canellos, and Manuel Roig-Franzia, great readers as well as great writers. I was also lucky to have the support of Peter Baker, Susan Glasser, and my other good friends at the *Post,* as well as Jon Gross, Jed Kolko, Ron Mitchell, Mark Wiedman, and the South Beach posse. And I would have been lost without my most sympathetic reader, Cristina Dominguez, who put up with me when I was a *jodón,* and stole my *corazón.*

Finally, I thank my loving family, including my ninety-eight-year-old grandmother, Lotte Grunwald, my brother, Dave, and my sister, Judy, and her husband, Steve, who shamed me by producing Allie and Zach in less time than I took to produce this book. I've dedicated this book to my amazing parents, Doris and Hans Grunwald. It's a totally inadequate gesture of my love, and all things considered, they'd rather have a couple more grandchildren. But this will have to do for now.

Miami Beach, June 2005

Index

About the Author

Michael Grunwald is a reporter for *The Washington Post.*
He has won the George Polk Award for national reporting,
the Worth Bingham Prize for investigative reporting, and
numerous other prizes, including the Society of Environmental Journalists award for his reporting on the Everglades.
He lives in Washington, D.C.